Galileo Galilei

Two New Sciences

Galileo Galilei
Two New Sciences

Including Centers of Gravity
&
Force of Percussion

Translated, with
Introduction and Notes, by
Stillman Drake

*

The University of Wisconsin Press

Published 1974
The University of Wisconsin Press
Box 1379, Madison, Wisconsin 53701

The University of Wisconsin Press, Ltd.
70 Great Russell Street, London

Copyright © 1974
The Regents of the University of Wisconsin System

First printing

Printed in the United States of America
For LC CIP information see the colophon

ISBN 0–299–06400–X cloth, 0–299–06404–2 paper

Contents

*This work is respectfully dedicated
to the memory of*

Antonio Favaro

(1847–1922)

*whose lifelong devotion to the study of
Galileo's books and manuscripts
alone made it possible*

Preface

This work, Galileo's last and scientifically his most important, has been translated into Spanish, Russian, Japanese, and French during the past quarter-century, and perhaps also into other languages. Earlier translations into German and English have been reissued. A compendious Italian edition has appeared, in which the Latin sections and Galileo's manuscript notes on motion are fully translated. This marked revival of interest in Galileo's physics reflects the maturing of history of science as a discipline in its own right, and leaves no doubt of the significance of Galileo in the transition from ancient and medieval to recognizably modern physics. Those same circumstances have rendered obsolete the English translation of this work that has remained standard since its first appearance in 1914.

The present volume contains a complete and entirely new translation. Its purpose is to provide a reliable and readable version of the main text and certain supplemental material from Galileo's hand, omitted from the 1914 English translation. The text presented is essentially that of the first edition (1638) with an important addition posthumously published in the second edition (1655), and a further dialogue intended for the 1638 edition but not completed to Galileo's satisfaction. Some of the other corrections and additions dictated by Galileo, who had become blind before he received the printed book, are included in footnotes. Numerous minor variants of text, essentially grammatical in character, are ignored. These may be found in the edition of Galileo's works superintended by Antonio Favaro (*Le Opere di Galileo Galilei*, Edizione Nazionale, Florence, 1898) to which page references are supplied throughout.

A further volume, now in preparation, will include documents, fragments, and correspondence relating to Galileo's new science of motion. This will also include a dialogue on Euclid's definition of proportion, often called "The Fifth Day," though that seems to have been composed as the first dialogue of another book, interrupted by Galileo's death early in 1642. Chronological arrangement of the documents, a task not previously attempted on a similar scale, reveals greater coherence in Galileo's work on motion and related topics than has hitherto been suspected. Detailed notes to the present translation as well as to the documents and correspondence, and a general biographical and historical appraisal of Galileo's physics, are projected to follow as a separate volume. Some such notes will necessarily be very long, and unsuited to appearance as footnotes here, while if placed at the end of this volume they would be difficult to compare conveniently with the text.

Two principal devices have been employed to hold down footnotes in this volume. First, a general introduction has been supplied, in which has been placed much of the material that would ordinarily go into notes, followed by a glossary of technical terms. Second, liberal use has been made of square brackets in the text. These may at first disturb the reader's eye, but in the translator's opinion they will be less distracting than the indication and inclusion of footnotes. Words or phrases given in Italian or Latin, alone or followed by the ordinary English equivalents, will permit the reader to decide for himself on some problems of meaning. Other English words in brackets are intended as glosses, phrasing the ideas (as understood by the translator) more completely than in the original. Bracketed material can usually be skipped over without disturbing the grammar of the sentence.

The translator is deeply grateful to the John Simon Guggenheim Memorial Foundation for a fellowship in 1971–72 which made possible the research behind this and the proposed forthcoming volumes. The University of Toronto obligingly granted a concurrent research year. The translation has been reviewed by Vittorio de Vecchi, James MacLachlan, and Michael Mahoney with many resulting suggestions of value to the translator, who nevertheless is fully responsible for the text as it stands. In the reading of typescript drafts against the original and against other translations, as in the reading of proofs, I have had immeasurable assistance from my wife, Florence Selvin Drake, a patient collaborator in this as in all my activities.

Introduction

Galileo's *Two New Sciences* was not published until 1638, when he was seventy-four years of age. Yet its principal conclusions had been worked out three decades earlier. Their novelty and their importance to the founding of modern physics make it almost incredible that Galileo should have held them so long unpublished. Accidental circumstances were largely responsible.

From the beginning of his professional career, Galileo's main interest was in problems of motion and of mechanics. His first treatise on motion (preserved in manuscript but left unpublished by him) belongs probably to the year 1590, when he held the chair of mathematics at the University of Pisa. It was followed by a treatise on mechanics begun at the University of Padua in 1593 and probably put into its posthumously published form around 1600. Galileo made some interesting discoveries concerning fall along arcs and chords of vertical circles in 1602, and two years later he hit upon the law of acceleration in free fall. With the aid of this law he developed a large number of theorems concerning motion along inclined planes, mainly in 1607–8. Toward the end of 1608 he confirmed by ingenious and precise experiments an idea he had long held: that horizontal motion would continue uniformly in the absence of external resistance. These experiments led him at once to the parabolic trajectory of projectiles. Meanwhile he had been at work on a theory of the breaking strength of beams, which seems also to have been virtually completed in 1608.

Galileo was organizing a book substantially along the lines of the Third and Fourth Days of the *Two New Sciences* when, in mid-1609, word reached him that a device had been invented in Holland that could make distant objects appear closer. Anxious to improve his salary and his position at the University of Padua, he put aside his projected book on motion and concentrated his attention on the telescope. He was able almost at once to produce an instrument about three times as powerful as its foreign rivals, for which he was rewarded in August 1609 by a doubled salary and life tenure as professor of mathematics at Padua.

But Galileo did not immediately return to his work on motion. He continued instead to improve his telescope, and early in 1610 he had an instrument of sufficient power for astronomical use. The fame of his resulting discoveries spread throughout Europe. This enabled him to obtain in mid-1610 a position at the court of Grand Duke Cosimo II of Tuscany, without teaching duties. His main interest now turned to astronomy. By 1613 he had become an open champion of the Copernican system, involving him in several controversies.

Early in 1616 Galileo was forbidden by order of the church to continue his campaign for Copernicus. Among the projects to which he then turned was the organization and completion of his long-neglected book on motion. At his direction, two pupil-assistants made copies of the main propositions he had previously worked out at Padua. But Galileo had added little to them when a new astronomical event distracted him. This was the appearance of three comets late in 1618, which again involved him in a long controversy. His concluding polemic, *Il Saggiatore*, appeared in 1623 and so impressed a new and friendly pope as to gain him permission to write about the rival systems of astronomy.

From 1624 to 1630, Galileo was mainly occupied in writing his famed *Dialogue Concerning the Two Chief World Systems, Ptolemaic and Copernican*, published in 1632. In that book he used some arguments drawn from his earlier studies of motion in order to refute the prevailing objections against any motion of the earth. It is doubtful that he returned seriously to the older project before he had finished the *Dialogue* in manuscript in 1630. He then began to add to his earlier theorems on motion, but was interrupted once more, this time by an order to stand trial in Rome for "vehement suspicion of heresy." In mid-1633 he was sentenced to life imprisonment and was forbidden to publish any more books on any subject whatever. Yet before the year was out, he had begun to assemble the *Two New Sciences* in its final dialogue form, writing most of the dialogue parts during the years 1633–35.

As the manuscript neared completion, Galileo made inquiries through friends at Venice, Vienna, and Paris with a view to its publication. It was ultimately sent to Leyden to be published by the celebrated printing establishment of the Elzevirs, who had recently printed Latin translations of Galileo's forbidden *Dialogue* and his *Letter to Christina* concerning freedom of scientific inquiry. A Leyden firm was then printing the *Discourse on Method* of René Descartes, together with its mathematical and scientific appendices.

Having found a publisher beyond the reach of the Roman Inquisition, Galileo proceeded to cover his tracks by formally dedicating his new book to a French ambassador who had recently returned to Paris from Rome. The dedication makes it appear that actual publication was not Galileo's own decision, but that of his distinguished admirer. Galileo had in fact given the ambassador a copy of the manuscript, or most of it, in 1636. By that time, however, another copy was already in the hands of the Elzevirs; hence Galileo's letter of dedication must be regarded as a polite fiction. The foreword, ostensibly written by the publishers, is believed to have been drafted by Galileo, though a reference in it to an opinion of Descartes shows that the Elzevirs were responsible for at least part of it.

Galileo was much displeased by the printed title of the book, which he considered undignified and ordinary. His own original title for it is not known, but from the fact that he usually referred to the work as "my dialogues on motion," we may surmise that his title began with the word *Dialogues* and that it went on to describe the contents and purposes of the work. Very likely it was phrased in such a way as to distinguish even more sharply than the printed title between the *discorsi*, by which Galileo meant oral discussions and verbal reasonings, and the *dimostrazioni matematiche*. The feminine form of the final adjective assures us of his intended distinction between traditional Aristotelian physics, heard by the Stagirite's pupils, and Galileo's physics formalized in rigorous mathematical proofs. The discussions were written entirely in Italian and were aimed at the general reader. The mathematical demonstrations, intended for scholars, fell into two groups, one written in Italian and the other in Latin, probably for reasons outlined below.

The appendix on centers of gravity includes Galileo's first mathematical studies, submitted in 1588 to support an application for a chair of mathematics at Bologna. It was probably expanded and brought to its present form in 1613, when it was intended to be published, but was put aside in deference to the wishes of Luca Valerio, an older mathematician who was revising his own earlier book on the same subject, a work much admired by Galileo.

The additional "day" here translated into English for the first time was planned to appear with the original book, but Galileo was unable to complete it to his own satisfaction. Still another dialogue (on the Eudoxian theory of proportion) was dictated by Galileo in his last days, but its opening passages show it to have been intended for another book.

II

The three interlocutors who appear in the *Two New Sciences* as published in 1638 bear the same names as those of the 1632 *Dialogue*. Two of them had in fact been close friends of Galileo while alive, and he wished to perpetuate their memory. Filippo Salviati, scion of a very distinguished Florentine family, was born in 1582 and died during a visit to Spain in 1614. Salviati was made Galileo's own spokesman in the *Dialogue*. Giovanni Francesco Sagredo, born in 1571, was a Venetian patrician, active in governmental and diplomatic affairs and a talented amateur of science. He had studied with Galileo at Padua and remained his close friend from 1600 to his death in 1620. Sagredo spoke for the intelligent layman in the *Dialogue*, playing the role of an uncommitted person for whose support the other two interlocutors were contending. Simplicio represented the old traditions; he was described in the preface to the *Dialogue* as a real person whose name was concealed for reasons of courtesy. It is generally agreed that Simplicio was essentially a composite of Galileo's Peripatetic opponents, but Galileo may have had in mind his personal friend and professional opponent Cesare Cremonini, professor of philosophy at Padua, or Lodovico delle Colombe, Galileo's later adversary at Florence. Both men, though poorly versed in mathematics, were masters of the discursive Aristotelian science then taught in all universities.

The names of the interlocutors remain the same here, but their roles are subtly shifted. In the earlier *Dialogue*, Galileo's own views are identifiable with certainty only when they are introduced by an interlocutor as those of a certain Academician whom they all know and admire. This device is not entirely absent from the present book, but here we have also a long written treatise, read aloud in full by Salviati and subjected to discussion. This treatise was printed in Latin, distinguishing it from the Italian dialogues. Avowedly Galileo's, it implies rather different roles for all the speakers from those they had played in the *Dialogue*.

The Latin treatise, set in distinctive type here, embodies Galileo's definitive thought in a language intelligible to all scholars of the period. I am inclined to believe that this is the only material in the book (except the Latin appendix) which Galileo considered to be valid beyond question, so that he wished to attach his own name directly to its exact formulation. Everything in Italian may be taken as to some degree tentative in Galileo's mind, even including the mathematical demonstrations that relate to the strength of materials which appear in the Second Day.

Thus Salviati remains a spokesman for Galileo, but no longer his sole or even his chief spokesman. Galileo speaks for himself in the Latin treatise, and Salviati is accordingly left free to comment or explain. On some occasions he phrases his remarks positively, while on others he expresses only a preference between conflicting opinions, and sometimes he even speaks paradoxically. Salviati reflects, on the whole, the thoughts of an aged, mellowing Galileo who is willing at the end of his long life to recognize merit in views different from his own.

Sagredo is no longer merely a representative of the educated layman. He still plays that role, but in this book there is no need (as in the *Dialogue*) for him to act the part of a foil for the arguments of two opposed experts. The physical topics dealt with here did not stand, like the systems of Ptolemy and Copernicus, at opposite poles in a highly controversial subject. Sagredo now speaks for Galileo at the middle stage of his development—from his move to Padua in 1592 to the essential completion of his mathematical physics a quarter-century later—a period that closely agrees with Sagredo's own active intellectual life. We find him raising questions that had puzzled Galileo earlier, and taking positions that Galileo had once considered and then rejected. Indeed, some of the questions Sagredo raised still puzzled Galileo, as is evidenced by Salviati's replies to them.

But it is above all the role of Simplicio that is changed from that of the *Dialogue*. Little remains of the stubborn and unshakeable loyalty to Aristotle and to every current dogma of the Peripatetics which had necessarily characterized Simplicio's earlier performance. Now he is often and easily

won over to Salviati's positions (though he has trouble with
Galileo's mathematics), quite in accordance with the theme
that truly new sciences are being introduced and expounded,
rather than that traditional issues are being debated as had
been the case in the *Dialogue*. Simplicio even laments his
past neglect of mathematics, and he shows bewilderment
when he cannot reconcile a Peripatetic position with facts
of observation (and, in some cases, not even with Aristotle's
logic). By no means is he a representative of that dogmatic
Aristotelianism which in fact continued to oppose Galileo's
new physics for many years.

Even Simplicio, I think, can be regarded in part as a spokesman for Galileo in this book. He speaks for the young
Galileo—an Aristotelian by schooling who, though not
unwilling to consider new views, yet demands that he first
be shown something better than the Peripatetic physics and
philosophy already known to him. For as Aristotle himself
wrote in *De caelo* (and this might well be taken as Galileo's
motto for his final book), "It is wrong to remove the foundations of a science unless you can replace them with others
more convincing." It is the young student Galileo that I hear
speaking when Simplicio rejects with withering sarcasm
the idea that all bodies of the same material, regardless
of weight, fall with the same speed; he says that in that case,
one would have to believe that a birdshot falls as swiftly
as a cannonball. And on first hearing the strange new idea,
advanced by G. B. Benedetti a generation earlier, what else
would Galileo himself have said before he had worked out
for himself its many implications for a new physics?

In the added "day," Simplicio is replaced by a new interlocutor, Paolo Aproino. He, too, was a real person, a pupil
of Galileo's at Padua in 1608, who lived to see a manuscript
of at least part of the *Two New Sciences* though he died in
the year it was published. Aproino replaces Simplicio not
as a Peripatetic; no such person was needed in a discussion
of the force of percussion, which had never been a topic of
Aristotelian physics. Neither did Aproino replace Simplicio
in the latter's new role as spokesman for the student Galileo;
for Galileo never mentioned percussion until 1594, when he
had already been a professor for five years. Aproino was
introduced as someone who could describe certain experiments of Galileo's, and discussions at which he had been
present; and indeed the very year 1608 has bequeathed us

the only other documentary evidence we have of Galileo's experimental care and proficiency. Aproino, then, speaks not for the student Galileo, but for Galileo's students—much as Simplicio had spoken in the *Dialogue* for Galileo's professors.

III

The Leyden edition of 1638 is the principal authority for the text. One partial manuscript survives which contains the First Day and nearly all of the Second. This copy, in the hand of a scribe with many corrections and additions in Galileo's own hand, will be referred to as the Pieroni manuscript. It was sent to Giovanni Pieroni, a Florentine engineer employed by Emperor Ferdinand II, with the idea of having it printed abroad. Failing in this, the good-hearted, inept, and incredibly verbose Pieroni eventually returned it to Galileo, among whose documents it is preserved. No other manscript of the book is known.

This translation follows the text established by Antonio Favaro with very few changes, duly noted. Basically it is the text of the 1638 edition after correction of misprints and with an additional proof which Galileo wished to have inserted in later editions. One long supplementary passage dictated by Galileo after printing is included in the text, set in smaller type, while some shorter ones appear in the notes. Changes and additions made by Favaro on the basis of the Pieroni manuscript have been adopted here, but only a few principal differences between Favaro's text and that of 1638 are noted. Most variant readings of the manuscript, recorded in Favaro's edition, are omitted. The pagination of that edition, identified above, is carried at the top of each page and in boldface type in the margins in this translation. In the footnotes, boldface type indicates the Favaro pagination rather than that of this book itself.

A brief account of previous translations will be not without interest, and will lead naturally into the reasons for the appearance of yet another English version at this time.

An abridged French translation and paraphrase by Marin Mersenne was published in 1639. The first English translation, made by Thomas Salusbury and printed at London about two decades before Isaac Newton's *Mathematical Principles of Natural Philosophy* (1687), was justified by the scientific value that Galileo's work possessed until Newton's great book rendered it obsolete. Salusbury's first complete translation had little influence in England, however, because all but about a dozen copies were destroyed by the Great Fire of London in 1666. A complete Latin version appeared at Leyden in 1699–1700 without identification of the translator. Just possibly it was made under Galileo's own direction, for the Elzevirs had planned to issue all his works in Latin, and a copy sent for that purpose may have remained at Leyden unpublished.

A second English translation was made by Thomas Weston, largely because of the difficulty of finding copies of the first, and was published posthumously in 1730 by Weston's brother John. The Salusbury translation had been made from the Bologna edition of 1655, and Weston made use of it in conjunction with the original text of 1638; many parallel passages and unusual English words show the connection. Weston's translation is of particular importance textually because of his detection of an editorial error in 1655, discussed below, that has survived in all other editions, including even that of Favaro. But except for a so-called second edition, which was merely an issue of the 1730 printed sheets with a title page dated 1734, Weston's translation suffered the same general neglect that befell Galileo's basic conception of physics during the next century and a half. Cartesian and Newtonian physics, reuniting science with philosophy, had quickly replaced Galileo's abortive attempt to isolate physics from metaphysical speculation.

Extremely rapid development of physical science during the nineteenth century awakened a new interest on the part of thoughtful physicists in the basic foundations of their science. Some of them pursued this interest in the form of philosophical and metaphysical investigations, while others turned to historical inquiries. Eminent among the latter was Ernst Mach, who called attention to the then neglected pioneer work of Galileo in his historical survey of mechanics first published in 1883. A complete German translation of the *Two New Sciences* and its supplements, with the first critical notes, was made by A. von Oettingen in 1890–91.

The great rarity of both the Salusbury and Weston translations, coupled with this renewed interest in Galileo's work, resulted in a third English translation by Henry Crew and Alfonso De Salvio, first published at New York in 1914. Their version has ever since been heavily relied on by historians of science in English-speaking lands, and by some others for whom the original mixture of Italian and Latin presented difficulties.

History of science was still something short of maturity as an independent discipline when the first German and the third English translations were made. This fact left its mark on both of them. The translators were primarily scientists and linguists rather than historians, and they tended to present Galileo's thought without due consideration of the nature of earlier physics and its special terminology. The American translators in particular have been justly criticized for permitting the intrusion of their historical preconceptions into Galileo's writings. Thus, near the beginning of the Third Day, they have Galileo say, "There is, in nature, perhaps nothing older than motion, concerning which books written by philosophers are neither few nor small; nevertheless I have discovered by experiment some properties of it which are worth knowing and which have not hitherto been either observed or demonstrated." Alexandre Koyré, the outstanding modern analyst of the scientific revolution, remonstrated long ago that the phrase "by experiment" was quite unjustifiably inserted, Galileo's own word having been simply *comperio*, "I find." This needless gloss was particularly offensive to Koyré because he held that Galileo had made no use of experiment in the modern sense, and attributed his success to simple faith in mathematical reasoning.

In defence of Professor Crew's phrase, it was replied that *comperio* necessarily implied experiment, considering that there was no other way for Galileo to have determined the actual law of free fall. The reply, though true in a very important sense, hardly constitutes a justification, for implications belong not in the text of a translation, but in notes or commentaries. Nevertheless, inaccuracies of the above kind hardly afforded grounds for a new translation. If Galileo's role in the evolution of experimental physics could be judged only from one isolated sentence, then nothing worth saying about it could ever be deduced; if not, then other information relating to Galileo's experiments would correct any misapprehensions thus created.

Another type of inaccuracy, however, is not so easily brushed aside. This is the excessive use in translation of modern terms having technical meanings in physics. Near

the beginning of the Fourth Day, the 1914 translation reads:
"Imagine any particle projected along a horizontal plane. . . .
This particle will move along this plane with a motion that
is uniform and perpetual." The word "particle" now has
a special sense in physics which has nothing in common
with Galileo's word, *mobile*. The modern physical particle
is essentially devoid of weight, whereas Galileo's whole
discussion at this place depended on the concept of heaviness.
In this instance, an inappropriate technical term was anachro-
nistically substituted for an obsolete technicality. In other
instances, words like "velocity" and "gravity" were left
unchanged, although their meanings have been very much
altered since Galileo's day.

Even these additional sources of misunderstanding still did
not seem to me sufficiently serious to require the effort of
making a new translation. The easy style of the existing
version, to say nothing of its beautiful format, went far (in my
opinion) to offset purely technical faults. What ultimately
changed my view was the detection of a seemingly trivial error,
which in the beginning amounted to no more than the use of
singulars in place of Galileo's plurals in one sentence. That
sentence, however, was one of crucial importance; it contained
Galileo's entire argument aginst the old belief that speeds in
free fall grow in proportion to the distances fallen from rest.
The grammatical change concealed the essential nature of his
reasoning, which in turn was related to an important mathematical insight. It made Galileo appear guilty of an elementary
logical blunder in the subject of his greatest expertise, and on
top of that it revived a long-forgotten debate among his
contemporaries.

Checking further, I found that this arbitrary change is not
limited to English; it occurs also in French, German, Russian,
and (I believe) in all modern translations except the Spanish.
Yet Galileo's words had been correctly given in the old Latin
and English translations. What kind of inaccuracy would be
newly introduced by modern translators, and adopted almost
universally by them? Clearly, the change must be one that
arises from a misunderstanding of Galileo's thought and is
related to some current theory about its relation to modern
physics on the one hand and to new historical information
about medieval physics on the other. This kind of inaccuracy
appears to me to be different in kind from the others previously
mentioned. When theories of what Galileo should have said
are allowed to alter his own words, it is time to call a halt.

Detection of a single widespread error by attending to Galileo's precise language shed new light in turn on other aspects of his thought on falling bodies. Mistranslations in related passages were also found, as might be expected. Some of these probably originated from inattention, while others seem to have been induced by a gratuitous assumption of medieval influence. Thus, the concept of "mean speed" was introduced into the statement of Galileo's basic first theorem on accelerated motion, in place of "one-half the final speed"; and in the proof, the word *totidem* was omitted in both its occurrences. A similar mistranslation of the same theorem was published by Koyré himself, while mistakes in dealing with *totidem* in the proof are found also in the German and French translations. This word means "as many as," and in ordinary speech that idea may not be precise; but in a mathematical proof it can only mean "in precisely the same number," which is in fact the basis of Galileo's proof. Not only that, but *totidem* signaled Galileo's formal introduction of one-to-one correspondence between members of infinite aggregates, a concept of great significance to modern mathematics.

also in Bradwardine 19

Discussion of this theorem with Professor I. B. Cohen and his graduate students at Harvard University while the present translation was in prospect led me to hypothesize a misprint in still another proposition of the Third Day. One of the students, Mr. Kenneth Manning, soon afterwards wrote to me that the fault did not occur in the 1638 edition. So plausible was this editorial meddling in the 1655 edition that even Favaro had accepted the supposed emendation, whence it has passed into every modern text and translation, and has infected most interpretations of Galileo's mathematical physics.

In brief, no previous English translation any longer meets the standards required by modern historians of physics. The importance of Galileo's science is such that not only specialists, but general historians, need an accurate and convenient text of this book. No doubt there will still be faults in this one, for "never hath book been printed but error hath affixed his sly imprimatur." But at least some pervasive and serious past inaccuracies have been eliminated, which is the first step toward proper understanding of Galileo's pioneer role in modern physics.

IV

The present translation is not strictly literal. English does not easily support the very long sentences that in Italian or

Latin are kept clear as to internal references by a wealth of
gender and case distinctions. Nor could literal translation
preserve much of the literary flavor of the original. Each interlocutor has idiosyncrasies of expression related to his pro-
fession and social position, yielding nuances which in literal
translation would hardly be recognized. Clauses must often be
rearranged in the interests of clarity. Some words are omitted
that are customary in Italian but superfluous in English, while
some that are needed in English but not Italian are supplied
without the usual square brackets when those would be pedantic. In general, an attempt has been made to stay as far away
from a literal as from a free translation.

Apart from the many problems already mentioned, two
others of particular importance arise in translating this book.
One is rooted in Galileo's adherence to Euclidean terminology
and to a form of geometrical demonstration long out of date.
The other problem relates to a vocabulary of physics that
enabled Galileo to address readers steeped in Aristotelian
philosophy and yet to present an analysis alien to that in many
ways. The translator can at best deal only with one side of such
problems, leaving the other side to be handled by the reader. A
glossary follows this introduction, one part of which gives the
English words and phrases adopted here for Euclidean tech-
nicalities and for Galileo's own mathematical expressions,
while the other gives his Italian or Latin physical terminology
with the English word or phrase used in its translation, usually
with some explanation of reasons for my choice. But a glossary
alone cannot supply the Euclidean and Aristotelian background against which the book was written and should be
read. Some discussion of this difficulty follows, directed mainly
to points that seem to me most likely to be generally over-
looked.

To begin with, Galileo wrote for a mixed readership. He had
good reason to despair of support from professors in the
universities, so he addressed the educated Italian public, using
a form much favored during the early days of the printed book
in didactic works; that is, a lively and plausible dialogue in
colloquial language. For the benefit of scholars outside Italy,
he included his Latin treatise on motion. From both classes of
readers in his day, Galileo could expect a good general understanding of Aristotelian science and Euclidean mathematics.
Neither group of prospective readers, on the other hand,
possessed our conceptions of elementary physics or of algebra.

Hence in order to understand Galileo, the modern reader should do his best to forget modern science and mathematics and to recall, or if necessary to acquire, something of the earlier background.

Next, it was no small part of Galileo's purpose to overcome a certain traditional resistance to the use of mathematics in the solution of physical problems. That resistance, quite unknown today, existed on both sides. Professors of physics were invariably philosophers who seldom had any special training in mathematics. Galileo strove to show them, as well as the general public, the potential utility of mathematics in physical investigations. Professors of mathematics, on the other hand, seldom exhibited much interest in physical problems as such. Galileo wanted to excite their interest in these, and particularly in the ancient Archimedean approach to them, which at that time was not a part of the standard university curriculum in mathematics.

For both purposes, Galileo selected certain well-defined and very restricted areas of inquiry within which mathematical analysis could clearly advance physics. Though now characteristic of science, his procedure of adhering to limited objectives and dealing with specific problems having precise solutions had little appeal to men schooled in abstract principles of great generality. It still has little appeal to most historians of science, who strive to look behind Galileo's work in physics and find there some grand speculative scheme of the universe, though it was precisely the absence of anything of the sort that was to cause Galileo to be so quickly forgotten. If Galileo ever believed that all phenomena of nature could be explained, or even fully analyzed, he failed to say so; rather, he publicly expressed doubt that any phenomenon in nature, even the very least that existed, could ever be completely understood by any theorist. The grand program which we know as "the mechanical philosophy" was due not to Galileo, but to Descartes, who flatly rejected Galileo's physics as having been built without foundation.

Within the restricted areas of inquiry dealt with by Galileo, a single mathematical theory was rigorously applied. This was the theory of proportion of magnitudes in general, first propounded by Eudoxus, as set forth in the fifth book of Euclid's *Elements*. The classic form of proof using that theory seems to us clumsy and verbose as compared with our easy manipulation of algebraic equations, a technique that was just acquiring a reasonably modern form in Galileo's time. He himself never used algebra, not even in his private papers. The question arises whether in a modern translation, when we alter the original language, we should not also translate the old geometry into modern algebra.

In my opinion it would be a mistake to paraphrase Galileo's proofs algebraically, even in footnotes. Such a procedure does not translate Galileo's thought into modern idiom, but instead transforms his mathematical ideas into others that are alien to his and that he himself mistrusted, especially in physics. The reasons for this have solid grounds in the history of mathematics, as will be explained in part below. Readers who wish to see algebraic paraphrases of Galileo's proofs will find them in the recent French translation by my respected friend, Professor Maurice Clavelin. Consideration was duly given to their repetition here, but I have decided against it. Those readers who wish only to know what conclusions Galileo reached that are still acceptable today need read only the statements of his theorems, without bothering to study the proofs he offered. Those who wish also to know why Galileo believed his proofs to be both mathematically rigorous and physically sound must take the trouble to learn the Eudoxian theory of proportion, and there is no royal road to Euclid's geometry, least of all a highway through algebra. Lest my point of view be thought merely eccentric (as it is indeed idiosyncratic today), I shall try to explain it, citing first from the article "Geometry" in the eleventh edition of the *Encyclopaedia Britannica*, which reflects a view regarded as authoritative about the year I was born: "The fifth book of Euclid is not exclusively geometrical. It contains the theory of ratios and proportion of quantites in general. The treatment, as here given, is admirable, and in every respect superior to the algebraical method by which Euclid's theory is now generally replaced. We shall treat the subject in order to show why the usual algebraical treatment is not really sound."

Those words were written well after the rigorous analysis of the real number system had been carried out by Richard Dedekind, and new foundations for function theory had been laid by Karl Weierstrass. The *Encylopaedia* writer was acutely conscious that the fully arithmetized continuum is not exhausted by all possible algebraic numbers. Both Dedekind and Weierstrass had appealed to the Eudoxian definition of "same ratio" in order to give rigor to the concept of the continuum. That definition was not only available to Galileo, but he made it the subject of special lectures at Padua, and wrote a little dialogue on the definitions in Euclid's fifth book.

Thus Galileo was able to deal rigorously with continuous magnitude (on which he wrote a treatise, now lost) in a sense in which many people today, educated in elementary algebra but not in Euclidean geometry, are not; nor are they generally aware of the fundamental assumptions lying behind algebraic manipulations that they perform mechanically. Galileo, on the contrary, was perfectly clear about the meaning of "same ratio" because he was obliged to make repeated conscious use of its Eudoxian definition. The same was not true of his medieval predecessors, who had been obliged to create an elaborate arithmetical theory of proportion based ultimately on Euclid, Book VII, because the definitions of Book V came to them in defective form through an Arabic version and were further garbled by incorrect commentaries in the standard Latin version of the Middle Ages. Not until the mid-sixteenth century were these things specifically cleared up. I believe that the essential difference between medieval and Galilean mathematical physics depended largely on that neglected fact.

The price Galileo paid for rigor in the avoidance of algebra and the use of Eudoxian proportion theory was that his mathematical physics was restricted to comparisons of ratios. A common failure to recognize this relational—and hence relative—feature of Galileo's physical laws has delayed a proper understanding of his work, both theoretical and experimental, especially as compared and contrasted with the achievements of the Middle Ages and with those that came soon after Galileo's death. This same misunderstanding, widely prevalent today, would in my opinion be only aggravated by any paraphrasing of Galileo's theorems, let alone his proofs, in algebraic form. An example may make the point clearer.

Algebraically, speed is now represented by a "ratio" of space traversed to time elapsed. For Euclid and Galileo, however, no true ratio could exist at all except between two magnitudes of the same kind. Whatever space and time may be, they are not magnitudes of the same kind—or if they are, then that is thanks to Einstein, and it is not something that Galileo would have regarded as capable of rigorous deduction. No longer bound by Euclid's definition of ratio, we can write $\bar{v} = s/t$ as a definition of average speed, and we can give rules such as $s=kt$ for uniform motion and $s=kt^2$ for uniformly accelerated motion. Such expressions entail a metaphysics in which we can calculate individual speeds from times and distances, whereas Galileo could only compare them in pairs alike in kind. The above rules do imply for us all the laws of motion that took Galileo many pages to state and demonstrate; but they also imply things that he not only did not establish, but would have rejected as rash assumptions. That is why our equations are by no means conceptually equivalent to Galileo's proportionalities.

Algebraic notation provides no way to insure the maintenance of Euclid's restrictions on ratios. The very essence of an equation is that any term can be transposed from one side to the other, whereas it is essential to a proportionality that this cannot be done unless all terms happen to represent magnitudes of the same kind (as, for example, pure numbers). Equations thus imply that a given physical magnitude can be precisely measured by measuring some other *kind* of physical magnitude. This may be true, and I am not trying to refute modern physics: all I wish to do is to emphasize that it is very different from Galilean physics. He was content to stop when he had established a proportionality between different physical magnitudes, without attempting any kind of identification of them or any explanation of the relation. So far as he was concerned, any explanation why the world is as we find it could come only from philosophers and theologians. This view was built into Galileo's mathematics and constitutes one reason that his physics aroused much opposition. It was soon supplanted by the Cartesian and then the Newtonian systems of the universe, each equipped with elaborate philosophical and theological components.

Galileo did, however, employ one device that is not found in Euclid (except in one spurious definition, never applied) but that had been utilized by Archimedes and Ptolemy. This was the formation of "compound ratios" by a process that we would call the multiplication of a ratio by some different ratio (our phrasing would have been meaningless to Euclid). Used without restriction, this procedure would have permitted the creation of "ratios" between physical magnitudes of different kinds. It is therefore important to note its use by Galileo; for example, in the Third Day, Book I, Theorem IV. This theorem states that for two different uniform motions, $s_1 : s_2 :: v_1 t_1 : v_2 t_2$. The compounded ratio on the right is made up of two proper ratios, v_1/v_2 and t_1/t_2. Without some restriction it might be assumed that the same compounded ratio could equally well be divided into v_1/t_2 and v_2/t_1, as indeed we are free to do algebraically. But under the Euclidean definition of ratio, Galileo took care not to admit any decomposition of compound ratios into components different from those that originally went into their composition. In this way he was able to express certain functional relationships without assuming or implying that any particular distance was the product of a particular speed and a particular time. This may seem to you a dubious gain, or even an outright disadvantage; if so, you have not yet perceived what Galileo and the ancients demanded of mathematical rigor, or what he was willing to pay for it in terms of suspended judgment about physical entities. The *ratio* of two speeds, which Galileo sought, was the same as the *ratio* of two such products, for whatever kind of magnitude v_1t_1 and v_2t_2 may be, those two expressions certainly represent magnitudes of the same kind. To equate them with speeds was, to Galileo, merely bad metaphysics, even though their ratio was demonstrably a ratio of uniform speeds.

Finally, Galileo's treatment of the infinite and the infinitely small in mathematical terms is of particular interest. Aristotle had dealt with Zeno's paradoxes of motion by attempting to resolve them. Galileo's attitude was different; he saw paradoxes of the infinite not as puzzles to be resolved, but as realities to be recognized. His discussions of them are neither trivial nor mystical, as is often charged, and they are best appreciated in terms of the sharp distinction in classical mathematics between the discrete and the continuous. This is again related to the differences between Book VII and Book V of Euclid, and to the fundamental difference between medieval and Galilean mathematical physics. Galileo's repeated insistence that in order for a body to accelerate from rest to any given speed, it must pass through every lesser speed, was utterly rejected from physics by Descartes, who remarked that though this might sometimes happen, it could not be so for most bodies. Honoré Fabri, his contemporary, argued at length that no mathematical instant could be treated as a physical entity. The great generality of modern mathematical methods blurs for us the nature of such disputes in which Galileo's continuity physics warred with a sort of quantum physics of motion rooted in traditions carried forward from the medieval Euclid borrowed from the Arabs.

Readers interested in understanding the nature of Galileo's mathematics and its applications by him to problems of physics should begin, not with algebraic versions of his proofs, but by reading T. L. Heath's introductions to his English editions of Euclid and Archimedes. This should be followed by reading the definitions in the fifth book of Euclid, with Heath's commentaries on them, and then the postulates and opening theorems of the Archimedean treaties *On Plane Equilibrium*, *On Floating Bodies*, and *On Spiral Lines*.

v

Galileo's terminology in physics includes that of Aristotle, supplemented by words taken from mechanics or borrowed from ordinary speech and given new technical senses. Mechanics was then not regarded as a part of physics proper, which was itself a part of philosophy. Physics was traditionally the study of natural motion (or more correctly, of change) in terms of its causes. Mechanics was considered to be a mixed science, partaking of both physics and mathematics but belonging to neither. The earliest known treatise on mechanics, questionably ascribed to Aristotle but certainly in the tradition of his school, said in its introductory section:

> As the great poet Antiphon wrote, "We gain skill by mastery over things in which we are conquered by nature (*physei*)." Of this kind are those [problems] in which the lesser masters the greater, and things possessing little weight move heavy weights, and all similar artifices, which we call mechanical problems. These are not altogether identical with physical problems, nor are they entirely separate from them; but they have a share in both mathematical and physical speculations; for the method is demonstrated by mathematics, while the [fact of] practical application belongs to physics.

Aristotle himself was a competent and well-informed mathematician, as evidenced (among other things) by his treatment of Zeno's paradoxes. But Aristotle did not encourage the pursuit of physics (that is, the study of natural phenomena) along mathematical lines. His main reason was that the mathematician and the physicist never consider the same aspects of anything that they may both happen to study. Aristotle's fundamental purpose in physics was to determine the causes of change; since in his view mathematics did not deal with physical causes, problems of the physicist were regarded by him as left untouched by mathematics. In his own words, near the end of the first book of his *Metaphysics*, "The minute accuracy of mathematics is not to be demanded in all cases, but only in the case of things which have no matter. Hence its method is not the method of physical science, for presumably the whole of nature has matter." Einstein was to say much the same thing in a famous lecture on geometry and experience: "As far as the laws of mathematics refer to reality, they are not certain; and as far as they are certain, they do not refer to reality."

At the very beginning of *Two New Sciences* the reader will see how different Galileo's view was from Aristotle's with regard to the applicability of mathematics to things that have matter. Later, in the Third Day, will be found his outright rejection of the quest for causes in the study of naturally accelerated motion; that is, in free fall. Galileo's youthful studies of motion had begun with the usual attempt at an analysis by causes, and that concept continued to occur in his later books on physical science. His use of the word "cause," however, became progressively less philosophical and more objective as he undertook to improve physical science by giving greater certainty to the study of motion through more use of mathematics than Aristotle had deemed justifiable. An example, taken from the *Dialogue*, is Galileo's mature treatment of a question about the cause of equilibrium on a steelyard between a huge bale and a small counterweight: "One can see no other cause than the disparity in the movements which must be made by each of them." Behind this effect, the writer of the Aristotelian *Questions of Mechanics* had seen circularity of motion as the cause. With Galileo, the cause had become simply a correlated phenomenon.

Galileo's mature refusal to enter into debates over physical causes epitomizes his basic challenge to Aristotelian physics. The most intelligent explanation of acceleration in free fall had been offered by Jean Buridan in the fourteenth century; it was still popular among Galileo's contemporaries G. B. Benedetti, Isaac Beeckman, G. B. Baliani, and Honoré Fabri. Yet it was not so much as mentioned, let alone adopted, by Galileo. Buridan's explanation was that impetus accumulates in the falling body after the initial motion is begun by its heaviness alone, and that those two causes acting together account for the increase in speed. The introduction of a second cause *after* the first obviously entails a temporal discontinuity, unacceptable to Galileo. Other causal explanations had also been given, and in rejecting all such "fantasies," as Galileo called them, he described his alternative program: to investigate the implications of a mathematically defined motion, and then to see whether they fitted any actual observations. Certainty arises neither from mathematics alone nor from experience alone, but only from their agreement.

Aristotle's search for causal principles was prosecuted characteristically through techniques of exhaustive classification. Fundamental to this was a dichotomy between natural and violent motion. Natural motion was that undertaken by a body freed from all constraint; under those conditions, it proceeded to its natural place in the universe. All other motions were called violent, or forced. Since it is those latter motions with which modern physics is mainly concerned, we tend to overlook the fact that force played a very minor role in Aristotle's physics. This being the science of nature, it was concerned primarily with natural motions, which by definition excluded force. A special concern with force entered into physics during the Middle Ages, but force as a physical entity is nearly as inconspicuous in Galileo as in Aristotle. To both of them, acceleration in free fall was a natural event and did not need explanation. To Aristotle, it did not even merit quantification.

The essential point is perhaps best phrased by saying that where earlier physicists had habitually regarded force as a *cause* of motion, Galileo saw it mainly as an *effect* of motion. In his first published book on a physical topic, the *Discourse on Bodies in Water* (1612), he introduced the word *momento* in the double sense of "static moment" and of "momentum", and when his critics objected to this as a neologism, he wrote, in a second edition of the same year:

Moment, among mechanics, signifies that force [*virtù*], that power [*forza*], that efficacy with which the mover moves and the moved body resists, which force depends not only on simple heaviness, but also on the speed of motion, and on the varying inclinations of the space over which the motion takes place, as a descending body has more impetus in a very steep descent than in one less steep. And in sum, whatever the cause [*cagione*, reason] of such force, it still retains the name of "moment." Nor does it appear to me that his sense should be linguistically novel, for if I am not mistaken we often say, "This is a weighty matter, and that other is of little moment," and "Let us turn to light things, and put aside those of moment"—metaphors that are, I believe, taken from mechanics. . . .

The power of heaviness is increased by speed of motion, so that weights absolutely equal, but associated with unequal speeds, are of unequal power, moment, or force; and the swifter is more powerful in the ratio of its speed to the speed of the other. . . . And such is the power and force conferred by speed of motion on the moveable that receives it, that this can exactly compensate a proportional weight by which the other, slower, movable may be increased. . . . This relation between heaviness and speed is found in every mechanical device, and it was taken as a principle by Aristotle in his *Questions of Mechanics*. Whence we may take it as truly assumed that weights absolutely unequal are reciprocally counterpoised and rendered of equal momenta whenever their weights are inversely proportional to their speeds of motion.

The reader's first impression from this may be that Galileo thought of force as a cause of motion, much as we do now. Anyone having that impression should reread the passage, noting this time that Galileo says, "The force depends on . . . ," "Whatever the cause of such force . . . ," and "The power is increased by . . . ," while he does not say that force or power causes anything, or that anything depends on it. Inveterate habits of thought associated with words like "force" often distract our attention from what Galileo actually wrote. And thus historians often attribute to him things he did not say; by making him modern on the one hand, or medieval on the other, they imply for him goals of physics he did not pursue.

Galileo's mature position—that causal inquiries might be well abandoned in physics—had no immediate appeal to his contemporaries, just as it had had no obvious source in earlier physics. It was pretty much ignored for a long time, except for occasional lip service. Thus Christian Huygens stated it in the preface to his *Treatise on Light*, but reverted to the mechanical philosophy early in the text. Finally it reappeared toward the end of the last century when Heinrich Hertz wrote, in the introduction to his *Principles of Mechanics*:

We form for ourselves images or symbols of external objects, and the form which we give them is such that the necessary consequents of the images in thought are always the images of the necessary consequents in nature of the things pictured. In order that the requirement may be satisfied, there must be a certain conformity between nature and thought. Experience teaches us that the requirement can be satisfied, and hence that such a conformity does exist.

All the above has been said for the purpose of suggesting certain frameworks within which the reader may consider what Galileo wrote, different from the ordinary framework of physics immediately before and shortly after the *Two New Sciences* appeared. The glossary of physical terms will provide information concerning their handling in this translation. But the best efforts of a translator cannot prevent the intrusion of anachronistic ideas suggested by modern English words. All that can be done is to make this danger known, "as we erect a beacon to denote the presence of a shoal that we cannot remove," in the striking metaphor of Alexander Bryan Johnson's *Treatise on Language* (1836). Johnson's advice to readers of his pioneering work is also applicable to readers of Galileo's *Two New Sciences*, which was no less pioneering in character:

As, however, the following sheets are the painful elaboration of many years, when my language or positions shall, in a casual perusal, seem absurd (and such cases may be frequent), I request the reader to seek some more creditable interpretation. The best which he can conceive should be assumed to be my intention; as on an escutcheon, when a figure resembles both an eagle and a buzzard, heraldry decides that the bird which is most creditable to the bearer shall be deemed to be the one intended by the blazon.

This is a very good rule for reading any book worth reading at all, and especially a pioneering work. Not only does it assure a fair hearing to the author in return for his pains—and Galileo's pains were plentiful, both those he took to make his ideas clear, and those he received for doing so—but it assures the reader the maximum reward for his trouble. For how can a reader gain more from another's words than by forcing himself to arrive at the best which he can conceive?

Glossary

*English terms used for
Galileo's mathematical expressions*

composition (of a ratio)—the formation of one ratio from
    another by adding its terms and relating their sum to the
    second term; thus $a:b$ becomes $(a+b):b$ "by com-
    position."

compound(ed) ratio—the product of two ratios, in modern
    language, although multiplication of one ratio by another
    is not a Euclidean concept, and is not an easy one to
    define except for purely numerical ratios. To signal this
    special operation, the preposition "from" has been used
    before each component ratio. But where a magnitude
    (rather than a ratio) is said to be compounded from, or of,
    other magnitudes (as, for example, an impetus from or of
    two impetuses), then the preposition chosen by Galileo
    is usually retained.

conversion (of a ratio)—taking, in place of $a:b$, the ratio
    $a:(a-b)$.

division (of a ratio)—the formation of one ratio from another
    by taking its first term in relation to the excess of that term
    over the second; thus $a:b$ becomes $(a-b):b$ "by
    division." (Heath's word is "separation," but Galileo
    calls this operation *divisio*.)

duplicate(d) ratio—the squared ratio, sometimes called
    "doubled ratio," and occasionally even "double ratio,"
    though the latter normally means merely the ratio $2:1$.
    The distinction in terminology (doubled vs. double) was
    not always duly observed even in antiquity. When a
    possibility of confusion occurs in Galileo's text, the
    translation used here is "squared ratio" or "square of
    the ratio."

equal in the square—said of a magnitude whose square is
    equal to the sum of the squares of two other named
    magnitudes. Galileo's phrases are *potentia aequale,
    potentia aequipollet, eguale in potenza*, and the like. What
    is meant is the rule of vector addition. The phrase derives
    from Euclid, Book X, with regard to magnitudes
    "commensurable [only] in the square." Euclid's word

dynamis (which became *potentia* in Latin) applied only to the square, and not to powers in general. Aristotle remarked that "It is by a change in meaning that a 'power' in geometry is so called" (*Metaphysics* 1019b. 33–34). Nevertheless, the idea that irrationals are "potentially" commensurable is not entirely unrelated to potentiality in its important philosophical sense in Aristotelian physics.

equidistance of ratios—when two sets of magnitudes having the same number of terms are in the same proportion, the first and last terms of the first set are in the same ratio as the first and last terms of the second set; this equality of ratio is said to be proportionality "by equidistance of ratios." Thus, if A, B, C, D, ... X are related to a, b, c, d, ... x in such a way that $a:b::A:B$, $b:c::B:C$, and so on, then $a:x::A:X$ "by equidistance of ratios," being equally separated in order. The old expression for this was proportionality *ex aequali*.

inverse ratio—the reciprocal of the ratio referred to. (Heath's expression for this is "alternate ratio.")

inverse proportion(ality)—If $a:b::c:d$, then $a:b$ is said to be inversely proportional to $d:c$. The relationship is variously expressed by Galileo, usually by applying to the description of a proportionality the words "taken in inverse (or contrary) order," this phrase applying only to the second ratio named. The concept, perfectly familiar today, is not Euclidean, though it was used by Archimedes, Ptolemy, Pappus, and many early writers on mechanics, usually in clumsy expressions because the idea is as unnatural in geometry as it is convenient in physics.

mean proportional—synonymous with geometric mean; as we think of it, the square root of the product of the extremes in a numerical proportionality. Galileo often writes simply "mean"; in order to avoid confusion with the arithmetic mean (half the sum of two terms), the expression is completed here without square brackets.

number—multitude of units. The "least" number for Aristotle and Euclid was two, since one is not a multitude. What we call "the number one" was then often referred to as "unity." When Galileo suggested unity as in a sense "the infinite number," he was neither contradicting any current notion, nor indulging in mysticism; he was in effect suggesting a place in the number system for something that at the time was not included in it.

permutation (of a proportionality)—interchange of mean terms, strictly permissible only when all four magnitudes are of the same kind. Not a Euclidean expression. (Heath uses the word "alternately" in Euclid V. 16, but no corresponding word occurs in the Greek text, and to me Heath's choice would be appropriate only if we also used his choice for "inverse," q.v.)

perturbed equidistance of ratios—when two sets of three magnitudes each happen to be such that $a:b::B:C$ and $b:c::A:B$, then $a:c::A:C$ "by perturbed equidistance"; cf. "equidistance of ratios," above.

ratio—a quantitative relationship between two magnitudes of the same kind, of which either can be made to exceed the other by multiplication. In Galileo's day, a ratio was never confused with the numbers by which it is expressed, or with a fraction; still less, with any quantity or magnitude. A ratio was strictly a *relation* of two magnitudes. We ignore the ancient distinctions because our concept of "number" (q.v.) is utterly different from Euclid's; our real number system includes irrationals and transcendentals, though those are certainly not "multitudes of units" any more than zero, or one, is a multitude of units.

triplicate(d) ratio—the cube of the ratio named; cf. "duplicate(d) ratio," above.

Galileo's physical terms and their English translations

braccio—a measure of length, here left untranslated. Pronounced bráh-cho; plural *braccia*, bráh-cha. Literally, an arm; the Florentine braccio of Galileo's time was 58.4 cm., or about an inch less than two feet. "Foot" is used here for a half-braccio. Other measures are rendered in English by using familiar approximate equivalents.

conficere—traverse. In classical Latin, this word of Galileo's for moving through a distance did not have that sense, but rather the sense of "make" or "complete" or (by transference) of "diminish," because the action of a thing often reduces or destroys it. *Conficere* was not in common use for motion at Galileo's time, though it is found in some of his earliest notes on motion. There, as in his published books, it alternates with *peragere; conficere*

seems to be relatively more frequent in his later fragments
on motion. In order that the reader may know which term
Galileo chose in each instance for the idea of traversing,
peragere is translated as "run through," having perhaps
a slightly more active sense. Other words, in more
common usage at the time, are *transire* and *pertransire*,
though these are seldom found in the *Two New Sciences*.
When they do occur, they are translated as "go through"
and "pass through." All the above words are associated
with a distance or space; when the subject is motion
itself, or time, Galileo's usual word is *absolvere*, translated
here as "finish," or, when related to space, "cover."

equabile—equable. This is synonymous with "uniform," the
word commonly used by medieval as by modern physicists. For some reason Galileo seldom employed
"uniform" except conjoined with, and as a further
explanation of, "equable." When "uniform" occurs in
the text, it is literally translated here.

gravità—heaviness. This means the tendency of a body having
weight (*peso*) to move downwards. Heaviness, the
tendency, was not regarded as identical with weight, the
property, though it was measured thereby. Sometimes
both words appear in the same sentence. Occasionally
gravità, which persists during free fall, is distinguished
grom *peso*, which disappears during free fall, in somewhat the way we now distinguish mass from weight,
allowing for the fact that Galileo's physics deals only
with bodies near the earth's surface. When the plural is
used, *gravità* is translated "weights" for convenience of
reading. It is translated "gravity" only when opposed to
"levity," a quality imputed by Aristotle (but rejected by
Galileo) to things that seem to go naturally upward.

impeto—impetus. This word is usually treated by Galileo as
freely interchangeable with *momento*, discussed below. In
one instance Galileo speaks of the impetus of a weight
that is merely laid on a stake, suggesting that impetus
could be considered as existing in virtual as well as in
actual motion. Impetus is sometimes treated as if synonymous with "speed," but only when two motions of the
same body are being compared, so that the weight
component of *momento* is the same in both cases and can
be neglected. In the earlier *Dialogue*, impetus was spoken
of as "impressed force," but Salviati refuses to endorse

this usage in the present book. Impetus here, like force, is more an effect than a cause of motion. It is usually not distinguished from the motion itself when the motion in question is that of a heavy body.

infiniti—infinitely many. In a few instances, "infinitely great," when the context refers to magnitude rather than quantity.

latio—movement. This translation enables the reader to know when Galileo has departed from the much more usual *motus*, "motion," which is synonymous. The word is uncommon in this sense, but occurs in the first published Latin translation (1544) of Archimedes' *On Spiral Lines*, as pointed out to me by Dr. Winifred Wisan.

mobile—moveable. This unattractive English noun, so spelled here in order to distinguish it from the adjective "movable" which also occur sometimes, is in my opinion required because of technical and philosophical implications that would be introduced by the free translation, "movable body." That term was used by Albertus Magnus, whose pupil Thomas Aquinas disputed its propriety and preferred "movable entity," because the nature of body is not the proper subject of physics, but of metaphysics. It was Descartes, and not Galileo, who introduced the word body (*cor(p)s, corpus*) into modern physics. In earlier physics, the moveable (*mobile*) was always distinguished from the mover (*movens, motor*), and "moving body" as a translation of *mobile* might imply to the modern reader a body that causes another to move. Galileo's "moveable" is always to be thought of as a tangible heavy object near the earth's surface. It is, moreover, a thing that is acted on rather than one that acts, unless the context shows it to be both.

momento—moment (pl. moments) or momentum (pl. momenta.) The latter translation is used when the context implies motion. Static moment is conceived by Galileo as the effective downward tendency of a weight acting through a lever arm, and he treats it as the product of weight and distance, in that any change in one is exactly compensated by an inversely proportional change in the other. Momentum, on the other hand, is the combined tendency of weight and speed; near the earth, an equivalent to our normal concept of momentum expressed as *mv*. Galileo also uses the phrase *momento di* . . . , translated "moment of" and meaning roughly "effectiveness of" speed, or heaviness, or the like—an idea sometimes also conveyed by *grado di . . .*, or "degree of."

mutatio—displacement. This term is infrequently used by Galileo, and seems not to have been meant as essentially different from *latio* or *motus*. In Aristotle, it distinguished overall change between *termini* from actual motion through the intervening interval, whence a "mutation" could be truly instantaneous, whereas motion could not. The translation "displacement" enables the reader to tell when this word was used; "mutation" was avoided because of its ordinary English connotation of change of quality.

parti non quante—unquantifiable parts. Cf. *parti quante*, below. Salusbury and Weston used the seemingly more logical term, "unquantified parts," but this creates confusion with a third classification that Galileo uses: "parts neither quantified nor unquantifiable, but corresponding to any assigned number"—that is, not infinitely numerous, but indefinitely many. The essential characteristic of "unquantifiable parts" is that they are uncountable and are devoid of size; they can exist *only* in infinite aggregates, which aggregates are necessarily of finite size, yet not of unique size. As individual parts, or in aggregates necessarily finite in number (if such a thing were conceivable), unquantifiable parts would have no size and would in every way be equivalent to mathematical points, except as to a physical distinction that Galileo occasionally makes between "filled" and "void" points.

parti quante—quantified parts. This is the expression adopted by both Salusbury and Weston. The concept is a technical one, and it deserves a recognizably technical term in translation. Quantified parts are capable of being counted; hence they must have dimensions and cannot be mathematical points. Usually it is the idea of countability that is emphasized by Galileo, so that *quante* has chiefly the sense of "so many." Sometimes, however, the idea of size is emphasized, and then *quante* has the sense of "so big." Quantified parts are always capable of being divided, and hence they are not in general to be identified with the *minima naturalia* so important in debates of medieval Aristotelians. *Minima naturalia* would make up that special class of *parti quante* which happened to be incapable of division for physical, rather than for mathematical, reasons. Galileo appears to accept the existence in nature of such particles (atoms), incapable of physical division without transformation into something else (such as fire, or light); but he does not discuss this concept in detail in this book.

resistente—resistent. This spelling distinguishes the noun from the adjective "resistant." It is applied to the medium, or to some other body offering resistance to the motion of the moveable under discussion.

vacuo—void. The translation "vacuum" would probably be misleading even in those passages in which the exclusion of air alone is meant, and even though Galileo does occasionally use the word *voto*, which means literally "void," and is so translated. Likewise, *vacuo* when used adjectivally has been translated as "void" rather than as "empty," because the question debated is always the possibility of void spaces in nature existing in the normal physical world, and not as the result of force or artifice. Two kinds of void were customarily debated, distinguished by their sizes. Most writers felt less repugnance against interstitial voids than against macroscopic voids. Aristotle rejected both; but, as Galileo pointed out, Aristotle's main argument had been directed against any void within which some motion was conceivable. For that reason, Galileo implied in at least one place that his dimensionless interstitial point-voids might not have been rejected even by Aristotle.

velocità—speed. The scalar quantity, without regard to direction. In the Fourth Day, Galileo does give rules for vector addition, but there speaks of *impeto* rather than of *velocità*.

Short Titles Used in Footnotes

Assayer	Galileo, *The Assayer*, in S. Drake and C. D. O'Malley, *The Controversy on the Comets of 1618*. Philadelphia, 1960.
Bodies in Water	Galileo, *Discourse on Bodies in Water*, tr. T. Salusbury. Urbana, 1960.
Dialogue	Galileo, *Dialogue Concerning the Two Chief World Systems*, tr. S. Drake. Berkeley and Los Angeles, 1953 or 1967.
Mechanics in Italy	S. Drake and I. E. Drabkin, *Mechanics in Sixteenth-Century Italy*. Madison, 1969.
On Mechanics *On Motion*	I. E. Drabkin and S. Drake, *Galileo: On Motion* and *On Mechanics*. Madison, 1960.
Opere	*Opere di Galileo Galilei*, ed. A. Favaro. Edizione Nazionale. Florence, 1890–1910, or reprint editions (see item 14, in Bibliography). Citations without volume number refer to Vol. VIII, for which paginations are shown in the present volume.

Two New Sciences

DISCORSI
E
DIMOSTRAZIONI
MATEMATICHE,

intorno à due nuoue scienze

Attenenti alla

MECANICA & I MOVIMENTI LOCALI,

del Signor

GALILEO GALILEI LINCEO,

Filofofo e Matematico primario del Sereniſſimo
Grand Duca di Tofcana.

Con vna Appendice del centro di grauità d'alcuni Solidi.

IN LEIDA,

Appreſſo gli Elfevirii. M. D. C. XXXVIII.

Galileo Galilei

Lincean Academician

Chief Philosopher and Mathematician to the
Most Serene Grand Duke of Tuscany

Discourses
&
Mathematical Demonstrations
Concerning

Two New Sciences

Pertaining to
Mechanics & Local Motions

*With an Appendix
On Centers of Gravity of Solids*

Leyden

At the Elzevirs, 1638

*   *   *

To which is added a further dialogue
On the Force of Percussion

To the very illustrious nobleman,
my Lord the

Count de Noailles

Councilor to his Most Christian Majesty;

Knight of the Holy Ghost;

Field Marshal of the Armies; Sensechal and

Governor of Rovergue; His Majesty's Lieutenant at

Auvergne; my lord and supreme patron

Most illustrious Sir:

I recognize as resulting from your excellency's magnanimity the disposition you have been pleased to make of this work of mine, notwithstanding the fact that I myself, as you know, being confused and dismayed by the ill fortune of my other works, had resolved not to put before the public any more of my labors. Yet in order that they might not remain completely buried, I was persuaded to leave a manuscript copy in some place, that it might be known at least to those who understand the subjects of which I treat. And thus having chosen, as the best and loftiest such place, to put this into your excellency's hands, I felt certain that you, out of your special affection for me, would take to heart the preservation of my studies and labors. Hence, during your passage through this place on your return from your Roman embassy, when I was privileged to greet you in person (as I had so often greeted you before by letters), I had occasion to present to you the copy that I then had ready of these two works. You benignly showed yourself very much pleased to have them, to be willing to keep them securely, and by sharing them in France with any friend of yours who is apt in these sciences, to show that although I remain silent, I do not therefore pass my life in entire idleness.

I was later preparing some other copies to send to Germany, Flanders, England, Spain, and perhaps also to some place in Italy, when I was notified by the Elzevirs that they had these works of mine in press, and that I must therefore decide about the dedication and send them promptly my thought on that subject. From this unexpected and astonishing news, I concluded that it had been your excellency's wish to elevate and spread my name, by sharing various of my writings, that accounted for their having come into the hands of those printers who, being engaged in the publication of other works of mine, wished to honor me by bringing these also to light at their handsome and elaborate press. Thus these writings of mine are to be revived through their having had the good fortune to fall under the award of so great a judge. In that marvelous combination of many virtues that render your excellency admirable to all, with incomparable magnanimity and out of zeal for the public good, to which it seemed to you these writings of mine should contribute, you have desired to widen the limits and boundaries of their honor.

Now that matters have arrived at this stage, it is certainly reasonable that, in some conspicuous way, I should show myself grateful by recognizing your excellency's generous affection. For it is you who have thought to increase my fame by having these works spread their wings freely under an open sky, when it appeared to me that my reputation must surely remain confined within narrower spaces. Hence to your name, illustrious Lord, it is right that I dedicate and consecrate this offspring of mine. To this action I am impelled not only by the accumulation of my obligations, but by self-interest as well, for if I may be permitted to say so, you are now obliged to defend my reputation against anyone who attacks it, you having entered me in the lists against all adversaries. Wherefore, advancing under your banner and your protection, I humbly make obeisance to you, and wish you, as the reward of these graces, the summit of all happiness and greatness. From Arcetri, the sixth of March 1638.

From your Excellency's

Most devoted servitor
GALILEO GALILEI

Civil life being maintained through the mutual and growing aid of men to one another, and this end being served principally by the employment of arts and sciences, their inventors have always been held in great esteem and much revered by wise antiquity; and the more excellent or useful an invention has been, the greater the praise and honor given to its inventors, even to the point of deifying them, mankind having by common consent wished to perpetuate the memory of the authors of their well-being through that sign of supreme honor. Similarly, those are worthy of great praise and admiration who by the acuity of their minds have improved things previously discovered, revealing the fallacies and errors of many propositions put forth by distinguished men and received as truth for many ages. For such exposure is praiseworthy even if the discoverers themselves have but removed something false without introducing the truth, which is hard to acquire. Thus the prince of orators declares: "Oh, that we could get at truth as easily as we refute falsehood!"[1]

And indeed such praise has been earned by these last centuries of ours in which the arts and sciences discovered by the ancients, through the work of perspicacious talents and by many tests and proofs, have been brought to great and ever-increasing perfection. This is particularly evident in the mathematical sciences, in which (omitting many others who have worked in them with great success) our Signor Galileo Galilei, Lincean Academician, has earned the highest place rightly and beyond any doubt, and with the applause and **46** approval of all experts, both by his having shown the inconclusiveness of many arguments concerning various conclusions, confirming this through sound demonstrations with which his previously published works are filled, and also through things discovered by means of the telescope—which device first appeared in these our lands, but was brought to much higher perfection by him. For he gave us news before anyone else of those four companions of Jupiter, of the true and certain nature of the Milky Way, of sunspots, of the rough surface and dark spots of the moon, of three-bodied Saturn,

---

1. Cicero, *De natura deorum* I.91.

hornèd Venus, and of the nature and location of comets—all of these being things never known by ancient astronomers or philosophers, so that it may be said that he has restored astronomy and presented it to the world in a new light. Inasmuch as it is in the skies and heavenly bodies that the power, wisdom, and goodness of the supreme Creator appear more evidently and admirably than in the rest of created things, all this enhances the greatness and merit of a man who has opened up this knowledge and rendered such bodies distinctly visible despite their great and almost infinite distance from us. For it is commonly said that seeing teaches more in a single day, and with greater certainty, than can instruction however many times repeated. And as another says, intuitive knowledge is on a level with definition.

The grace conceded to this man by God and nature (though only through many labors and vigils) is still more evident in the present work, wherein he is seen to be the discoverer of two whole new sciences, which he has conclusively—that is, geometrically—demonstrated from their first principles and foundations. What renders this work even more remarkable is that one of these two sciences concerns an age-old subject, among the most important in nature, which has been the subject of speculation by all great philosophers, and upon which many volumes have been written. I speak of local motion, a matter containing an infinitude of wonderful properties, none of which has been previously discovered or demonstrated by anyone. The other science that he has demonstrated concerns the resistance which solid bodies make against separation by force, a subject of great utility, especially in the mechanical arts and sciences, and likewise full of phenomena and theorems not previously noticed.

Of these two new sciences, full of propositions that will be boundlessly increased in the course of time by ingenious theorists, the outer gates are opened in this book, wherein with many demonstrated propositions the way and path is shown to an infinitude of others, as men of understanding will easily see and acknowledge.

Table of the Principal Matters That Are Treated in the Present Work[1]

1. This table of contents reversing the essential content of the two first days, was prepared by the Elzevirs.
2. Sometimes called the Sixth Day, this incomplete dialogue was first published in 1718, as part of the second collected edition of Galileo's works. A so-called Fifth Day, first published by Vincenzio Viviani (1622–1703) in 1674, does not belong to this book.

First Day

*Interlocutors: Salviati, Sagredo
and Simplicio*

Salviati. Frequent experience of your famous arsenal, my Venetian friends, seems to me to open a large field to speculative minds for philosophizing, and particularly in that area which is called mechanics, inasmuch as every sort of instrument and machine is continually put in operation there. And among its great number of artisans there must be some who, through observations handed down by their predecessors as well as those which they attentively and continually make for themselves, are truly expert and whose reasoning is of the finest.

Sagredo. You are quite right. And since I am by nature curious, I frequent the place for my own diversion and to watch the activity of those whom we call "key men" [*Proti*] by reason of a certain preëminence that they have over the rest of the workmen. Talking with them has helped me many times in the investigation of the reason for effects that are not only remarkable, but also abstruse, and almost unthinkable. Indeed, I have sometimes been thrown into confusion and have despaired of understanding how some things can happen that are shown to be true by my own eyes, things remote from any conception of mine. Nevertheless, what we were told a little while ago by that venerable workman is something commonly said and believed, despite which I hold it to be completely idle, as are many other things that come from the lips of persons of little learning, put forth, I believe, just to show they can say something **50** concerning that which they don't understand.

Salv. You mean, perhaps, that last remark that he offered when we were trying to comprehend the reason why they make the sustaining apparatus, supports, blocks, and other strengthening devices so much larger around that huge galley that is about to be launched than around smaller vessels. He replied that this is done in order to avoid the peril of its splitting under the weight of its own vast bulk, a trouble to which smaller boats are not subject.

11

Sagr. I mean that, and particularly the finishing touch that he added, which I have always considered to be an idle notion of the common people. This is that in these and similar frameworks one cannot reason from the small to the large, because many mechanical devices succeed on a small scale that cannot exist in great size. Now, all reasonings about mechanics have their foundations in geometry, in which I do not see that largeness and smallness make large circles, triangles, cylinders, cones, or any other figures [or] solids subject to properties different from those of small ones; hence if the large scaffolding is built with every member proportional to its counterpart in the smaller one, and if the smaller is sound and stable under the use for which it is designed, I fail to see why the larger should not also be proof against adverse and destructive shocks that it may encounter.

Salv. The common notion is indeed an idle one, so much so that with equal truth its contrary may be asserted; one may say that many machines can be made to work more perfectly on a large scale than on a small one. For example, take a clock that is both to show the hours and to strike; one of a certain size will run more accurately than any smaller one. The common idea is adopted on better grounds by some persons of good understanding when, to explain the occurrence in large machines of effects not in agreement with pure and abstract geometrical demonstrations, they assign the cause of this to the imperfection of matter, which is subject to many variations and defects.

51   Here I do not know whether I can declare, without risking reproach for arrogance, that even recourse to imperfections of matter, capable of contaminating the purest mathematical demonstrations, still does not suffice to excuse the misbehavior of machines in the concrete as compared with their abstract ideal counterparts. Nevertheless I do say just that, and I affirm that abstracting all imperfections of matter, and assuming it to be quite perfect and inalterable and free from all accidental change, still the mere fact that it is material makes the larger framework, fabricated from the same material and in the same proportions as the smaller, correspond in every way to it except in strength and resistance against violent shocks [*invasioni*]; and the larger the structure is, the weaker in proportion it will be. And since I am assuming matter to be inalterable—that is, always the same—it is evident that for this [condition] as for any

other eternal and necessary property, purely mathematical demonstrations can be produced that are no less rigorous than any others.

Therefore, Sagredo, give up this opinion you have held, perhaps along with many other people who have studied mechanics, that machines and structures composed of the same materials and having exactly the same proportions among their parts must be equally (or rather, proportionally) disposed to resist (or yield to) external forces and blows [*impeti*]. For it can be demonstrated geometrically that the larger ones are always proportionately less resistant than the smaller. And finally, not only artificial machines and structures, but natural ones as well, have limits necessarily placed on them beyond which neither art nor nature can go while maintaining always the same proportions and the same material.

Sagr. Already I feel my brain reeling, and like a cloud suddenly cleft by lightning, it is troubled. First a sudden and unfamiliar light beckons to me from afar, and then immediately my mind becomes confused, and hides its strange and undigested fancies.

From what you have said, it seems to me, must follow the impossibility of constructing two similar and unequal structures of the same material that would have proportionate resistance. But if that is so, it will be impossible even to find two sticks of the same wood that differ in size and are nevertheless similar in strength and stability.

52

Salv. So it is, Sagredo. And the better to make sure that we both have the same idea, I say that if we shape a wooden rod to a length and thickness that will fit into a wall at right angles, horizontally, and the rod is of the greatest length that can support itself, so that if it were a hairbreadth longer, it would break of its own weight, then that rod will be absolutely unique [in shape and size]. For example, if its length is one hundred times its thickness, then no different rod of the same material can be found which has, like this, a length one hundred times its thickness, and is just able to sustain its own weight and no more; for longer bars will break, and shorter ones will be able to sustain something more than their own weights. And what I have said about the state of self-support, assume to be said about any other constituents [*constituzione*]; thus if a scantling can bear the weight of ten like scantlings, a [geometrically] similar beam will by no means be able to bear the weight of ten like beams.

Here you and Simplicio must note how conclusions that are true may seem improbable at a first glance, and yet when only some small thing is pointed out, they cast off their concealing cloaks and, thus naked and simple, gladly show off their secrets. For who does not see that a horse falling from a height of three or four braccia will break its bones, while a dog falling from the same height, or a cat from eight or ten, or even more, will suffer no harm? Thus a cricket might fall without damage from a tower, or an ant from the moon. Small children remain unhurt in falls that would break the legs, or the heads, of their elders. And just as smaller animals are proportionately stronger or more robust than larger ones, so smaller plants will sustain themselves better. I think you both know that if an oak were two hundred feet high, it could not support branches spread out similarly to those of an oak of average size. Only by a miracle could **53** nature form a horse the size of twenty horses, or a giant ten times the height of a man—unless she greatly altered the proportions of the members, especially those of the skeleton, thickening the bones far beyond their ordinary symmetry.

Similarly, to believe that in artificial machines the large and small are equally practicable and durable is a manifest error. Thus, for example, small spires, little columns, and other solid shapes can be safely extended or heightened without risk of breaking them, whereas very large ones will go to pieces at any adverse accident, or for no more cause than that of their own weight.

Here I must tell you of a case really worth hearing about, as are all events beyond expectation, especially when some precaution taken to prevent trouble turns out to be a powerful cause thereof. A very large column of marble was laid down, and its two ends were rested on sections of a beam. After some time had elapsed, it occurred to a mechanic that in order to insure against its breaking of its own weight in the middle, it would be wise to place a third similar support there as well. This suggestion seemed opportune to most people, but the result showed quite the contrary. Not many months passed before the column was found cracked and broken, directly over the new support at the center.

Simp. A truly remarkable event, and most unexpected, if indeed this was due to the addition of the new support in the middle.

Salv. It surely did result from that, and to recognize the cause of the effect removes the marvel of it. For the two pieces of the column being placed flat on the ground, it was seen that the beam-section on which one end had been supported had rotted and settled over a long period of time, while the support at the middle remained solid and strong. This had caused one half of the column to remain suspended in the air; and, abandoned by the support at the other end, its excessive weight made it do what it would not have done had it been supported only on the two original [beams], for if one of them had settled, the column would simply have gone along with it. And doubtless no such accident would have happened to a small column of the same stone, if its length bore to its thickness the same ratio as that of   **54** the length to the thickness of the large column.

Sagr. Thus far I am convinced of the truth of the effect, but stop short of the reason why any material, in becoming larger, should not by that very accumulation [of size] multiply its resistance and its strength. I am the more puzzled by seeing other cases in which there is a much greater increase in hardiness and resistance to rupture than there is in size of material. For example, if two nails are driven into a wall, and one is twice as thick as the other, it will hold not only twice the weight, but three or four times as much.

Salv. Say eight times, and you will not be far from the truth. But this effect is not contrary to that other, although superficially it seems to be.

Sagr. Then smooth out for us these rough spots, Salviati, and clear up these obscurities, if you have any way of doing so, for indeed I am beginning to think that this subject of resistance is a field full of beautiful and useful considerations. And if you are willing that it be made the subject of our discussions today, that will be most welcome to me, and I believe to Simplicio.

Salv. I cannot refuse to be of service, provided that memory serves me in bringing back what I once learned from our Academician [Galileo] who made many speculations about this subject, all geometrically demonstrated, according to his custom, in such a way that not without reason this could be called a new science. For though some of the conclusions have been noted by others, and first of all by Aristotle, those are not the prettiest; and what is more important, they were not proved by necessary demonstrations from their primary and unquestionable foundations.

Since, as I say, I want to prove these to you demonstratively, and not just persuade you of them by probable arguments, I assume that you have that knowledge of the basic mechanical conclusions that have been treated by others up to the present which will be necessary for our purpose.

First of all, we must consider what effect is at work in the breaking of a stick, or of some other solid whose parts are firmly attached together; for this is the primary concept, and it contains the first simple principle that must be assumed as known. To clarify this, let us draw the cylinder or prism **55** *AB*, of wood or other solid and coherent material, fastened above at *A*, and hanging plumb; at the other end, *B*, let the weight *C* be attached. It is manifest that whatever may be the tenacity and mutual coherence of the parts of this solid, provided only that that is not infinite[ly strong], it can be overcome by the force of the pulling weight *C*, of which the heaviness [*gravità*] can be increased as much as we please, and that this solid will finally break, just like a rope. And just as we understand that the resistance of a rope is derived from the multitude of hempen fibers that compose it, so in wood there are seen fibers and filaments stretched out length-wise which render it even more resistant to breakage than hemp of the same length would be. In a stone or metal cylinder, the coherence of parts seems still greater, and depends on some other cement than that of filaments or fibers. Yet even these [cylinders] are broken by a sufficient pull.

Simp. If this business proceeds as you say, I understand how the filaments in wood, which are as long as the wood itself, can render it strong and resistant to the great force that is applied to break it. But how can a rope, composed of threads of hemp no longer than two or three braccia each, be made one hundred braccia long and still remain so strong? I should also like to hear your opinion concerning that attachment between the parts of metals, stones, and other materials devoid of such filaments but nevertheless, if I am not mistaken, still more tenacious.

Salv. It would be necessary to diverge to new speculations, not very relevant to our purpose, if we wanted to find the solutions of the difficulties mentioned.

Sagr. If digressions can bring us knowledge of new truths, why should they trouble us? We are not committed to any closed and concise method, but meet only for our own pleasure. If we digress now, it is in order not to lose information; who knows, if we let this occasion pass, that we shall meet with it again some other time? In fact, how do we know that **56** we shall not discover curious things that are more interesting than the answers we originally sought? Hence I beg you to give Simplicio satisfaction; nor am I less curious about this than he is, or less desirous of knowing what that cement may be that so tenaciously holds together the parts of solids, which are nevertheless ultimately sundered. Moreover, this knowledge is necessary for an understanding of the coherence between the parts of those very filaments of which some solids are composed.

Salv. I am here to be of service, so let it be as you please. The first difficulty is how the filaments of a rope one hundred braccia long, each of these extending no more than two or three braccia, can be so solidly connected together that great force [*violenza*] is needed to part them. Well, Simplicio, tell me: can you not hold one end of a single thread of hemp between your fingers to tightly that I, pulling at the other end, will break it before freeing it from your hand? Surely you can. Now, if the threads of hemp were not just held strongly at one end, but were tightly held throughout their lengths by the threads surrounding them, is it not evident that it would be much harder for a person who pulled them to tear them away from one another than to break them? But the very act of twisting [in making] rope binds the threads mutually in such a way that later, when the rope is pulled with great force, its filaments will break rather than separate from one another. This is manifestly known by seeing that the filaments at the broken ends are very short, and not one braccio or more in length, as would be seen if the parting of the rope were made not by a breaking of its filaments, but by their mere separation one from another, and their slipping.

Sagr. In confirmation of this it may be added that sometimes rope is broken not by pulling it lengthwise, but merely by excessive twisting of it. This seems to me to argue conclusively that its threads have been mutually compressed among themselves in such a way that those pressing do not permit those pressed to move even that little bit that would allow the [outer] turns to stretch sufficiently to encircle the rope, which in being twisted is shortened and consequently somewhat thickened.

Salv. Right you are; and now see how one truth draws
57 another in its train. That tightly held thread, which does not
obey the person who pulls on it with some force and tries
to draw it out from between the fingers, resists because it
is retained by a double compression, for the upper finger
presses against it no less than the lower, the one pressing
against the other. Doubtless if those two pressures could be
separated, one alone would produce one-half the resistance
that depends on the two conjoined. But since we cannot,
by raising the upper finger for example, take away its pressure
without also removing the rest, we need some new device
to preserve [just] one pressure, finding a way in which the
thread shall press itself against the finger or some other
solid body on which it is situated. We have to arrange things
so that the same force, pulling to free the filament, presses
it only the harder, the more strongly it pulls; and this is
done by winding the thread spirally around the solid.

For better understanding, I shall draw a diagram. Let *AB*
and *CD* be two cylinders, between which is the thread *EF*,
which for greater clarity we shall draw as a small cord.
There is no doubt that if the two cylinders are pressed strongly
one against the other, the cord *FE*, pulled at end *F*, will
withstand considerable force before it will move between
the two pressing solids, though if we remove one of these
while the cord continues to touch the other, it will not be
kept by that [single] contact from running freely. But if

we hold it lightly against the top of cylinder *A* and wind
it round in the spiral *AFLOTR*, then when we pull this by
the end *R*, it will obviously begin to bind on the cylinder.
If the turns of the spiral are numerous, and we pull hard,
the cord will be always more compressed against the cylinder;
and if the contact is extended by multiplying the [turns of
the] spiral, it will be less capable of being overcome, and it
will be harder and harder to move the cord in compliance
58 with the pulling force. Now who does not see that such is
the resistance of those filaments which, together with thousands of like windings, make up the thick rope? Indeed,
such binding by twisting cements things so tenaciously that
from a few rushes, and not very long ones, woven with but
few turns, very strong cord is made that I believe is called
pack twine.

Sagr. As a result of your train of reasoning, my mind
pauses at the marvels of two effects for which the reasons

were never before well understood by me. One is the effect
of two or three turns of hemp around the drum of a winch;
this will not only hold firmly, but will not give way to slipping
even though pulled with immense force by a weight it sustains.
Moreover, the winch being turned, its drum can lift and draw
up great stones with successive revolutions, merely by the
contact of the rope that binds the drum; and the arms of
a mere boy can hold that rope and draw in the slack.

The other effect is that of a simple but clever device invented
by a young kinsman of mine, to enable him to descend from
a window by means of a rope without cruelly cutting the
palms of his hands, as had happened a short time before to
his considerable injury. For easy understanding I shall
make a little sketch. Around a wooden cylinder AB, as
thick as a cane and four inches long, he carved a spiral channel
of one and one-half turns, no more, just wide enough to
fit the cord he wished to use; this entered the channel at
A and emerged at B. Then he enclosed the cylinder and cord
with a tube of wood or sheet metal, slit lengthwise and hinged
so that it could be freely opened and closed. Holding this
tube and pressing with both hands, the cord having been
tied to a fixed object above, he hung by his arms, and with
pressure on the cord between tube and cylinder he could
at will hold himself without dropping, by pressing his hands
strongly, or by relaxing his grip somewhat, drop slowly
at his pleasure.

Salv. Truly an ingenious invention, though for a complete
explanation of its nature I can already see dimly that some   **59**
additional theory needs to be added. But I do not wish now
to digress on this subject, especially since you want to hear
my thoughts about resistance to breakage on the part of
other bodies, whose texture is not of filaments, as is that of
ropes and most kinds of wood, but whose parts cohere by
reason of other causes. These, in my opinion, may be reduced
to two kinds, one of which is the celebrated repugnance
that nature has against allowing a void to exist. The other,
when this of the void is deemed insufficient, requires the
introduction of some sticky, viscous, or gluey substance
that shall tenaciously connect the particles of which the body
is composed.

I shall speak first of the void, showing by clear experiences
the nature and extent of its force. To begin with, we may see
whenever we wish that two slabs of marble, metal, or glass,

exquisitely smoothed, cleaned, and polished and placed one
on the other, move effortlessly by sliding, a sure argument
that nothing gluey joins them. But if we want to separate
them while keeping them parallel, we meet with resistance;
for the upper slab in being raised draws the other with it,
and holds it permanently even if it is large and heavy. This
clearly shows nature's horror at being forced to allow, even
for a brief time, the void space that must exist between the
slabs before the running together of parts of the surrounding
air shall occupy and fill that space. It is also observed that
if the two surfaces are not perfectly clean, so that their contact
is not everywhere perfect, and we want to separate them
slowly, the only resistance we feel is that of the heaviness
[of the upper slab], whereas in rapid separation the lower
stone is also lifted and immediately falls back, following
the upper one only during the brief time that suffices for the
expansion [*distrazzione*] of the small amount of air between
the imperfectly fitting surfaces, and for entrance of surrounding air. Doubtless the resistance that is so sensibly
perceived between the two surfaces likewise resides between
the parts of a solid, and enters into their attachment at least
to some extent, and as a concomitant cause.

    Sagr. Please pause here, and allow me to mention a certain
60 idea that has just now come to mind. Seeing the lower slab
follow the upper when that is lifted with swift motion assures
us that motion in the void would not be instantaneous
despite the opinion of many philosophers, and perhaps of
Aristotle himself.[1] For if it were, the two surfaces would
be separated without any resistance whatever, the same
instant of time sufficing for their separation and for the
running together of the surrounding air to fill the void that
might [otherwise] remain between them. Thus, from the
following of the upper slab by the lower, it is deduced that
motion in a void would not be instantaneous. It is then further
deduced that some void indeed does remain between the
surfaces, at least for a very brief time; that is, for as long as
the time consumed by the ambient air in running to fill this
void. For if no void existed there, neither would there be
any need on the part of the ambient air of running together,

---

1. Aristotle had argued against the existence of a void by a *reductio ad
absurdum* which invoked speed of motion and the principle that there can
be no ratio between the finite and the non-existent or the infinite. Cf. *Physica*,
215a.24–216a.26, and see note 26, below.

or of any other motion. Hence we must say that by force (or contrary to nature) a void is sometimes to be admitted—though in my opinion nothing is contrary to nature save the impossible, and that never happens.

But here another difficulty arises, and this is that although experience assures me of the truth of the conclusion, my mind is still not entirely satisfied about the cause to which the effect is to be attributed. For the effect of separating the two surfaces occurs prior to the [existence of this] void, which consequently follows the separation. Now, it seems to me that the cause should precede the effect, in time at least, if not in physical existence [*natura*]; also, that for a positive effect, there should be a positive cause. Hence I cannot see how the cause of adherence of the two slabs and their repugnance to being separated—effects that are actual—can be a void that does not exist [first], but which must follow. And there can be no action by things that do not exist, according to the definite statement of the Philosopher.[2]

Simp. Since you concede this axiom to Aristotle, I don't think you will ignore another that is elegant and true; namely, that nature does not undertake to do that which refuses [*repugna*] to be done; from this pronouncement it seems to me that there follows a solution of your problem.[3] Since void space is self-refusing, nature prohibits any action in consequence of which a void would follow, and such is the separation of the two surfaces.

Sagr. Well, assuming that what Simplicio adduces is an **61** adequate resolution of my doubt, it seems to me that this same refusal of a void should be sufficient to hold together the parts of a stone or metal solid, or of things [even] more firmly joined and resistant to division, should any exist. If for one effect there is only one cause, as I have always understood and believed (or if many are assigned, they are reducible to one), then why won't this one of the void, which surely does exist, suffice also for all resistances [to separation]?

Salv. I do not wish at present to enter into a contest as to whether the void is in itself enough, without any other retainer, to hold united the separable parts of coherent

2. Cf. *Physica* 225a.25 26; *De anima* 217a.17

3. Cf. *De caelo* 311b.33; see also *Dialogue*, p. 19 (*Opere*, VII, 56), where the axiom ascribed by Galileo to Aristotle is the same, but does not exactly agree with the reference cited above. In this sentence, the 1638 edition reads "our problem." Favaro adopted the better reading of the Pieroni MS.

[*consistenti*] bodies. But I will say that the void which fights and is vanquished between two plates is not in itself enough reason for the firm bonding [*collegamento*] of the parts of a solid marble or metal cylinder which, strained [*violentate*] by strong forces pulling them directly, are finally separated and divided. Now, if I can find a way to distinguish this known resistance, that depends on the void, from any other resistance, whatever it may be, that joins with this in strengthening the attachment, and if I make you see that the former one alone is far from sufficient for the whole effect, will you not then grant me that another [resistance] must be introduced?——Help him, Simplicio, since he is hesitating about what to reply.

Simp. Sagredo's hesitation must be for some other reason, there being no room for doubt about such a clear and necessary consequence.

Sagr. You guess right, Simplicio. I was wondering whether, since it takes more than a million in Spanish gold every year to pay the army, something besides small coins must be provided for the soldiers' pay.[4] But go on, Salviati; assume that I grant your argument, and show us how to separate the operation of the void from all other [actions]; then, measuring this, make us see that it is inadequate for the effect we are discussing.

Salv. Your daemon is guiding you. I shall tell you first how to separate the force of the void from other [forces], and then how to measure it. To separate it, let us take some continuous material whose parts lack any resistance to separation other than that of the void. Water has been demonstrated at length, in a certain treatise of our Academician, to be such a material.[5] Thus when a cylinder of water is displaced [within a tube], and in drawing it, a resistance is felt against the detachment of its parts, no other cause can be recognized for this than repugnance to

62

4. Sagredo means to hint that the other (much greater) resistance may turn out in the end to be made up of a myriad of small resistances not differing in kind from that of the void, just as the million of gold is made up of small coins; see further at p. **66** and note 7, below. (References in bold face type are to pagination of Vol. VIII of the *Opere*, given in the margins and running heads of the present text.)

5. Galileo had argued in a previous book that no internal resistance to separation existed in water, as shown by the settling of fine dust from cloudy water. For this and other arguments, see *Bodies in Water*, pp. 40 ff. (*Opere*, IV, 103–8).

a void. In order to make the experiment, I have imagined an artifice which I can better explain by a diagram than by mere words. Consider *CABD* here to be the profile of a cylinder of metal, or better of glass, empty within and very accurately turned, into the hollow of which there enters, with the smoothest contact, a wooden cylinder which can be driven up and down, of profile *EGFH*. This is drilled through the center so that through the hole there passes an iron wire, hooked at end *K*, while the other end, *I*, is broadened out in the shape of a conical screwhead. Things are so arranged that the upper part of the hole through the wood is indented in the form of a conical surface, shaped exactly to receive the conical extremity *I* of the iron [wire] *IK* when pulled down in the direction of *K*. Insert the wood, which we may call the piston [*zaffo*, a stopper] *EH*, in the cylinder-hole *AD*, not so as to reach the upper surface of the cylinder, but to remain two or three inches away. This space is first filled with water, poured in while the vessel is held with its mouth *CD* upward, the piston *EH* then being replaced while the screwhead *I* is kept a little way from the indentation in the wood in order to allow the escape of air pressing against the piston, which will get out through the hole in the wood, this having been drilled a little larger than the stem of the iron *IK*. All the air having escaped, the wire is drawn back again, sealing the piston with its screwhead *I*, and the whole vessel is rotated to bring it with the mouth [*CD*] down.

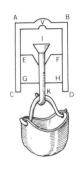

A container is now attached to the hook *K*, into which sand or some other heavy material is put, loading it until finally the upper surface *EF* of the piston is detached from the lower surface of the water, to which nothing held it joined except repugnance to the void. Then, by weighing the piston together with the iron, the container, and whatever it contains, **63** we shall have the amount of the force of the void.

Next, to a marble or glass cylinder of the same size as the cylinder of water we attach a weight which, together with the weight of the marble or glass itself, balances the weight of all the things weighed before. If this breaks the cylinder, we can unquestionably affirm that the void alone is cause enough to hold the parts of the marble or crystal together. But if it is not sufficient, and in order to break [the cylinder] we must add four times the above weight again, then we must say that the void offers one-fifth of the resistance, while the other [resistance] is four times that of the void.

Simp. It cannot be denied that the invention is ingenious; yet I consider it to be subject to many difficulties that leave me in doubt. For who will assure us that air may not penetrate between the glass and the piston, even though this is well packed with tow or some other yielding material? And in order that the cone *I* be well fitted to the hole, it may not be enough to treat the latter with wax or turpentine. Besides, why might not the parts of water expand or rarefy? Why should not air, or exhalations, or other more subtle substances, penetrate through porositites of the wood, or even of the glass itself?

Salv. Simplicio very cleverly arrays his difficulties against us, and in part, as to the penetration of air through the wood or between the wood and the glass, he administers remedies. Beyond this, I note, we can discover for ourselves whether the difficulties advanced are valid, and at the same time acquire new knowledge. First, if it is the nature of water to suffer expansion, though [only] by force, as happens with air, then the piston will be seen to drop. Next, if in the upper part of the glass we make a small protruding indentation, as at *V*, then air or any more tenuous and spiritous material, penetrating through the substance or the porosity of glass or wood, will be seen to collect in the indentation *V*, the water giving way to it. But if those things are not observed, we may be assured that the experiment has been tried with all the proper precautions, and we shall know that water is not expansible, nor glass penetrable by any material however subtle.

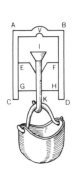

Sagr. Thanks to this reasoning, I find the cause of an effect that has for a long time kept my mind full of marvel and empty of understanding. I once observed a cistern in which a pump had been installed to draw water, perhaps by someone who vainly believed that more water can be drawn [thus], or as much with less labor, than by means of an ordinary bucket. This pump had its piston and valve up above, so that the water was pumped by suction and not by impulsion, as in pumps that have the apparatus down below. As long as the water was up to a certain height in the cistern, this [pump] drew it admirably; but when the water went down below a certain mark, the pump would no longer work. The first time I noticed this event, I thought that the apparatus was worn out; but when I found a master [mechanic] to repair it, he

64

told me that there was nothing at all wrong except the water [level] which, having gone down too far, did not allow itself to be lifted to such a height. He added that neither with pumps nor with any other device that lift water by suction is it possible to make this rise a hairbreadth more than eighteen braccia; whether pumps are [of] large [bore] or small, that is the measure of this absolutely limited height.[6]

Well, up to now I have been so dull-witted that although I understood that a rope, a wooden staff, or an iron rod can be lengthened until its own weight breaks it when attached from above, it never occurred to me that the same thing will happen, and much more easily, with a rope or rod of water. And that which is drawn up in a pump is nothing else than a cylinder of water which, having its attachment above and being lengthened more and more, finally arrives at that boundary beyond which it breaks, just as if it were a rope.

Salv. That is exactly how the matter goes, and since the same height of eighteen braccia is the predetermined limit of height to which any quantity of water can be sustained, whether the [cylinder of the] pump is wide, or narrow, or thin as a straw, then if we weigh the water contained in eighteen braccia of tube, whether broad or narrow, we shall have the value of the resistance of the void for cylinders of any solid material as large as the hollows of those tubes. Having said this, let us show how one may easily find, for all metals, stones, wood, glass, etc., the lengths up to which cylinders, threads, or rods of any thickness may be brought, and beyond which they cannot sustain themselves but will **65** break of their own weight.

Take, for example, a copper wire of any thickness and length, fix one of its end on high, and to the other end add greater and greater weight until it finally breaks. Let the maximum weight that it can sustain be, for example, fifty pounds. It is obvious that fifty pounds of copper, over and above the weight of this wire, say one-eighth of an ounce, drawn into a wire of the same thickness, would be the maximum length of wire that could maintain itself. Next, measure the length of the wire that broke, and let this be one braccio; since this weighed one-eighth of an ounce and sustained itself plus fifty pounds, which is 4,800 times one-eighth of

6. Galileo did not accept the suggestion made to him in 1630 by G. B. Baliani (1582–1666) that failure of siphons and suction pumps above thirty feet should be ascribed to atmospheric pressure; cf. *Opere*, XIV, 158–60.

an ounce, we shall say that all copper wires, of whatever thickness, can sustain themselves up to a length of 4,801 braccia, and no more. A copper rod that is able to sustain itself up to a length of 4,801 braccia encounters a resistance dependent on the void that, in comparison with its other resistances, is as much as the weight of a rod of water eighteen braccia long and as thick as the copper; and if we find, for example, that copper is nine times as heavy as water, then the resistance to breakage of any copper rod, so far as the void is concerned, will be as the weight of two braccia of the same rod. By similar reasoning and procedures we can find the maximum lengths of wires or rods of all solid materials that can sustain themselves, as well as the part played by the void in their resistance.

Sagr. It remains now for you to tell us what it is that the balance of this resistance consists in; that is, what that gluey or viscous thing is that holds the parts of solids attached, in addition to the resistance which derives from the void. I cannot imagine any cement that cannot be burned and consumed in a very hot furnace over a period of two, three, or four months, let alone in ten, or a hundred. Yet silver, gold, or liquefied glass may remain in a furnace that long, and when taken out again and cooled, the parts of these become reunited and attached together as before. Moreover, the difficulty that I feel about the attachment between the parts of the glass, I shall feel about that of the parts of the cement; that is, what it can be that holds them so firmly joined.

66   *Salv.* A little while ago I told you that your daemon was guiding you; now I find myself in the same straits. Seeing clearly that a repugnance to the void is undoubtedly what prevents the separation of two slabs except by great force, and that still more force is required to separate the two parts of the marble or bronze column, I cannot see why this [repugnance to the void] must not likewise exist and be the cause of coherence between smaller parts, right on down to the minimum ultimate [particles] of the same material. And since for any effect there is one unique and true and most potent cause, if I can find no other glue, why should I not try to see whether this cause, the void, already found, may suffice?

Simp. If you have already demonstrated that in the separation of two large pieces of a solid, the resistance of the large void is very small in comparison with that which holds together the minimum particles, then why do you not wish to admit it as certain that the latter [resistance] has a cause quite different from that of the former?

Salv. To this, Sagredo replies that every individual soldier was paid with pennies and farthings collected by general levies, although a million in gold was not enough to pay the whole army.[7] Who knows that there are not other tiny voids operating on the most minute particles, so that the same coinage as that with which the parts are joined is used throughout? I shall tell you what has sometimes passed through my mind on this; I do this not as the true solution, but rather as a kind of fantasy full of undigested things that I subject to your higher reflections. Take what you will from it, and judge the rest as suits you best.

Sometimes, in considering how heat [*fuoco*, fire] goes snaking among the minimum particles of this or that metal, so firmly joined together, and finally separates and disunites them; and how then, the heat departing, they return to reunite with the same tenacity as before, without the quantity of gold being diminished at all, and that of other metals very little, even though these remain disunited for a long time, I have thought that this may come about because of very subtle fire-particles. Penetrating through the tiny pores of the metal, between which (on account of their tightness) the minimum [particles] of air and other fluids could not pass, these [fire-particles] might, by filling the minimum voids distributed between these minimum particles [of metal],[8] free them from that force with which those voids attract one [particle] against another, forbidding their separation. And being thus able to move freely, their mass [*massa*] would become fluid, and remain so until the fire-particles between them depart. But when these go, leaving the pristine voids, the usual attraction returns, and consequently the attachment of the parts.

And to Simplicio's objection it seems to me that one may reply that although such voids are very tiny, and as a result

67

<hr />

7. This completes Sagredo's metaphorical remark on p. **61**; cf. note 4, above.

8. The coherence of material atoms is here ascribed to the presence of interstitial void points (or perhaps very minute spaces) rather than to any property inherent in material atoms as such. In this way nature's horror of the void (an Aristotelian principle) is preserved, but only for points (or vanishingly small intervals). See further, pp. **93** ff., and notes 10, 32, and 37, below.

each one is easily overpowered, still the innumerable multitude
of them multiplies the resistances innumerably, so to speak.
The character and extent of the force resulting from an
immense number of very weak momenta conjoined may be
most evidently argued from our seeing a weight of millions
of pounds, sustained by very thick ropes, ultimately yield
and allow itself to be lifted by the assault of innumerable
atoms of water, which, driven by the south wind or extended
in a thin fog, go moving through the air to be driven between
the fibers of the ropes; the immense force of the hanging
weight being unable to prevent their entrance, they penetrate
through narrow pores into the ropes, swelling and hence
shortening them, by which means the enormous bulk is
raised.[9]

Sagr. There is no doubt that as long as a resistance is not
infinite, it can be overcome by the sheer multitude of minimal
forces. Thus a number of ants might bring to land a ship
loaded with grain, for our eyes daily show us that an ant can
readily transport a grain, and it is clear that in the ship there
are not infinitely many grains, but some limited number.
We can take a number several times as great, and put that
number of ants to work; and they will bring to land not only
the grain, but the ship along with it. It is true that the number
would have to be large, but in my opinion so is that of the
voids that hold together the minimum particles of a metal.

Salv. But if an infinitude were required, you would perhaps
hold this to be impossible?

Sagr. No, not if the metal were infinite in bulk, [but]
otherwise . . .[10]

68   *Salv.* Otherwise, what? Well, since paradoxes are at hand,
let us see how it might be demonstrated that in a finite continuous extension it is not impossible for infinitely many voids
to be found. At the same time we shall see, if nothing else, at

9. A popular story at the time was that during the raising of the Vatican
obelisk in 1586, stretching of the ropes at a crucial moment was countered
by pouring water on them.

10. No dots are present in the 1638 edition to indicate that Sagredo has
paused here, at a loss to go on. Viviani added in his copy the words ". . .
if they were infinitely many, they could have no size; but a while ago they
were given size." What had actually been said (p. **86**) was that fire-particles
could fill them, but not specifically that such particles had dimensions;
cf. note 8, above. What Sagredo had in mind when he paused was more
probably ". . . there could be no room for infinitely many voids," as seen
from the ensuing discussion.

least a solution of the most admirable problem put by Aristotle among those that he himself called admirable; I mean among his *Mechanical Questions*. And its solution may perhaps be no less enlightening and conclusive than that which he himself alleges, and yet different from that which the learned Monsignor di Guevara very acutely considers.[11]

But first it is necessary to explain a proposition not touched on by others, upon which the solution of this question depends; and if I am not mistaken, this [proposition] will later entail other new and admirable things. To understand this, let us draw the diagram with attention. We are to think of an equilateral and equiangular polygon of any number of sides described around the center G. For the present, let this be a hexagon *ABCDEF*, similar to and concentric with which we shall draw a smaller hexagon marked *HIKLMN*, and extend one side of the larger, *AB*, indefinitely in the direction S. The corresponding side of the smaller, *HI*, is extended in the same direction by line *HT* parallel to *AS*, and through the center we draw *GV* parallel to both these.

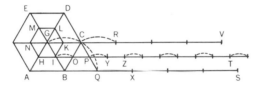

This done, we suppose the larger polygon to rotate along the line *AS*, carrying with it the smaller polygon. It is clear that the point *B*, one end of side *AB*, remains fixed. When revolution begins, the corner *A* rises and the point *C* drops, describing the arc *CQ*, so that side *BC* fits the equal line *BQ*. In this revolution, the corner *I* of the smaller polygon is lifted above line *IT*, because *IB* is oblique to *AS*; and point *I* does not return to the parallel *IT* until point *C* gets to *Q*, when point *I* will have dropped to *O* after describing arc *IO*, outside the line *HT*, the side *IK* having then passed to *OP*. During all this time, the center *G* will have been moving along outside the line *GV*, to which it does not return until it has described the entire arc *GC*.

This first step having been taken, the larger polygon is

69

11. Giovanni di Guevara (1561–1641), Bishop of Teano, had discussed this problem with Galileo but took a different approach in his *In Aristotelis Mechanicas comentarii* (Rome, 1627).

now situated with its side *BC* on line *BQ*; side *IK* of the smaller one is on line *OP*, having jumped over the part *IO* without touching it; and the center *G* has come to *C*, tracing its whole path outside the parallel *GV*. The entire figure is again at a place similar to its first position. Commencing the second turn and coming to the second place, side *DC* of the larger polygon will fit on the part *QX*; *KL* of the smaller, having first skipped the arc *PY*, falls on *YZ*; and the center, still moving outside *GV*, falls on it only at *R* after the big jump *CR*. And eventually, when one entire revolution has been made, the larger polygon will have touched, along *AS*, six lines equal [in all] to its perimeter, with nothing interposed [between them]; the smaller polygon will likewise have impressed six lines equal to its circumference but interrupted [*discontinuate*] by the interposition of five arcs, under which there are stretches which are parts of *HT* not touched by this polygon; and the center *G* has never met the parallel *GV* except at six points. From this, it is understood that the space passed over by the smaller polygon is almost equal to that passed by the larger one; that is, line *HT* [nearly equals] *AS*, being smaller only by the chord of one of these arcs, if we understand line *HT* to include the spaces of the five [skipped] arcs.[12]

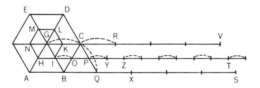

Now, what I have here set forth and explained by the example of these hexagons, I wish to be understood as happening with all other polygons, of as many sides as you please, provided that they are similar, concentric, and joined so that the turning of the larger governs that of the smaller, no matter how much smaller it may be. Understand, I say, that the lines passed over by these are approximately equal, when we count as space passed over by the smaller those intervals under the little arcs, which are not touched by any

70

12. Whether there are exactly five skipped spaces, or something more than that, is crucial to the paradox. This difficulty does not really vanish when Galileo passes to the circle "at one fell swoop" (pp. **92–93**), any more than *n* ever becomes *n* + 1 or some intermediate quantity; see also notes 13 and 14, below.

part of the perimeter of this smaller polygon. Therefore a
larger polygon having a thousand sides passes over and
measures a straight line equal to its perimeter, while at the
same time the smaller one passes an approximately equal
line, but one interruptedly composed of a thousand little
particles equal to its thousand sides with a thousand void
spaces interposed—for we may call these "void" in relation
to the thousand linelets touched by the sides of the polygon.
And what has been said thus far presents no difficulty or
question.

But now tell me: if around some center, say this point *A*,
we describe two concentric, joined circles, and from the points
C and *B* on their radii we draw the tangents *CE* and *BF*, with
the parallel *AD* to these [passing] through the center *A*;

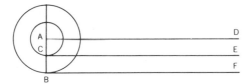

and if we suppose the greater circle to be turned on the line
BF, equal to its circumference as are likewise lines *CE* and
AD; then, when the greater circle has completed one revolution, what will the smaller circle have done, and the center?
The center will certainly have run over and touched the whole
line *AD*; and the circumference of the smaller [circle] will
with its contact have measured the whole of *CE*, behaving
like the polygons considered above. The only difference is
that there, the line *HT* was not touched in all its parts by the
perimeter of the smaller polygon, for it left untouched, by
the interposition of the voids skipped over, as many parts
as those touched by the sides.[13] But here, in the circles, the
circumference of the smaller circle is never separated from
the line *CE* in such a way that any part of *CE* is not touched;
nor is that in this circumference which touches ever less than
that which is touched in the straight line [*CE*]. How then,
without skipping, can the smaller circle run through a line
so much longer than its circumference?

Sagr. I was wondering whether one might say that just as
the center of the circle, all alone, being but a single point

13. Here Galileo says that for the hexagon there would be not five, but
six skipped spaces; cf. note 12, above.

drawn along on *AD*, touches the whole of that line, so the
points of the smaller circumference, driven by the larger
circumference, might be dragged through some particles of
the line *CE*.

71   *Salv.* This cannot be, for two reasons. First, because there
would be no more reason that some of the contacts analogous
to *C*, rather than others, should be dragged along some parts
of the line *CE*. If this were the case, and such contacts being
infinitely many by reason of their being points, the draggings
along *CE* would be infinitely many; and being quantified
[*quanti*], these would form an infinite line; but *CE* is finite.
The second reason is that since the larger circle in its revolution
continually changes its [point of] contact, the smaller circle
cannot avoid likewise [continually] changing its contact, as
it is only through the point *B* that a line can be drawn to the
center *A* and still pass through the point *C*. So whenever the
larger circle changes contact, the smaller does also; nor does
any point of the smaller [circle] touch more than one point
of the straight line *CE*.

Besides, even in the revolution of the polygons, no point
of the perimeter of the smaller is fitted to more than one
point of the line that is measured by that same perimeter.
This may easily be understood by considering the line *IK* as
parallel to *BC*, so that until *BC* falls on *BQ*, *IK* remains
lifted above *IP*, nor does it fall [flat] before that very instant
in which *BC* is united with *BQ*. But at that instant *IK* as a
whole unites with *OP*, and later on it is just as suddenly lifted
above it.

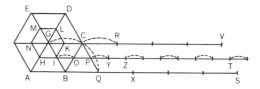

Sagr. This business is truly very intricate, and no solution
at all occurs to me; so tell us what occurs to you.

Salv. I return to the consideration of the polygons discussed
earlier, the effect of which is intelligible and already understood. I say that in polygons of one hundred thousand sides,
the line passed over and measured by the perimeter of the
larger—that is, by the hundred thousand sides extended
[straight and] continuously—is equal to that measured by

the hundred thousand sides of the smaller, but with the interposition [among these] of one hundred thousand void
spaces.[14] And just so, I shall say, in the circles (which are
polygons of infinitely many sides), the line passed over by
the infinitely many sides of the large circle, arranged continuously [in a straight line], is equal in length to the line
passed over by the infinitely many sides of the smaller, but
in the latter case with the interposition of as many voids
between them. And just as the "sides" [of circles] are not
quantified, but are infinitely many, so the interposed voids
are not quantified, but are infinitely many; that is, for the
former [line touched by the larger circle there are] infinitely
many points, all filled [*tutti pieni*], and for the latter [line
touched by the smaller circle], infinitely many points, part
of them filled points and part voids.

Here I want you to note how, if a line is resolved and divided
into parts that are quantified and consequently numbered   **72**
[*numerate*], we cannot then arrange these into a greater
extension than that which they occupied when they were
continuous and joined, without the interposition of as many
void [finite] spaces. But imagining the line resolved into
unquantifiable parts—that is, into its infinitely many indivisibles—we can conceive it immensely expanded without
the interposition of any quantified void spaces, though not
without infinitely many indivisible voids.

What is thus said of simple lines is to be understood also
of surfaces and of solid bodies, considering those as composed
of infinitely many unquantifiable atoms; for when we wish
to divide them into quantifiable parts, doubtless we cannot
arrange those in a larger space than that originally occupied
by the solid unless quantified voids are interposed—void,
I mean, at least of the material of the solid. But if we take the
highest and ultimate resolution [of surfaces and bodies] into
the prime components, unquantifiable and infinitely many,
then we can conceive such components as being expanded
into immense space without the interposition of any quantified
void spaces, but only of infinitely many unquantifiable voids.
In this way there would be no contradiction in expanding,
for instance, a little globe of gold into a very great space
without introducing quantifiable void spaces—provided,

14. Strictly speaking, it is necessary to add ". . . of which 99,999 are each
equal to a side of the smaller polygon, while one is the excess over that of
a side of the larger polygon."

however, that gold is assumed to be composed of infinitely many indivisibles.[15]

Simp. It seems to me that you are traveling along the road of those voids scattered around by a certain ancient philosopher.[16]

Salv. At least you do not add, "who denied Divine Providence," as in a similar instance a certain antagonist of our Academician very inappropriately did add.[17]

Simp. Indeed I perceived, not without disgust, the hatred in that malicious opponent; yet I shall not touch on that, not only by reason of the bounds of good taste, but because I know how far such ideas are from the temperate and orderly mind of such a man as you, who are not only religious and pious, but Catholic and devout.

Getting back to the point, I feel many difficulties that are born of the reasoning just heard; doubts from which I really don't know how to free myself. For one, I advance this: if the circumferences of the two circles are equal to the two straight lines *CE* and *BF*, the latter taken as continuous and the former with the interposition of infinitely many void points, how can *AD*, described by a center that is one point only, be called equal to this point, of which [entities] it contains infinitely many? Also, this composing the line of points, the divisible of indivisibles, the quantified of unquantifiables— these reefs seem to me to be hard to pass. And not absent from my difficulties is the necessity of assuming the void, so conclusively refuted by Aristotle.

Salv. There are these [difficulties] indeed, and others; but let us remember that we are among infinites and indivisibles, the former incomprehensible to our finite understanding by reason of their largeness, and the latter by their smallness. Yet we see that human reason does not want to abstain from giddying itself about them. Taking some liberties on that account, I am going to produce a fantastic idea of

15. Cf. note 10, above; the reference here seems to be to point-atoms of gold, since *minima naturalia*, indivisible physically but divisible mathematically, could not be infinitely many in a finite bulk.

16. The reference is probably to Epicurus (341–270 B.C.) as expounded by Lucretius (98–55 B.C.), since the scattered voids are interstitial, as well as for reasons implied in note 17, below.

17. Orazio Grassi (1583–1654), in his *Ratio ponderum librae et simbellae* (Paris, 1626), which was an attack on Galileo's *Il Saggiatore* of 1623. Cf. *Opere*, VI, 475–76, where it is the Epicureans that are named though Democritus (460–357 B.C.), is usually considered as the chief atomist of antiquity. Cf. note 16, above.

mine which, if it concludes nothing necessarily, will at least by its novelty occasion some wonder. Or perhaps it will seem to you inopportune to digress at length from the road that we started on, and hence will be distasteful.

Sagr. Please let us enjoy the benefit and privilege that comes from speaking with the living and among friends, about things of our own choice and not by necessity, which is very different from dealing with dead books that excite a thousand doubts and resolve none of them. So make us partners in whatever reflections suggest themselves to you in the course of our discussions. We do not lack time to continue and resolve the other matters we have undertaken, thanks to our present freedom from necessary occupations. In particular, the doubts raised by Simplicio are by no means to be skipped over.

Salv. Be it so, since that is the way you wish it. Let us begin from the first—how a single point can ever be understood to be equal to a line. The most that can be done at present is for me to try to put at rest, or at any rate to moderate, this improbability with an equal or greater one, as a marvel is sometimes put to rest by a miracle. I shall do this by showing you two equal surfaces, and two bodies, also equal, with the said surfaces as their bases. These will [all] go continually and equally diminishing during the same time, their remaining parts always being equal, until finally the surfaces and the solids terminate their preceding perpetual equality by one solid and one surface becoming a very long line, while the other solid and the other surface become a single point; that is, the latter two become a single point, and the former two, infinitely many points.

Sagr. This seems to me a truly remarkable proposal; let **74** me hear its explanation and demonstration.

Salv. We must draw a diagram for it, since the proof is purely geometrical.[18] Take the semicircle *AFB* whose center is *C*, and around it the rectangular parallelogram *ADEB*; from the center to points *D* and *E*, draw the straight lines *CD* and *CE*. Next imagine the whole figure rotated around

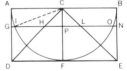

18. The ensuing paradox had been hinted at in the *Dialogue*, p. 247 (*Opere*, VII, 271–72). Galileo had previously sent it to Buonaventura Cavalieri (1598?–1647) to caution him regarding the perils of the "method of indivisibles" in geometry. The paradox has a double purpose here: to illustrate the nature of mathematical definitions, and to show the pitfalls of analogy in transferring the word "equal" from entities of *n* dimensions to their supposed counterparts of *n* − 1 dimensions. Cf. notes 19 and 22, below.

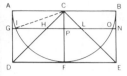

the fixed radius *CF*, perpendicular to the straight lines *AB* and *DE*. It is manifest that a cylinder will be described by the rectangle *ADEB*, a hemisphere by the semicircle *AFB*, and a cone by the triangle *CDE*. We now suppose the hemisphere removed, leaving [intact] the cone and those remains of the cylinder which in shape resemble a soupdish, for which reason we shall call it by that name.

First, we shall prove this soupdish to be equal [in volume] to the cone. Then, drawing a plane parallel to the circle at the base of the soupdish, of diameter *DE* and with center *F*, we shall prove that this plane, passing for example through the line *GN*, and cutting the soupdish at points *G, I, O* and *N*, and the cone at points *H* and *L*, leaves the part of the cone *CHL* always equal to the part of the soupdish whose cross section is represented by the "triangles" *GAI* and *BON*. Moreover, we shall prove that any base of the cone, say the circle whose diameter is *HL*, is equal to that circular surface which is the base for that part of the soupdish; this is, as it were, a washer [*nastro*, ribbon] of breadth *GI*.

Note here what sort of things mathematical definitions are; that is, the mere imposition of names, or we might say abbreviations of speech, arranged and introduced in order to remove the tedious drudgery that you and I felt before we agreed to call one surface the "washer," and presently feel until we call the [upper section of the] soupdish the "cylindrical razor." Now, call these what you will, it suffices to understand that the plane at any level, provided that it is **75** parallel to the base, or circle of diameter *DE*, always makes the two solids equal; that is, the part of the cone *CHL*, and the upper part of the soupdish [i.e., the cylindrical razor]. Likewise it makes equal the two surfaces that are the bases of those solids; that is, the washer and the circle *HL*.

From this follows the marvel previously mentioned; namely, that if we understand the cutting plane to be gradually raised toward the line *AB*, the parts of the solids it cuts are always equal, as likewise are the surfaces that form their bases. Lifting it more and more, the two always-equal solids, as well as their always-equal bases, finally vanish—the one pair in the circumference of a circle, and the other pair in a single point, such being the upper rim of the soupdish and the summit of the cone. Now, during the diminution of the two solids, their equality was maintained right up to the end; hence it seems consistent to say that the highest and last boundaries of the reductions are still equal, rather than that one is infinitely greater than the other, and so it appears that the circumference of an immense circle may be called equal to a single point!

What happens in the solids likewise happens in the surfaces that are their bases. These also maintain equality throughout the diminution in which they share; and at the end, in the instant of their ultimate diminution, the washer reaches its limit in the circumference of a circle, and the base of the cone in a single point. Now, why should these not be called equal, if they are the last remnants and vestiges left by equal magnitudes?[19]

Note next that if these vessels were as large as the immense celestial hemispheres, and the ultimate edges and the points of the contained cones always preserved their equality, those edges would terminate in circumferences equal to great circles of the celestial orbs, and the cones [would terminate] in single points. Hence, along the line in which such speculations lead us, the circumferences of all circles, however unequal [in size], may be called equal to one another, and each of them [may be called] equal to a single point!

Sagr. The speculation appears to me so delicate and wonderful that I should not oppose it even if I could. To me it would seem a sort of sacrilege to mar so fine a structure, trampling on it with some pedantic attack. Still, for our full satisfaction, **76** let us have that proof, which you call geometrical, of the constant maintenance of equality between those solids, and between their bases. I think this must be very clever, since the philosophical meditation stemming from this conclusion is so subtle.

Salv. The demonstration is also brief and easy. In the diagram drawn, angle *IPC* being a right angle, the square of the radius *IC* is equal to the two squares of the sides *IP* and *PC*. But the radius *IC* is equal to *AC*, and this to *GP*; and *CP* is equal to *PH*. Therefore the square of the line *GP* is equal to the two squares on *IP* and *PH*, and four times the former equals four times the sum of the latter; that is, the square

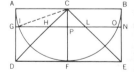

19. Use of the word "equal" in this way violates Berkeley's axiom that conclusions reached from the behavior of given entities and dependent on their existence cannot be rigorously applied to other entities deprived of them when they vanish. Sagredo's reply bears out the view of Abraham Kästner (1719–1800) and Bernard Bolzano (1781–1848) that Galileo inserted this paradox not as a conclusion to be accepted, but solely to stimulate careful thought.

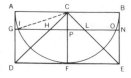

of the diameter *GN* is equal to the two squares *IO* and *HL*. And since circles are to each other as the squares of their diameters, the circle of diameter *GN* will be equal to the two circles of diameters *IO* and *HL*; hence removing the common circle whose diameter is *IO*, the remaining circle *GN* will be equal to the circle whose diameter is *HL*.

So much for the first part [areas]. As to the other part [volumes], let us skip that proof for the present; if we wish to see it, we shall find it in the twelfth proposition of the second book of *De centro gravitatis solidorum* by Signor Luca Valerio, the new Archimedes of our age, who makes use of it for another proposition of his.[20] [We omit the proof] also because in our case it is enough to have seen how the two surfaces described are always equal, and that in diminishing always equally, they tend to end, the one in a single point, and the other in the circumference of a circle of any size whatever; for our marvel turns on this consequence alone.

Sagr. The proof is as ingenious as the reflection based on it is remarkable. Now let us hear something about the second difficulty advanced by Simplicio, if you have anything new to say about it, which I believe may not be the case, since the controversy has been so widely agitated.

Salv. I shall give you my own special thought on it, first repeating what I said a while ago; that is, that the infinite is inherently incomprehensible to us, as indivisibles are likewise; so just think what they will be when taken together!

77   If we want to compose a line of indivisible points, we shall have to make these infinitely many, and so it is necessary [here] to understand simultaneously the infinite and the indivisible. Many indeed are the things I have on many occasions turned over in my mind on this matter. Some of them, perhaps the most important, I may not recall offhand; but in the progress of the argument I may happen to awaken objections and difficulties in you, and especially in Simplicio, in meeting which I shall remember things that without such stimulus would remain asleep in my imagination. So, with our customary freedom, let it be agreed that we bring in our human caprices, as we may well call them in contrast with those theological [*sopranaturale*] doctrines that are the only

20. Luca Valerio (1552–1618) met Galileo at Pisa about 1590 and later corresponded with him at Padua. The book cited was first published at Rome in 1603/04; Bk. II, Prop. XII, includes a denonstration from which Galileo derived the foregoing paradox.

true and sure judges of our controversies and the unerring guides through our obscure and dubious, or rather labyrinthine, opinions.[21]

One of the first objections usually produced against those who compound the continuum out of indivisibles is that one indivisible joined to another indivisible does not produce a divisible thing, since if it did, it would follow that even the indivisible was divisible; because if two indivisibles, say two points, made a quantity when joined, which would be a divisible line, then this would be even better composed of three, or five, or seven, or some other odd number [of indivisibles]. But these lines would then be capable of bisection, making the middle indivisible capable of being cut. In this, and other objections of the kind, satisfaction is given to its partisans by telling them that not only two indivisibles, but ten, or a hundred, or a thousand do not compose a divisible and quantifiable magnitude; yet infinitely many may do so.[22]

Simp. From this immediately arises a doubt that seems to me unresolvable. It is that we certainly do find lines of which one may say that one is greater than another; whence, if both contained infinitely many points, there would have to be admitted to be found in the same category a thing greater than an infinite, since the infinitude of points of the greater line will exceed the infinitude of points of the lesser. Now, the occurrence of an infinite greater than the infinite seems to me a concept not to be understood in any sense.

Salv. These are some of those difficulties that derive from

21. The ensuing argument required very tactful treatment, since the proposition that a line might be composed of indivisibles, strongly opposed by Aristotle, had been condemned as heretical in 1415 by the Council of Constance. John Wyclif was exhumed and his body burned for this and other Epicurean doctrines. Cf. note 17, above, and see Aristotle, *Physica*, Bk. VI; *De caelo*, 299a.10 ff., as well as the pseudo-Aristotelian treatise *On Indivisible Lines*.

22. The meaning here is not that the adversaries are literally satisfied, but that this is the proper reply to them. Galileo's position is quite unrelated to the older discussions cited in note 21, above, or to that of Thomas Bradwardine (1290?–1349) in his *De continuo*, all of which debates concerned "indivisibles" that differed only in size from whatever they were supposed to compose. Galileo speaks here of elements having one less dimension than the aggregates in which they are supposed to exist; it was of such "indivisibles" that Cavalieri (note 18, above) made use in his geometry. In evaluating the role of Cavalieri's work in the development of the calculus, his "indivisibles" have frequently been confused by historians with infinitesimal magnitudes having the same dimensionality as the continuum to be analyzed, an approach studiously avoided by Cavalieri himself.

reasoning about infinites with our finite understanding, giving
78 to them those attributes that we give to finite and bounded
things. This, I think, is inconsistent, for I consider that the
attributes of greater, lesser, and equal do not suit infinities,
of which it cannot be said that one is greater, or less than,
or equal to, another.[23] In proof of this a certain argument
once occurred to me, which for clearer explanation I shall
propound by interrogating Simplicio, who raised the difficulty.
I assume that you know quite well which are square numbers,
and which are not squares.

Simp. I know well enough that a square number is that
which comes from the multiplication of a number into itself;
thus four and nine and so on are square numbers, the first
arising from two, and the second from three, each multiplied
by itself.

Salv. Very good. And you must also know that just as
these products are called squares, those which thus produce
them (that is, those which are multiplied) are called sides,
or roots. And other [numbers] that do not arise from numbers
multiplied by themselves are not squares at all. Whence if
I say that all numbers, including squares and non-squares,
are more [numerous] than the squares alone, I shall be
saying a perfectly true proposition; is that not so?

Simp. One cannot say otherwise.

Salv. Next, I ask how many are the square numbers; and
it may be truly answered that they are just as many as are
their own roots, since every square has its root, and every
root its square; nor is there any square that has more than
just one root, or any root that has more than just one square.[24]

Simp. Precisely so.

Salv. But if I were to ask how many *roots* there are, it
could not be denied that those are as numerous as all the
numbers, because there is no number that is not the root of
some square. That being the case, it must be said that square
numbers are as numerous as all numbers, because they are
as many as their roots, and all numbers are roots. Yet at the

23. Having previously warned against the dangers in applying the word
"equal" to infinites in the same sense as to finite magnitudes (notes 18 and
19, above), Galileo next turns to the positive integers to introduce the idea
of one-to-one correspondence. His conclusions are valid and consistent,
though by a much later extension of the concept of number we are
now permitted to speak of different orders of infinite aggregates.
24. Negative roots were excluded under the Euclidean definition of number;
see Glossary of mathematical terms.

outset we said that all the numbers were many *more* than all the squares, the majority being non-squares. Indeed, the multitude of squares diminishes in ever-greater ratio as one moves on to greater numbers, for up to one hundred there are **79** ten squares, which is to say that one-tenth are squares; in ten thousand, only one one-hundredth part are squares; in one million, only one one-thousandth. Yet in the infinite number, if one can conceive that, it must be said that there are as many squares as all numbers together.

Sagr. Well then, what must be decided about this matter?

Salv. I don't see how any other decision can be reached than to say that all the numbers are infinitely many; all squares infinitely many; all their roots infinitely many; that the multitude of squares is not less than that of all numbers, nor is the latter greater than the former. And in final conclusion, the attributes of equal, greater, and less have no place in infinite, but only in bounded quantities. So when Simplicio proposes to me several unequal lines, and asks me how it can be that there are not more points in the greater than in the lesser, I reply to him that there are neither more, nor less, nor the same number [*altrettanti*, just as many], but in each there are infinitely many. Or truly, might I not reply to him that the points in one are as many as the square numbers; in another and greater line, as many as all numbers; and in some tiny little [line], only as many as the cube numbers—in that way giving him satisfaction by putting more of them in one than in another, and yet infinitely many in each?[25] So much for the first difficulty.

Sagr. Hold on a minute, and allow me to add to what has been said a thought that has just struck me. If matters stand as has been said up to this point, it seems to me that not only may one infinite not be said to be greater than another infinite, but it may not even be said that an infinite is greater than a finite. For if the infinite number is greater than one million, say, it would follow that in passing from one million to other continually larger numbers, one would be traveling toward the infinite [number], which is not so; rather, the opposite is so, and the larger the numbers to which we pass, the farther we get from the infinite number. For with numbers,

25. The word "more" should be read as if placed in quotation marks. Galileo had already shown that the word had no meaning in this context, but offered this ingenious rationalization to anyone who might still think differently.

the larger they are taken, the scarcer become the square numbers contained within these, whereas in the infinite number, the squares cannot be less than all the numbers, as was just now concluded. Hence to go to ever and ever larger numbers is to move away from the infinite number.

80   *Salv.* And so, from your ingenious reasoning, it is concluded that the attributes of greater, less, or equal are out of place not only between infinites, but even between infinites and finites.[26]

Now let us pass to another consideration, which is that the line, and every continuum, being divisible into ever-divisibles, I do not see how to escape their composition from infinitely many indivisibles; for division and sub-division that can be carried on forever assumes that the parts are infinitely many. Otherwise the subdivision would come to an end. And the existence of infinitely many parts has as a consequence their being unquantifiable, since infinitely many quantified [parts] make up an infinite extension. And thus we have the continuum composed of infinitely many indivisibles.

Simp. But if we can continue forever the division into quantified parts, what need have we, in this respect, to introduce the unquantifiable?

Salv. The very ability to continue forever division into quantifiable parts implies the necessity of composition from infinitely many unquantifiables. For, getting down to the real trouble, I ask you to tell me boldly whether in your opinion the quantified parts of the continuum are finite, or infinitely many?

Simp. I reply to you that they are both infinitely many and finite; infinitely many potentially [*in potenze*]; and finite actually [*in atto*]; that is, potentially infinitely many before division, but actually finite [in number] after they are divided. For parts are not understood to be *actually* in their whole until after [they are] divided, or at least marked. Otherwise they are said to be *potentially* there.[27]

---

26. A cardinal principle of Aristotle's was that there can be no ratio between finite and infinite; cf. note 1, above, and *De caelo* 274a.10; 274b.12. The principle is probably put into Sagredo's mouth because the mature Galileo neither entirely rejected nor fully accepted it.

27. Aristotle's distinction of potentiality and actuality was fundamental to his physics, since it entered into his very definition of motion; see *Physica* 201a.10. With respect to potentiality and the infinite, see *Physica* 206a.15–206b.25. Here Galileo proceeds to show that the distinction is meaningless mathematically unless it affects quantity or magnitude.

Salv. So that a line twenty spans long, for instance, is not said to contain twenty lines of one span each, actually, until after its division into twenty equal parts. Before this, it is said to contain these only potentially. Well, have this as you please, and tell me whether, the actual division of such parts having been made, that original whole has increased, diminished, or remains still of the same magnitude?

Simp. It neither increases nor diminishes.

Salv. So I think, too. Therefore the quantified parts in the continuum, whether potentially or actually there, do not make its quantity greater or less. But it is clear that quantified parts actually contained in their whole, if they are infinitely many, make it of infinite magnitude; whence infinitely many quantified parts cannot be contained even potentially except in an infinite magnitude. Thus in the finite, infinitely many quantified parts cannot be contained either actually or potentially.   **81**

Sagr. Then how can it be true that the continuum may be unceasingly divided into parts always capable of new division?

Salv. That distinction between act and potency seems to make feasible in one way what would be impossible in another, but I expect to balance the accounts better by different bookkeeping. To the question which asks whether the quantified parts in the bounded continuum are finite or infinitely many, I shall reply exactly the opposite of what Simplicio has replied; that is, [I shall say] "neither finite nor infinite."

Simp. I could never have said that, not believing that any middle ground [*termine mezzano*, mean term] is to be found between the finite and the infinite, as if the dichotomy or distinction that makes a thing finite or else infinite were somehow wanting and defective.

Salv. It seems to me to be so. Speaking of discrete quantity, it appears to me that between the finite and the infinite there is a third, or middle, term; it is that of answering to every [*ogni*] designated number. Thus in the present case, if asked whether the quantified parts in the continuum are finite or infinitely many, the most suitable reply is to say "neither finite nor infinitely many, but so many as to correspond to every specified number." To do that, it is necessary that these be not included within a limited number, because then they would not answer to a greater [number]; yet it is not necessary that they be infinitely many, since no specified number is infinite. And thus at the choice of the questioner we may cut a given line into a hundred quantified parts, into a thousand, and into a hundred thousand, according to whatever number he likes, but not into infinitely many [quantified parts]. So I concede to the distinguished philosophers that the continuum contains as many quantified parts as they please; and I grant that it contains them actually or potentially at the pleasure and to the satisfaction of those gentlemen. But I then tell them further that in whichever way there are contained in a ten-fathom line ten lines of one fathom each, and forty of one braccio each, and eighty of one-half braccio, and so on, then in that same way it contains infinitely many points. You may call this "actually" or "potentially" as you choose, Simplicio, for on this particular I submit myself to your choice and judgment.

Simp. I cannot but praise your reasoning, yet I greatly fear that this parity between containing points and [containing] quantified parts does not quite work, and that it will not be so easy for you to divide a given line into infinitely many points as it is for those philosophers [to divide it] into ten fathoms or forty braccia. In fact I hold it to be quite impossible to put your division into practice, whence it will remain one of those potentialities [*potenze*] that are never reduced to act.

82

Salv. That a thing can be done only with labor and care, or over a long period of time, does not make it impossible. I think that you likewise cannot easily escape from labor and care in a division that is to be made of a line into a thousand parts, and still less if you have to divide it into 937, or some other large prime number [of parts]. But as to that division which you perhaps deem impossible, if I can make this as easy for you as it would be for someone else to cut the line into forty [equal parts], will you be content to admit this into our conversation more tranquilly?

Simp. I enjoy your way of sometimes dealing with things so pleasantly. To your question I reply that it seems to me more than sufficient if the case of resolution into points shall be no more laborious than its division into a thousand parts.

Salv. Here I want to say something that will perhaps astonish you concerning the possibility of resolving a line into its infinitely many [points] by following the procedure that others use in dividing into forty, sixty, or a hundred parts; that is, by dividing it into two, and then four, and so

on. Pursuing that method, anyone who believes he can find
its infinitely many points is badly mistaken, for with such
a procedure he will never achieve the division of the line
into all its quantified parts, even if he goes on forever; and
as to its indivisibles, he would be so far from arriving at the
desired end by that path that instead, he would be traveling
away from it. If anyone thinks that by continuing division
and by increasing the multitude of parts he is approaching
infinity, I believe that he is always receding farther from that.

My reason is this. In our discussion a little while ago,
we concluded that in the infinite number, there must be as
many squares or cubes as all the numbers, because both
[squares and cubes] are as numerous as their roots, and all
numbers are roots. Next we saw that the larger the numbers
taken, the scarcer became the squares to be found among
them, and still rarer, the cubes. Hence it is manifest that to   **83**
the extent that we go to greater numbers, by that much and
more do we depart from the infinite number. From this it
follows that turning back (since our direction took us always
farther from our desired goal), if any number may be called
infinite, it is unity. And truly, in unity are those conditions
and necessary requisites of the infinite number. I refer to
those [conditions] of containing in itself as many squares
as cubes, and as many as all the numbers [contained].

Simp. I don't quite see how this business should be understood.

Salv. The business has in it no room for doubt, because
unity is a square, and a cube, and a fourth power, and all the
other powers. There is no essential property belonging to
squares, cubes, and so on that does not belong to [the number]
one. For instance, a property of two square numbers is that of
having between them a number [that is their] mean proportional. Take as one extreme any square number, and as the
other, unity; there will always be found a numerical mean
proportional; thus let the two square numbers be 9 and 4;
between 9 and 1 the mean proportional is 3, and between 4
and 1 it is 2; between the two squares 9 and 4 we find 6, the
middle [term in geometric proportion]. A property of cubes
is that between them there are necessarily two mean proportionals; given 8 and 27, between them lie the [geometric]
means 12 and 18; between 1 and 8 are 2 and 4; and between
1 and 27 are 3 and 9. Thus we conclude that there is no
infinite number other than unity.

These are among the marvels that surpass the bounds of our imagination, and that must warn us how gravely one errs in trying to reason about infinites by using the same attributes that we apply to finites; for the natures of these have no necessary relation [*convenienza*] between them. Apropos of this, I do not wish to pass by in silence a remarkable event that just now occurs to me, illuminating the infinite difference and even the repugnance and contrariety of nature encountered by a bounded quantity in passing over to the infinite. Let us take this straight line *AB*, of any length whatever, and in it take some point *C* that divides it into unequal parts. I say that pairs of lines leaving from the points *A* and *B*, and preserving between themselves the same ratio as that of the parts *AC* and *BC*, will intersect in points that all fall on the circumference of the same circle. For example, *AL* and *BL*, coming from points *A* and *B* and having the same ratio as parts *AC* and *BC*, meet in a point *L*; with the same ratio, another pair *AK* and *BK* meet in *K*; others [are] *AI* and *BI*, *AH* and *HB*, *AG* and *GB*, *AF* and *FB*, *AE* and *EB*. I say that the meeting-points *L*, *K*, *I*, *H*, *G*, *F*, and *E* all fall on the circumference of the same circle. Thus if we imagine the point *C* moving continuously, under the rule that the lines produced from it to the fixed limits *A* and *B* shall maintain always the same ratio as that of the original parts *AC* and *CB*, then that point *C* will describe the circumference of a circle, as I shall next prove. And the circle described in this way will be ever greater, infinitely, according as the point *C* is taken closer to the midpoint *O* [of *AB*], while the circle will be smaller [which is] described by a point closer to the end *B*. Thus, following the above rule, circles will be described by motion of the infinitely many points that can be taken in the line *OB*, and the circles are of every size—less than the pupil of a flea's eye, or greater than the equator of the celestial sphere [*primo mobile*].

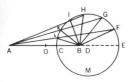

Now, if a circle is described by any point lying between the limits *O* and *B*, and immense circles by moving points close to *O*, then by moving the point *O* itself and continuing to do so in observance of the same law (that is, so that lines produced from *O* to the ends *A* and *B* shall keep the ratio of the original lines *AO* and *OB*), what line will point *O* trace? It will trace the circumference of a circle, but that of a circle greater than any other great circle, and therefore of an infinite circle. But [in fact] it traces a straight line perpendicular to *BA*, rising from point *O* and extending *in infinitum* without ever returning to join its last end with its first, as all the others do return. For that [which was] traced by the limited motion of point *C*, after marking the upper **85** semicircle *CHE*, went on to trace the lower, *EMC*, rejoining its extreme ends at point *C*. But (because points taken in the other part of *OA* also describe circles, the points near *O* the greatest ones) point *O*, being moved like all the others of line *AB*, in tracing its circle so that it will be made greatest of all, and consequently infinite, can never return to its original extreme; and in brief, it describes an infinite straight line as circumference of its infinite circle.[28]

Consider, then, what a difference there is [in moving] from a finite to an infinite circle. The latter changes its being so completely as to lose its existence and its possibility of being [a circle]. For we understand well that there cannot be an infinite circle, from which it follows as a consequence that still less can a sphere be infinite; nor can any other solid or surface having a shape be infinite. What shall we say of this metamorphosis in passing from finite to infinite? And why must we feel greater repugnance when, seeking the infinite in numbers, we come to conclude that it is in [the number] one? We break a solid into many parts, and go on to reduce it to very fine powder; if it were resolved into its infinitely many atoms, no longer divisible, why should we not say that it had returned to a single continuum, fluid perhaps, like water or mercury, or the original metal liquefied? Do we not see stones liquefied into glass, and glass under great heat made more liquid than water?

Sagr. Must we therefore believe that fluids are what they are because they are resolved into indivisibles, infinitely many, [as] their prime components?

Salv. I cannot find any better expedient for solving some of the sensible appearances, among which is this. When I take a hard body of stone or metal, and with hammer or file I proceed to divide it as finely as possible into impalpable powder, clearly its minimum [particles], though imperceptible individually to sight and touch, are still quantified, have shape, and are countable. That is why they support themselves cumulatively in a heap, a dent in this, up to a certain point, remaining a dent, without the surrounding parts rushing

28. Cf. *Dialogue*, p. 377 (*Opere*, VII, 404).

in to fill it. Agitated and stirred, these particles stop as soon
as the external mover abandons them. All these effects happen
86 also in all aggregates of larger corpuscles of every shape, even
spherical, as we see in mounds of flour, grain, lead shot,
and other materials. But if we seek these phenomena in water,
none are to be found. When raised, water immediately
smooths flat unless sustained by a vessel or other external
restraint; dented, it immediately runs to fill the cavity;
agitated, it goes on fluctuating for a long time, and its waves
extend through great distances.

From this, I think it is reasonable to argue that the minimum
[particles] into which water seems to be resolved, since it
has less consistency than the finest powder (or rather, has
no consistency at all), are quite different from quantified
and divisible minimum [particles], and I cannot find any
other difference here besides that of their being indivisible.
It also seems to me that their perfect transparency strengthens
this conjecture. If we take the most transparent crystal
that exists and begin to pound and break it, it loses its transparency when reduced to powder, and the more so the more
finely it is broken. But water, which is broken to the highest
degree, is yet diaphanous to the highest degree. Gold and
silver, pulverized by aqua fortis more finely than by any file,
still remain in powder and do not become fluid; nor do they
liquefy until the indivisibles of fire or of the sun's rays dissolve
them (as I think) into their first and highest components,
infinitely many, and indivisible.

Sagr. What you have said of [the sun's] light, I have often
observed with wonder. I have seen lead instantly liquefied
by a concave mirror three spans in diameter, and am of
the opinion that if the mirror were very large, smooth, and
of parabolic shape, it would liquefy any metal in short time.
For we see that a spherically concave mirror, neither very
large nor well polished, liquefies lead with great power
and burns every combustible material—effects that give
credibility to the wonders of the mirrors of Archimedes.[29]

Salv. As to Archimedes and the effects of his mirrors,
all the miracles that are read in other authors are rendered
credible to me by reading the books of Archimedes himself,

29. The story that Archimedes burned enemy ships by means of powerful
mirrors is not found before the twelfth century. It probably grew out of
accounts of the burning of enemy ships in the defence of Syracuse which
failed to mention the catapulting of incendiary material as the means.

long ago studied by me with infinite astonishment. And if any doubt lingered, the book lately published about the burning glass [*Specchio ustorio*] by Father Buonaventura Cavalieri, which I read with admiration, is enough to put **87** a stop to all difficulties for me.[30]

Sagr. I also saw that treatise and read it with pleasure and wonder; I was already acquainted with the author, and this confirmed the idea that I had formed of him—that he would turn out to be one of the chief mathematicians of our age. But returning to the remarkable effect of the sun's rays in liquefying metals, should we believe that so vehement an operation takes place without motion, or that it does so with the most rapid motion?

Salv. We see other fires and dissolutions to be made with motion, and very swift motion; behold the operations of lightning, and of gunpowder in mines and bombs. We see how much the use of bellows speeds the flames of coals mixed with gross and impure vapors, increasing their power to liquefy metals. So I cannot believe that the action of light, however pure, can be without motion, and indeed the swiftest.

Sagr. But what and how great should we take the speed of light to be? Is it instantaneous perhaps, and momentary? Or does it require time, like other movements? Could we assure ourselves by experiment which it may be?

Simp. Daily experience shows the expansion of light to be instantaneous. When we see artillery fired far away, the brightness of the flames reaches our eyes without lapse of time, but the sound comes to our ears only after a noticeable interval of time.

Sagr. What? From this well-known experience, Simplicio, no more can be deduced than that the sound is conducted to our hearing in a time less brief than that in which the light is conducted to us. It does not assure me whether the light is instantaneous, or time-consuming but very rapid. Your observation is no more conclusive than it would be to say: "Immediately on the sun's reaching the horizon, its splendor reaches our eyes." For who will assure me that the rays did

30. The book mentioned was published by Cavalieri at Bologna in 1632; it included a derivation of the parabolic trajectory of projectiles, which Galileo had discovered late in 1608 but had not yet published; cf. note 5 to Fourth Day. He was indignant on first hearing of this publication, but when he saw the book with its acknowledgment to him and learned that Cavalieri believed him to have published it earlier, he was appeased. Cavalieri's teacher had been Galileo's pupil, Benedetto Castelli (1578–1643).

not reach the horizon before [reaching] our vision?

Salv. The inconclusiveness of these and like observations caused me once to think of some way in which we could determine without error whether illumination (that is, the expansion of light) is really instantaneous.[31] The rapid motion of sound assures us that that of light must be very swift indeed, and the experiment that occurred to me was this. I would have two men each take one light, inside a dark lantern or other covering, which each could conceal and reveal by interposing his hand, directing this toward the vision of the other. Facing each other at a distance of a few braccia, they could practice revealing and concealing the light from each other's view, so that when either man saw a light from the other, he would at once uncover his own. After some mutual exchanges, this signaling would become so adjusted that without any sensible variation, either would immediately reply to the other's signal, so that when one man uncovered his light he would instantly see the other man's light.

This practice having been perfected at a short distance, the same two companions could place themselves with similar lights at a distance of two or three miles and resume the experiment at night, observing carefully whether the replies to their showings and hidings followed in the same manner as near at hand. If so, they could surely conclude that the expansion of light is instantaneous, for if light required any time at a distance of three miles, which amounts to six miles for the going of one light and the coming of the other, the interval ought to be quite noticeable. And if it were desired to make such observations at yet greater distances, of eight or ten miles, we could make use of the telescope, focusing one for each observer at the places where the lights were to be put into use at night. Lights easy to cover and uncover are not very large, and hence are hardly visible to the naked eye at such distance, but by the aid of telescopes previously fixed and focused they could be comfortably seen.

Sagr. The experiment seems to me both sure and ingenious. But tell us what you concluded from its trial.

Salv. Actually, I have not tried it except at a small distance, less than one mile, from which [trial] I was unable to make sure whether the facing light appeared instantaneously. But if not

---

31. In *Il Saggiatore* (1623), Galileo had spoken of light as propagated instantaneously by the ultimate subdivision of matter into its true indivisibles; cf. *Assayer*, p. 313 (*Opere*, VI, 352).

instantaneous, light is very swift and, I may say, momentary;
at present I should liken it to that motion made by the
brightness of lightning seen between clouds eight or ten miles   **89**
away. In this, we distinguish the beginning and fountainhead
of light at a particular place among the clouds, followed
immediately by its very wide expansion through surrounding
clouds. This seems to me to be an argument that the stroke
of lightning takes some little time, because if the illumination
were made all together and not by parts, it appears that we
should not be able to distinguish its place of origin and its
center from its extreme streamers and dilatations.

But in what seas are we inadvertently engulfing ourselves, bit
by bit? Among voids, infinites, indivisibles, and instantaneous
movements, shall we ever be able to reach harbor even after
a thousand discussions?

Sagr. These things are truly quite ill-adapted to our purpose.
The infinite, sought among numbers, seems to end at unity;
from indivisibles is born the ever-divisible; the void seems to
exist only by being indivisibly mixed into the plenum; in a
word, the nature of each of these things alters from our
common understanding of it, until the circumference of a circle
is replaced by a straight line. If I recall correctly, Salviati,
that is the proposition that you were to make clear to us by
geometrical demonstration. So it will be good if, without
further digression, you will produce it.

Salv. I am at your service; and for complete understanding,
I shall demonstrate the following problem:

> Given a straight line divided into unequal parts in any
> ratio, to describe a circle such that to any point of its
> circumference, two straight lines being drawn from the
> ends of the given line, [these lines] will retain the same
> ratio as that of the said parts of the given line, so that
> all [pairs] leaving from the same extremities will be
> homologous.

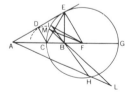

Let the given line be *AB*, divided in any way into unequal
parts at the point *C*; it is required to describe the circle such
that at any point on its circumference, two lines drawn from
A and *B* will meet, and will have between them the same
ratio as that of the parts *AC* and *BC*, making homologous
those which leave from the same endpoints [*A* and *B*]. Draw
a circle with center *C* and radius equal to the smaller part *CB*,
to which line *AD* will be tangent, this being indefinitely   **90**
prolonged in the direction of *E* from point *A*, the point of

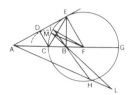

tangency being *D*. Join *C* and *D*; [*CD*] will be perpendicular to *AE*, and *BA* will be perpendicular to *BE*, which produced will meet *AE*, since angle *A* is acute. From this intersection *E*, drop a line perpendicular to *AE*, which produced will meet *AB*, indefinitely prolonged, at *F*.

I say first that the two lines *FE* and *FC* are equal, since if *EC* is drawn, we have in the two triangles *DEC* and *BEC* two sides of one, *DE* and *EC*, equal to two sides of the other, *BE* and *EC*. And *DE* and *EB* being tangent to the [dotted] circle *DB*, the bases *DC* and *CB* are equal, whence angles *DEC* and *BEC* are equal. Since angles *BCE* and *CEB* are complementary, as are angles *CEF* and *CED*, angles *FCE* and *FEC* are equal; hence also sides *FE* and *FC*. Now taking point *F* as center, with radius *FE* describe a circle *CEG* passing through point *C*. I say that this is the circle sought, at any point of the circumference of which, every pair of lines intersecting and passing through the extremities *A* and *B* will have the same ratio as that of the two parts *AC* and *BC*, which are joined at the point *C*. This is obvious of the two lines *AE* and *BE* that meet at point *E*, for the angle *E* of triangle *AEB* being bisected by *CE*, whatever ratio *AC* has to *CB* will be that of *AE* to *BE*. We prove the same of *AG* and *BG*, meeting at point *G*. By the similarity of triangles *AFE* and *EFB*, as *AF* is to *FE*, so *EF* is to *FB*, that is, as *AF* is to *FC*, so *CF* is to *FB*; and by division, as *AC* is to *CF* (that is, to *FG*), so *CB* is to *BF*, and the whole *AB* is to the whole *BG* as the part *CB* is to the part *BF*. And by composition, as *AG* is to *GB*, so *CF* is to *FB*, that is, as *EF* to *FB*, that is, *AE* to *EB*, and *AC* to *CB*, which was to be proved.

91   Now take any point *H* in the circumference, at which the two lines *AH* and *BH* meet. I say that as *AC* is to *CB*, so *AH* is to *HB*. Extend *HB* to the circumference at *I*, and join *I* and *F*; since (as seen before) *CB* is to *BF* as *AB* is to *BG*, the rectangle *AB–BF* is equal to the rectangle *CB–BG*, that is, *IB–BH*. Hence as *AB* is to *BH*, so *IB* is to *BF*, and the angles at *B* are equal; therefore *AH* is to *HB* as *IF* (that is, *EF*) is to *FB*, and *AE* to *EB*.

Besides this, I say that it is quite impossible for lines in this ratio and passing through ends *A* and *B* to meet at any point outside or inside the circle *CEG*. For if possible, let two lines *AL* and *BL* meet at *L*, outside [the circle]. Extend *LB* to the circumference at *M*, and draw *MF*. Then if *AL* is to *BL* as *AC* is to *BC* (that is, as *MF* to *FB*), we have two triangles

ALB and *MFB* which have their sides proportional around
the angles *ALB* and *MFB*, while the angles at the apex *B*
are equal, and the two remaining angles *FMB* and *LAB* are
less than a right angle, because the right angle at point *M*
has for its base the whole diameter *CG* and not just the part *BF*,
while the angle at point *A* is acute, since line *AL*, homologous
to *AC*, is greater than *BL*, homologous to *BC*. Therefore
triangles *ABL* and *MBF* are similar, and as *AB* is to *BL*,
MB is to *BF*, whence rectangle *AB–BF* will be equal to
rectangle *MB–BL*. But rectangle *AB–BF* has been shown
equal to *CB–BG*; therefore rectangle *MB–BL* is equal to
rectangle *CB–BG*, which is impossible; hence the intersection
[*L*] cannot fall outside the circle. In the same way it may be
demonstrated that it cannot fall inside, wherefore all intersections fall on the circumference itself.

But it is time now to give satisfaction to Simplicio by
showing him that it is not impossible to resolve a line into
its infinitely many points, and not only that, but that this
presents no greater difficulty than to distinguish its quantified
parts. First, one assumption; I do not think, Simplicio, that
you will deny this to me. I assume that you do not require
me to separate the points from one another and show them
to you distinctly one by one on this paper. In return, I shall   **92**
be content if without your detaching four or six parts of a
line one from another, you show me these divisions marked,
or even just bent at angles so as to form a square or a hexagon.
I am persuaded that you, too, will call such [parts] sufficiently
distinguished and actualized.

Simp. Of course.

Salv. Now, if bending of a line at angles, forming now a
square, now an octagon, now a polygon of forty or one
hundred or one thousand angles, is suffcient change
[*mutazione*] to reduce to act those four, eight, forty, one
hundred or one thousand parts that were previously in the
line "potentially," as you put it, then when I form of this
line a polygon of infinitely many sides—that is, when I bend
it into the circumference of a circle—may I not, with the same
license, say that I have reduced to act its infinitely many parts,
since you conceded that while it was straight, these were
said to be contained in it potentially? That such a resolution
[of a line] is made into its infinitely many points cannot be
denied, any more than that [a resolution was made] into its
four parts in forming a square, or into its thousand [parts] in

forming a milligon, inasmuch as none of the conditions are lacking here that are found in the polygon of one thousand or one hundred thousand sides. This latter, applied to a straight line and placed thereon, touches it with one of its sides, that is, with one one-hundred-thousandth part of it. The circle, which is a polygon of infinitely many sides, touches the straight line with one of its sides, which is a single point, different from all its neighbors, and therefore divided and distinguished from them no less than is one side of the polygon from its adjacent [sides]. And as the polygon, rotated on a plane, stamps out with the successive contacts of its sides a straight line equal to its perimeter, so does the circle, when rolled on a plane, describe with its infinitely many successive contacts a straight line equal to its circumference.

I don't know, Simplicio, whether the learned Peripatetics, to whom I grant as quite true the concept that the continuum is divisible into ever-divisibles in such a way that in continuing such division and subdivision one would never reach an end, will be willing to concede to me that none of their divisions is the last—as indeed none is, since there always remains another—and yet that there indeed exists a last and highest, and it is that which resolves the line into infinitely many indivisibles. I admit that one will never arrive at this by **93** successively dividing [the line] into a greater and greater multitude of parts. But by employing the method I propose, that of distinguishing and resolving the whole infinitude at one fell swoop—an artifice that should not be denied to me—I believe that they should be satisfied, and should allow this composition of the continuum out of absolutely indivisible atoms. Especially since this is a road that is perhaps more direct than any other in extricating ourselves from many intricate labyrinths. One such, in addition to that already mentioned of the [problem of the] coherence of the parts of solids, is the understanding of rarefaction and condensation, without our stumbling into the inconsistency of being forced by the former [rarefaction] to admit void spaces[32], and by the latter [condensation, to admit] the [inter]penetration of

---

32. Empty spaces of finite dimensions would create physical problems, since in Galileo's view natural effects would continually destroy them (nature's horror of a void.) Empty points in the mathematical sense raised no physical problem; hence Galileo employed them to attract physically adjacent particles and to keep them joined up to the limit of resistance to fracture. Cf. notes 8 and 10 above.

bodies, both these involving contradictions that seem to me to be cleverly avoided by assuming the said composition of indivisibles.

Simp. I don't know what the Peripatetics would say, inasmuch as the considerations you have set forth would strike them, I believe, for the most part as novelties, and as such they would need to be examined. It may be that the Peripatetics would find replies and solutions capable of untying those knots that I, from the shortness of time and the frailty of my intellect, cannot at present resolve. So leaving aside for now that [Peripatetic] faction, I should indeed like to hear how the introduction of these indivisibles facilitates the comprehension of condensation and rarefaction, while at the same time it circumvents [both] the void and the [inter]penetration of bodies.

Sagr. I too will hear this with pleasure, as it is still obscure to my mind. Provided, that is, that I shall not be defrauded of hearing, in accordance with what you said to Simplicio a short time ago, the reasonings of Aristotle in refuting the void, and then the solutions thereof at which you arrive, as is only fitting if you assume that which he denies.

Salv. Both shall be done. As to the first, it is necessary that just as we shall make use, in regard to rarefaction, of the line described by the smaller circle when that is driven by the revolution of the larger, which line is longer than its own circumference; so, for an understanding of condensation, we must show how the larger [circle, driven] by the revolution of the lesser, describes a straight line shorter than its own circumference. For a clear explanation of this, let us consider what happens with the polygons.

In a diagram similar to the previous one, let there be two **94** hexagons, *ABC* and *HIK*, around the common center *L*, and the parallel lines *HOM* and *ABc* on which they must be revolved. Let the corner *I* of the smaller polygon be fixed, and turn this polygon until the side *IK* falls on the parallel [*MH*]. In this motion, point *K* will describe arc *KM*, and side *KI* will unite with part *IM*. Let us see what side *CB* of the larger polygon will do. Since the revolution is made about the point *I*, the line *IB* with end *B* will go back, describing the arc *Bb* below the parallel *cA*, so that when side *KI* is joined with line *MI*, side *BC* will unite with line *bc*, going forward only as much as the part *Bc*, and leaving behind the part subtended by the arc *Bb*, which comes to be superimposed

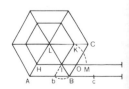

on the line *BA*. Assuming the rotation driven by the smaller polygon to continue in this way, it will trace and cover on its own parallel a line equal to its perimeter. But the larger [hexagon] will pass over a line shorter than its perimeter by one less line [of length] *bB* than [the number of] its sides, and this line will be approximately equal to that described by the lesser polygon, which it will exceed by only the length *bB*. Here, then, without any contradiction [*repugnanza*], is revealed the reason why the sides of the larger polygon, when driven by the smaller, do not cover a line greater than that traveled by the smaller; for a part of each side is superimposed on that which precedes and is adjacent to it.

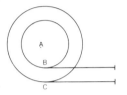

Now consider the two circles around center *A*, placed on their parallels so that the smaller touches one of these at point *B*, and the larger [touches] the other at point *C*. The smaller [circle] commencing to roll, its point *B* will not remain motionless for any time while the [imaginary] line *BC* goes backward carrying point *C*, as happened in the polygons, where point *I* remained fixed until side *KI* fell on line *IM*. There, line *IB* did carry *B* (one end of side *CB*) backward to *b* so that side *BC* fell on *bc*, superimposing part *Bb* on line *BA*, and advancing only by the part *Bc* equal to *IM*, or to one side of the smaller polygon. On account of these superpositions, equal to the excesses of the larger sides over the smaller, the residual advances made, equal to the sides of the lesser polygon, come to compose in one entire revolution the straight line equal to that marked and measured by the smaller polygon.

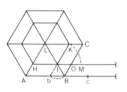

If we were to apply similar reasoning to the case of circles, we should have to say that where the sides of any polygon are contained within some number, the sides of any circle are infinitely many; the former are quantified and divisible, the latter unquantifiable and indivisible; either end of each side of the revolving polygon stays fixed for a time (that is, that fraction of the time of an entire revolution, which the side is of the entire perimeter), whereas in circles the delays of the ends of their infinitely many sides are momentary, because an instant in a finite time is a point in a line that contains infinitely many [points.] The backward turns made by sides of the larger polygon [each] advancing as far as the side of the lesser [polygon] are not those of a whole side, but only of its excess over a side of the smaller; in circles, the point (or "side") *C*, during the instantaneous rest of end *B*,

95

moves back as much as its excess over the "side" *B*, and advances by as much as [point] *B*.[33] To sum up, the infinitely many indivisible sides of the greater circle, with their infinitely many indivisible retrogressions, made in the infinitely many instantaneous rests of the infinitely many ends of the infinitely many sides of the lesser circle, together with their infinitely many advances, equal to the infinitely many sides of the lesser circle, compose and describe [*disegnano*, mark] a line, equal to that described by the lesser circle, which contains in itself infinitely many unquantifiable superpositions, making a compacting and condensation without any [inter]penetration of quantified parts.[34]

This is not to be understood as happening in the line divided into quantified parts, as in the perimeter of any polygon, **96** which extended into a straight line cannot be compressed into [any] shorter length except by superposition and interpenetration of its sides. The compacting of infinitely many unquantifiable parts without interpenetration of quantified parts, and the previously explained expansion of infinitely many indivisibles with the interposition of indivisible voids, I believe to be the most that can be said to explain the condensation and rarefaction of bodies without the necessity of introducing interpenetration of bodies and [appealing to] quantified void spaces. If anything in it pleases you, make capital of that; if not, ignore this as idle, and my reasoning along with it, and go search for some other explanation that will bring you more peace of mind. I repeat only this: we are among infinites and indivisibles.

Sagr. I freely confess that the idea is subtle, and to my ears novel and remarkable. Whether in fact nature proceeds in any such way, I cannot decide. The truth is that until I hear something that better satisfies me, I shall stick to this rather than remain completely dumb.

But Simplicio may have what I have not yet found—some

33. According to Berkeley's axiom (note 19, above), Galileo could not move from very small sides to point-sides for circles without losing the right to speak of any excess analogous to that existing between the sides of polygons of no matter how many sides. Being unable to argue consistently with his own views that points in one circle are somehow greater than those in another, he was left with no proper basis for the analogy here.

34. The purpose of this clearly deliberate complication in an ostensible summing up was to discourage all attempts to simplify paradoxes which in Galileo's opinion were genuine and needed to be thought about and thought through again and again.

way of explaining the explanation supported by his philosophers in this most abstruse matter. What I have read [in them]
up to the present concerning condensation is, for me, so
dense, and with regard to rarefaction what I have read is so
subtle, that my feeble vision neither takes hold of the latter,
nor penetrates the former.

Simp. I am filled with confusion, and find hard obstacles
in both opinions. Particularly in this new one; for according
to this rule, an ounce of gold might be rarefied and expanded
into a bulk greater than the whole earth, and all the earth
might be condensed and reduced into a bulk smaller than a
walnut. These things I do not believe, nor do I think that you
yourself believe them. The considerations and demonstrations
made by you up to this point, being mathematical things
abstracted and separated from sensible matter, I believe
would not work according to your rules if applied to physical
and natural materials.

Salv. I doubt that you want me to make you see the invisible,
nor am I able to do that; but so far as that which can be
understood by our senses is concerned, and since you mention
gold, do we not see immense expansions made of its parts?
I don't know whether you have thought of the way in which
artisans proceed in drawing gold for gilding, which is really
gold only on the surface, while the matter inside is silver. The
way they do this is to take a cylinder or rod of silver about
ten inches long and three or four fingers thick; this they cover
with beaten gold leaf, which as you know is so thin that it
goes floating through the air. They put on eight or ten such
leaves, no more; and they commence drawing it, thus gilded,
with great strength, passing it through the holes of a wire die
over and over again, drawing it successively through smaller
holes. After a great many passages they reduce it to the fineness
of a lady's hair or finer; yet it remains gilded on the surface.
I leave it to you to consider the thinness and expansion to
which the substance of gold is thus subjected.

Simp. I do not see that from this operation there comes in
consequence a thinning of the material of gold by performing
on it those marvels that you would have. First, the original
gilding was done with ten gold leaves, which are of perceptible
thickness; and second, though the silver grows in length with
the drawing and thinning, at the same time it diminishes just
as much in thickness, so that one dimension compensates the
other, and the surface is not increased in such a way that to

clothe the silver with gold must reduce it to greater subtlety than that of the original leaves.

Salv. You are very much mistaken, Simplicio, because the increase of surface is as the square root of the lengthening, as I can prove geometrically.

Sagr. For my sake as well as Simplicio's, I beg you to give us the demonstration, if you think we can follow it.

Salv. Let me see whether I can recall it offhand. It is manifest that the original thick cylinder of silver and the very long wire drawn from it are equal in volume, being of the same silver. So if I show the ratio that holds between the surfaces of the **98** equal cylinders, we shall have what we want. Hence I say that:

> The surfaces of equal cylinders, excluding [those of] their bases, are in the ratio of the square roots of their lengths.

Let there be two equal cylinders with heights *AB* and *CD*, and let line *E* be a mean proportional between them; then I say that the surface of cylinder *AB*, excluding its bases, has to the surface of cylinder *CD*, likewise without its bases, the same ratio that line *AB* has to line *E*, which is the square root of the ratio of *AB* to *CD*. Cut the cylinder *AB* at *F*, letting height *AF* equal *CD*. Now, since the bases of equal cylinders are inversely proportional to their heights, the circle at the base of cylinder *CD* will be to the circle at the base of cylinder *AB* as the height *BA* is to the height *DC*. And since circles are to one another as the squares of their diameters, these [two squares] have the same ratio as *BA* to *CD*. But as *BA* is to *CD*, so the square on *BA* is to the square on *E*, so that the four squares are proportional. Hence their sides are proportional, and as line *AB* is to *E*, so the diameter of circle *C* is to the diameter of circle *A*. And as the diameters, so are the circumferences; and as the circumferences, so also are the surfaces of cylinders equally high. Therefore as line *AB* is to *E*, so is the surface of cylinder *CD* to the surface of cylinder *AF*. Since height *AF* is to *AB* as surface *AF* is to surface *AB*, and as height *AB* is to line *E*, surface *CD* is to *AF*, then, by perturbed [proportion], surface *CD* will be to surface *AB* as height *AF* is to *E*; and by conversion, as the surface of cylinder *AB* is to the surface of cylinder *CD*, so is the line *E* to *AF* (or to *CD*), or as *AB* is to *E*, which is the square root of the ratio of *AB* to *CD*; and this is what was to be proved.

Now, if we apply what has thus been demonstrated to our original purpose, and assume that the silver cylinder which was gilded when it was no more than a foot [*mezzo braccio*] **99**

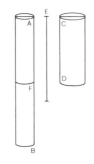

long and three or four fingers thick, then when drawn to the
fineness of a hair it is lengthened to forty thousand feet or
even more, and we shall find that its surface has grown two
hundred times over what it was originally. Hence those gold
leaves that were assumed to be ten in number, being extended
over a surface two hundred times as great, show us that the
gold that covers the surface of so many feet of wire can be no
thicker than one-twentieth [the thickness] of one ordinary
beaten gold leaf. Now consider whether this thinness is
possible to conceive without an enormous expansion of parts,
and judge whether this seems to you an experience that tends
in the direction of composition of infinitely many indivisibles
into physical materials—though for this, there are not lacking
other [experiences] stronger and more conclusive.

Sagr. The proof appears to me so elegant that even if it had
no power to persuade me of that original purpose for which
it was adduced (though indeed for me it has), the brief time
devoted to hearing it was well spent in any case.

Salv. Seeing how much you enjoy these geometrical demonstrations, the bearers to us of secure gains, I shall give you
the companion to this one, which settles a very curious
question. From the above, we know what happens with two
equal cylinders differing in height or length. It is good to
hear also what happens with cylinders equal in surface but of
unequal heights, meaning again the surrounding surfaces
without those of the upper and lower bases. I say that:

> Right cylinders of which the surfaces, excluding their
> bases, are equal, have [volumes in] the ratio of their
> heights taken inversely.

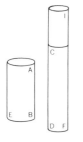

Let the surfaces of the two cylinders *AE* and *CF* be equal,
but let the height of *CD* be greater than that of *AB*; I say that
cylinder *AE* is to cylinder *CF* in the same ratio as is height
CD to *AB*. Since the surface *CF* is equal to *AE*, the volume
[*cilindro*] *CF* is less than *AE*; for if these were equal, then by
reason of the foregoing proposition, surface *CF* would be
100 greater than surface *AE*, and even more so if cylinder *CF*
were greater than *AE*. Take cylinder *ID* equal to *AE*; then,
by the foregoing, the surface of cylinder *ID* is to the surface
of *AE* as the height *IF* is to the [geometric] mean between *IF*
and *AB*. But it is given that surface *AE* is equal to *CF*; and
surface *ID* having to *CF* the same ratio as the height *IF* to
CD, it follows that *CD* is the mean proportional between
IF and *AB*. Furthermore, the cylinder *ID* being equal to

cylinder *AE*, both have the same ratio to cylinder *CF*. But *ID* is to *CF* as height *IF* is to *CD*; hence cylinder *AE* is to cylinder *CF* in the same ratio as line *IF* to *CD*; that is, as *CD* is to *AB*, which was the theorem.

From this we understand the reason for an event that is not heard without astonishment by most people; that is, that if the same piece of cloth, longer one way than the other, be made into a sack for holding grain, as is often done by placing a board at the bottom, it will hold more when we use for the height of the sack the smaller dimension of the cloth and wrap the longer around the board, than if made the other way. For example, if the cloth is six braccia one way and twelve the other, it will hold more when the length of twelve is wrapped around a board at the bottom and the sack is six braccia high, than if the enclosed circumference is six braccia and the height is twelve.

Now to this general information that there is greater capacity the former way than the latter, there is added from what has just been proved the specific and particular knowledge [*scienza*] of how much more is held. The sack will hold more to the extent that it is lower, and less to the extent that it is higher. In specific measures, if the cloth is twice as long as it is wide, then sewn lengthwise it will hold half as much as the other way. Similarly, using a straw mat to make a basket, **101** say [a mat] twenty-five braccia long and seven in width, then when rolled lengthwise it will hold only seven of those measures of which it will hold twenty-five when rolled the other way.

Sagr. And so, to our particular pleasure, we go on acquiring curious and useful knowledge. But in this last proposition, I doubt whether among people who lack knowledge of geometry you would find four in a hundred who would not be mistaken at first, [thinking] that bodies contained inside equal surfaces are equal in all respects. They make the same error when speaking of surfaces; for in determining the sizes of different cities, they often imagine that everything is known when the lengths [*quantità*] of the city boundaries are given, not knowing that one boundary may be equal to another, while the area contained by one may be much greater than that in the other. This happens not only with irregular surfaces, but also among regular ones, where those which have more sides are always more spacious [for the same perimeter] than those having fewer sides. Ultimately the circle, as a polygon of infinitely many sides, is the most capacious of all polygons of equal circumference. I recall having seen the proof of this with particular satisfaction when I was studying the *Sphere* of Sacrobosco and an added learned commentary.[35]

Salv. Very true: I also saw this passage, and had occasion to discover therefrom a unique and brief proof that the circle **102** may be concluded to be the greatest of all regular isoperimetric figures, while among the others, those with more sides are greater than those with fewer.

Sagr. I take such delight in proven propositions and selected demonstrations that depart from the trivial that I beg you to share this with me.

Salv. I hasten to prove to you briefly the following theorem:
The circle is the mean proportional between any two similar regular polygons of which one is circumscribed about it and the other is isoperimetric to it. Also, the circle being less than all circumscribed [figures], it is nevertheless the greatest of all isoperimetric [figures]. And among the circumscribed [polygons], those that have more angles are smaller than those that have fewer; on the other hand, among isoperimetric [polygons], those having more angles are the greater.

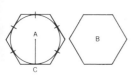

Take two similar polygons, *A* and *B*; let *A* be circumscribed about circle *A*, and let *B* be isoperimetric to this circle; I say that the circle is the mean proportional between them. Draw radius *AC*. The circle [*A*] is equal [in area] to that right triangle of which one side is the radius *AC* and the other is equal to the circumference; likewise, the polygon *A* is equal **103** to the right triangle that has one side equal to *AC* and the other to the perimeter of the polygon; hence it is evident that the circumscribed polygon has to the circle the same ratio that its perimeter has to the circumference of this circle, or to the perimeter of polygon *B*, which was assumed equal to that circumference. But polygon *A* is to *B* as the square of the ratio of its perimeter to the perimeter of *B*, these being similar figures; hence the circle *A* is the mean proportional between the polygons *A* and *B*. Since polygon *A* is greater than

35. The work mentioned, written in the thirteenth century, survives in many manuscript copies, and was repeatedly published; Galileo himself lectured on it at Padua as the standard elementary text on astronomy. The learned commentary mentioned here was written by Christopher Clavius (1537–1612), Jesuit mathematician at Rome and an early correspondent of Galileo's. See Clavius, *In sphaeram Ioannis De Sacro Bosco commentarius* (Rome, 1581), p. 81.

circle *A*, it is manifest that this circle *A* is greater [in area] than its isoperimetric polygon, *B*, and hence it is the greatest of all regular polygons to which it is isoperimetric.

As to the other part, we must prove that of polygons circumscribed around the same circle, that with fewer sides is greater than that with more sides; while on the other hand, of all isoperimetric polygons, that with more sides is greater than that with fewer sides. These [propositions] we prove thus.

To the circle of radius *OA* and center *O*, draw the tangent *AD*; let *AD* be the half-side of the circumscribed pentagon, and *AC* the half-side of the hectagon. Draw lines *OGC* and *OFD*, and taking *O* as center and *OC* as radius, describe the arc *ECI*. Since triangle *DOC* is greater than sector *EOC*, and sector *COI* is greater than triangle *COA*, triangle *DOC* will have a greater ratio to triangle *COA* than sector *EOC* has to sector *COI*, that is, than sector *FOG* to sector *GOA*. By composition and permuting, triangle *DOA* will have to sector *FOA* a greater ratio than triangle *COA* has to sector *GOA*, and ten triangles *DOA* will have to ten sectors *FOA* a greater ratio than fourteen triangles *COA* have to fourteen sectors *GOA*. Thus the circumscribed pentagon will have a greater ratio to the circle than will the hectagon, and hence the pentagon will be greater than the hectagon.

Now take a hectagon and a pentagon that are both isoperimetric to the same circle; I say that the hectagon is greater than the pentagon. For the same circle is a mean proportional between the circumscribed pentagon and the **104** isoperimetric pentagon, and is likewise the mean proportional between the circumscribed and the isoperimetric hectagon. It has been proved that the circumscribed pentagon is greater than the circumscribed hectagon; hence this pentagon will have a greater ratio to the circle than [will] the hectagon. That is, the circle will have a greater ratio to its isoperimetric pentagon than to its isoperimetric hectagon; therefore the pentagon is less than the isoperimetric hectagon, which was to be proved.

Sagr. A very refined proof, and most acute,[36] and one which at first glance seems to contain a sort of contradiction, since the reason for which the polygon of more sides is greater than its isoperimetric of fewer sides comes from the circumscribed of more sides being less than the circumscribed of

36. The remainder of this sentence was added in the margins of Galileo's copy of the printed book.

fewer sides. But where did we go astray, engulfing ourselves in geometry? We were about to consider the difficulties put forth by Simplicio, which truly need close attention—especially that of condensation, which seems very hard to me.

Salv. If condensation and rarefaction are opposing changes [*moti*], then wherever great rarefaction is found, condensation no less enormous cannot be denied. We see daily immense rarefaction; and what is still more remarkable, this is almost instantaneous. I refer to the boundless rarefaction of a small amount of gunpowder, when it is resolved into a vast bulk of fire. And what of the almost unlimited expansion of [its] light? If that fire and that light were to be reunited—which is not impossible, seeing that they previously took up so little space—what a condensation that would be! Reasoning thus, you will find thousands of like rarefactions, which are more readily observed than are condensations, since materials that are dense to begin with are more tractable and more [readily] subjected to our senses. We can handle wood, and see it

105   resolved into fire and light; but we do not thus see fire and light condensed to constitute wood. We see fruits, flowers, and a thousand other solid materials resolved (as a general rule) into odors; but we do not observe odorous atoms coming together in the constitution of scented solids.

But where we lack sensory observations, their place may be supplied by reasoning, which is able to make us no less capable of understanding the change [*moto*] of solids by rarefaction and resolution than [the change] of tenuous and rare substances by condensation. We shall investigate the possibility of condensation and rarefaction by theorizing how they can happen in bodies capable of being rarefied and condensed without [our] introducing the void and an interpenetration of bodies. This leaves open the possibility that materials may exist in nature that exclude those things, and hence do not involve events that you call contradictory and impossible. Finally, Simplicio, out of respect for you and the rest of your friends the most learned philosophers, I have worked out a theory of condensation and rarefaction in which these can be understood to take place without assuming interpenetration of bodies and [at the same time] without introducing void spaces, since those are effects that you despise and abhor—though if you would but concede them, I [for my part] should not oppose them so stubbornly.[37]

37. Here Galileo seems to incline toward the assumption of minute natural voids rather than point-voids; cf. notes 10, 15, and 32, above.

Hence you may either grant those contradictions, or welcome my theories, or find others that are more suitable.

Sagr. In the denial of interpenetration I am completely on the side of the Peripatetic philosophers. With regard to the void, I should like to hear judiciously weighed the demonstration with which Aristotle refutes it, and that with which you, [Salviati], oppose him. Do me the favor, Simplicio, of providing Aristotle's proof exactly; and you, Salviati, shall reply to it.

Simp. As I recall it, Aristotle does battle against some ancients who introduced the void as necessary for motion, saying that no motion could exist without it. Aristotle, opposing this [view], proves that on the contrary, the occurrence of motion, which we see, destroys the supposition of the void; and these are his steps.[38] He makes two assumptions; one concerning moveables differing in heaviness but moving in the same medium, and the other concerning a given moveable moved in different mediums. As to the first, he **106** assumes that moveables differing in heaviness are moved in the same medium with unequal speeds, which maintain to one another the same ratio as their weights [*gravità*]. Thus, for example, a moveable ten times as heavy as another, is moved ten times as fast. In the other supposition he takes it that the speeds of the same moveable through different mediums are in inverse ratio to the crassitudes or densities of the mediums. Assuming, for example, that the crassitude of water is ten times that of air, he would have it that the speed in air is ten times the speed in water.

From this second supposition he derives his proof [against the void] in this form: Since the tenuity of the void exceeds by an infinite interval the corpulence, though most rare [*sottilissima*], of any filled medium [*mezzo pieno*], every moveable that is moved through some space in some time through a filled medium must be moved through the void in a single instant; but for motion to be made instantaneously is impossible; therefore, thanks to motion, the void is impossible.

Salv. This argument is seen to be *ad hominem*; that is, it goes against those who would have the void as necessary for motion. Hence if I accept the argument as conclusive and grant that motion does not take place in the void, the supposition of the void taken absolutely, and not just in relation to motion, is not thereby destroyed.

---

38. See *Physica* 215a.24–216a.21; *De caelo* 301b.

But to say what those ancients [attacked by Aristotle] would perhaps reply, so that we may better judge the conclusiveness of Aristotle's argument, I think it possible to go against his assumptions and deny both of them. As to the first one, I seriously doubt that Aristotle ever tested [*sperimentasse*] whether it is true that two stones, one ten times as heavy as the other, both released at the same instant to fall from a height, say, of one hundred braccia, differed so much in their speeds that upon the arrival of the larger stone upon the ground, the other would be found to have descended no more than [*né anco*] ten braccia.

Simp. But it is seen from his words that he appears to have tested this, for he says "We see the heavier . . ." Now this "We see" suggests that he had made the experiment [*fatta l'esperienza*].

Sagr. But I, Simplicio, who have made the test, assure you **107** that a cannonball that weighs one hundred pounds (or two hundred, or even more) does not anticipate by even one span the arrival on the ground of a musket ball of no more than half [an ounce],[39] both coming from a height of two hundred braccia.

Salv. But without other experiences, by a short and conclusive demonstration, we can prove clearly that it is not true that a heavier moveable is moved more swiftly than another, less heavy, these being of the same material, and in a word, those of which Aristotle speaks. Tell me, Simplicio, whether you assume that for every heavy falling body there is a speed determined by nature such that this cannot 'be increased or diminished except by using force or opposing some impediment to it.

Simp. There can be no doubt that a given moveable in a given medium has an established speed determined by nature, which cannot be increased except by conferring on it some new impetus, nor diminished save by some impediment that retards it.[40]

Salv. Then if we had two moveables whose natural speeds

39. The text reads *mezza*, "one-half"; but since a musket ball could not weigh half as much as a cannonball, or even half a pound, it appears that the word "ounce" was inadvertently omitted.

40. This position is more extreme than the usual Peripatetic interpretation at the time. The essentials of Galileo's argument had been given in his early treatise *On Motion*, pp. 29–30 (*Opere*, I, 265–66). G.B. Benedetti (1530–90) had previously argued that two united bodies would not change speed if separated during free fall; see *Mechanics in Italy*, p. 206.

were unequal, it is evident that were we to connect the slower
to the faster, the latter would be partly retarded by the slower,
and this would be partly speeded up by the faster. Do you
not agree with me in this opinion?

Simp. It seems to me that this would undoubtedly follow.

Salv. But if this is so, and if it is also true that a large stone
is moved with eight degrees of speed, for example, and a
smaller one with four [degrees], then joining both together,
their composite will be moved with a speed less than eight
degrees. But the two stones joined together make a larger
stone than that first one which was moved with eight degrees
of speed;[41] therefore this greater stone is moved less swiftly
than the lesser one. But this is contrary to your assumption.   **108**
So you see how, from the supposition that the heavier body
is moved more swiftly than the less heavy, I conclude that
the heavier moves less swiftly.

Simp. I find myself in a tangle, because it still appears to
me that the smaller stone added to the larger adds weight to
it; and by adding weight, I don't see why it should not add
speed to it, or at least not diminish this [speed] in it.

Salv. Here you commit another error, Simplicio, because
it is not true that the smaller stone adds [*accresca*] weight to
the larger.

Simp. Well, that indeed is beyond my comprehension.

Salv. It will not be beyond it a bit, when I have made you
see the equivocation in which you are floundering. Note that
one must distinguish heavy bodies put in motion from the
same bodies in a state of rest. A large stone placed in a balance
acquires weight with the placement on it of another stone, and
not only that, but even the addition of a coil of hemp will
make it weigh more by the six or seven ounces that the hemp
weighs. But if you let the stone fall freely from a height with
the hemp tied to it, do you believe that in this motion the
hemp would weigh on the stone, and thus necessarily speed
up its motion? Or do you believe it would retard this by
partly sustaining the stone?

We feel weight on our shoulders when we try to oppose

41. A marginal addition in Galileo's copy of the book changes the rest
of this sentence to read "... therefore this composite (though it is greater
than that first [stone] alone) will be moved more slowly than the first alone,
which is lesser." He probably had noticed that it is not logically justified to
call two stones tied together "a greater stone" under the Aristotelian rule
as set forth by Simplicio; for this is not one stone, but still two, each endowed
with its own rule of motion governed by weight.

the motion that the burdening weight would make; but if
we descended with the same speed with which such a heavy
body would naturally fall, how would you have it press and
weigh on us? Do you not see that this would be like trying
to lance someone who was running ahead with as much
speed as that of his pursuer, or more? Infer, then, that in
free and natural fall the smaller stone does not weigh upon
the larger, and hence does not increase the weight as it does
at rest.

Simp. But what if the larger [stone] were placed on the
smaller?

109   *Salv.* It would increase the weight if its motion were faster.
But it was already concluded that if the smaller were slower,
it would partly retard the speed of the larger so that their
composite, though larger than before, would be moved less
swiftly, which is against your assumption. From this we
conclude that both great and small bodies, of the same specific
gravity, are moved with like speeds.[42]

Simp. Truly, your reasoning goes along very smoothly;
yet I find it hard to believe that a birdshot must move as swiftly
as a cannonball.

Salv. You should say "a grain of sand as [fast as] a mill-
stone." But I don't want you, Simplicio, to do what many
others do, and divert the argument from its principal purpose,
attacking something I said that departs by a hair from the
truth, and then trying to hide under this hair another's fault
that is a big as a ship's hawser. Aristotle says, "A hundred-
pound iron ball falling from the height of a hundred braccia
hits the ground before one of just one pound has descended
a single braccio." I say that they arrive at the same time.
You find, on making the experiment,[43] that the larger an-
ticipates the smaller by two inches; that is, when the larger
one strikes the ground, the other is two inches behind it.
And now you want to hide, behind those two inches, the
ninety-nine braccia of Aristotle, and speaking only of my
tiny error, remain silent about his enormous one.

Aristotle declares that moveables of different weight are
moved (to the extent this depends on heaviness) through

42. Mention of specific gravity appears superfluous here, but it is not;
the discussion thus far required comparison of bodies of the same material.
It was only after discussing resistance of the medium that an unqualified
statement could be made; see note 50, below.

43. The words "on making the experiment" had been inserted in Galileo's
own hand in the Pieroni MS, but were not printed in the 1638 edition.

the same medium with speeds proportional to their weights. He gives as an example moveables in which the pure and absolute effect of weight can be discerned, leaving aside those other considerations of shapes and of certain very tiny forces [*momenta*], which introduce great changes [*alterazione*] from the medium, and which alter the simple effect of heaviness alone. Thus one sees gold, which is most heavy, more so than any other material, reduced to a very thin leaf that goes floating through the air, as do rocks crushed into fine dust. If you wish to maintain your general pro-position, you must show that the ratio of speeds is observed **110** in all heavy bodies, and that a rock of twenty pounds is moved ten times as fast as a two-pound rock. I say this is false, and that in falling from a height of fifty or a hundred braccia, they will strike the ground at the same moment.

Simp. Perhaps from very great heights, of thousands of braccia, that would follow which is not seen at these lesser heights.

Salv. If that is what Aristotle meant, you saddle him with a further error that would be a lie. For no such vertical heights are found on earth, so it is clear that Aristotle could not have made that trial; yet you want to persuade us that he did so because he says that the effect "is seen."[44]

Simp. Well, the fact is that Aristotle did not make use of this rule [to refute the void], but of the other one, which, I believe, does not labor under these difficulties.

Salv. The other [rule] is no less false than this one, and I marvel that you yourself do not see through its fallacy and infer that if it were true that in mediums of different subtlety and rarity and yielding differently, such as water and air

---

44. A different version of this speech appears in the Pieroni MS, as follows: "*Salv*. If Aristotle had meant this, you would be burdening him with two more errors, whereas I remove two of the three because he did not actually commit them. One of the two [that you would add] would amount to a lie; for no such vertical heights are found on earth, so it is clear that Aristotle could not have made that trial. Yet (you say) he wants to persuade us that because he says the effect 'is seen,' he did make the experiment. The other error would be that if he introduced these considerations of ratios of speeds which hold for filled mediums, then in order to come to show the contradic-tions that would follow from their maintenance in void spaces [*mezzi vacui*, void mediums], and since such [ratios] are found only in mediums of immeasurable depths of thousands of braccia, he could not have concluded any more than that enormous void spaces cannot be found in nature—or at any rate [that they] are not found where heavy bodies do ordinarily move; a conclusion which, so far as I know, would be conceded to him by those ancients as well as by all modern philosophers."

for example, the same moveable were moved more swiftly
in air than in water, in the ratio of the rarity of air to that of
water, it would follow that every moveable falling in air
must also descend in water. But that is false, since many
bodies fall through air that do not descend in water, but
rise upward.

Simp. I fail to see how that must follow; and besides, I
say that Aristotle speaks of those heavy moveables that
descend in both mediums; not of those that fall in air and
rise in water.

Salv. You produce defences for the Philosopher that he
absolutely would not adduce, in order not to aggravate
111 his original mistake. Tell me whether the materiality
[*corpulenza*] of water, or whatever it may be that retards
motion, has some ratio to the materiality of air, that retards
it less; and if it does, assign that ratio at your pleasure.

Simp. It does have, and let us assume that the ratio is ten
to one, so that therefore the speed of a heavy body that
descends in both elements is ten times as slow in water as
in air.

Salv. Next, take one of those heavy bodies that go downward in air, but do not in water; say, a wooden ball. I ask
you to assign to this whatever speed you please for its descent
through air.

Simp. Let us assume that it moves with twenty degrees of
speed.

Salv. Very well. It is manifest that this speed has to some
lesser speed that ratio which the materiality of water has to
that of air; this speed is only two degrees. Thus, to go down
the line in agreement with Aristotle's assumption, one must
conclude that the wooden ball which, in its descent through
air that is ten times as yielding as water, is moved at twenty
degrees of speed, will descend through water with two [degrees
of speed], and not come floating up from the bottom as in
fact it does. Unless, of course, you mean that for wood,
to rise in water is the same thing as to fall with two degrees
of speed, which I do not believe. But since the wooden ball
does not sink to the bottom, I think you will grant that some
ball of other material can be found, different from wood,
that will descend through water with two degrees of speed.

Simp. No doubt something can be found, but of material
markedly heavier than wood.

Salv. That is what I sought. But this second ball, which

descends in water with two degrees of speed, will descend in air with what speed? You must reply, if you wish to use Aristotle's rule, that it will move at twenty degrees. But twenty degrees of speed is assigned by you yourself to the wooden ball, so that both it and the other (much heavier) ball will be moved at the same speed through air. Now, how does the Philosopher square this conclusion with that other of his, that moveables differing in heaviness are moved in the same medium with speeds differing in accordance with their weights [*gravità*]?

Without any deep thought, you cannot have failed to observe some frequent and palpable events, or to have   **112** noticed two bodies of which one will be moved in water a hundred times faster than the other, while in air, the faster of these does not outrun the other by even one part in a hundred. For instance, a marble egg will fall through water a hundred times as fast as a hen's egg, but through air it will not get four inches ahead in a distance of twenty braccia. One heavy body that takes three hours to get to the bottom in ten braccia of water will pass the same [ten] in air in a pulse beat or two.[45] From this experience it would follow that the density of water exceeds that of air by more than a thousand doubles. Yet on the other hand some other body, which might be a lead ball, will pass the same ten braccia through water in a time perhaps little more than double the time in which it will pass an equal space through air. From this second experience one would have to conclude that the density of water is little more than twice that of air!

Here, Simplicio, I know very well that you understand there is no room for any quibble [*distinzione*] or reply whatever. Let us conclude, then, that such an argument [as Aristotle's] proves nothing against the void, and if it did, it would destroy only [void] spaces of perceptible size. I neither suppose that the ancients assumed those to occur in nature, nor do I assume this myself, though indeed they may be created by force; this is deduced from various experiences that it would take too long to adduce now.

Sagr. Seeing that Simplicio remains silent, I'll take the field to say something. You have clearly demonstrated that it is not at all true that unequally heavy bodies, moved in

45. The remainder of this paragraph was added in the margin of Galileo's copy of the printed book, which concluded simply: "and one such (as for example, a ball of lead) will pass them in a time easily less than double."

the same medium, have speeds proportional to their weights [*gravità*], but rather have equal [speeds]. You assumed bodies of the same material (or rather, of the same specific gravity), and not (or so I think) of different density, because **113** I do not believe you mean us to conclude that a cork ball moves with the same speed as one of lead. Moreover, you have demonstrated quite clearly that it is not true that the same moveable, in mediums of differing resistances, maintains the same ratio in its speeds (or slownesses) as that of the resistances. It would now be a most satisfying thing to me to hear what ratios are observed in either case.

Salv. The questions are good, and I have often thought about them. I shall tell you my reasoning, and what I have ultimately deduced therefrom. It is certainly not true that the same moveable, in mediums of differing resistance, observes in its speed the ratio of the yieldings of these mediums; still less, that moveables of different heaviness, in the same medium, maintain in their speeds the ratio of the weights, meaning also [when they have] different specific gravities. After assuring myself of this, I began to combine these two phenomena together, noting what happened with moveables of different heaviness placed in mediums of different resistances, and I found that the inequality of speeds is always greater in the more resistant mediums, as compared with those more yielding. This difference is such that of two moveables descending in air and differing little in speed of motion, one of them will be moved in water ten times as fast as the other; or even such that one of them may swiftly descend in air, and not only fail to descend in water, but will remain quite still there, or even move upward. Thus sometimes one can find some kind of wood, or a knot or root, that remains at rest in water but will fall swiftly in air.

Sagr. I have tried many times, with great patience, to adjust a ball of wax so that it will not sink [or rise] by itself, adding grains of sand to it and seeking that degree of similarity with the weight [of an equal volume] of water that would hold it still in the midst of water. But with all my diligence I never did succeed in accomplishing this, so I do not know whether any solid material can be found that is so physically [*naturalmente*] similar to water in heaviness as, placed therein, to stay at any given place.

Salv. In this, as in a thousand other operations, there are many animals more skillful than we are. Fish are good evidence of this in the matter you mention, they being so expert in this exercise that they can at will equilibrate themselves not only with ordinary water, but with waters that are notably different, whether by nature or through the **114** advent of turbidity or saltiness, which makes a very great difference. They equilibrate so exactly, I say, that without the least movement they rest quietly at any place. This they do, I believe, by using the instrument given to them by nature for the purpose; that is, a little bladder that they have in their bodies which communicates with the mouth by a very fine tube. By means of this they can at will let out part of the air contained in this bladder; or, rising to the surface by swimming, they can draw in more [air], in this way rendering themselves heavier or less heavy than the water, and equilibrating themselves at will.

Sagr. By using a different artifice, I once fooled some friends to whom I had boasted of getting that ball of wax into exact equilibrium with water. Having put salt water into the lower part of a vessel, and fresh water above this, I showed them the ball at rest in the middle; pushed down or lifted up, it would not remain, but returned to the middle.

Salv. Nor is this experiment devoid of use; doctors in particular deal with the different qualities of water, and especially with comparisons of its lightness or heaviness, among other things. This they do with such a ball [as yours], prepared so that it cannot decide, so to speak, between sinking and rising in a given water. However small the difference in weight between two waters, if such a ball will descend in one, it will rise in the other. The experiment is so precise that the addition of just two grains of salt in six pounds of water will make a ball rise to the top that before would sink.

I want also to say something else, in confirmation of the delicacy of this experiment, and at the same time as a clear proof that water has no resistance to divison. Not only does mixture with some substance heavier than water make a noticeable difference in its heaviness, but merely heating or cooling slightly will produce the same effect. This operation is so subtle that the introduction of a few drops of water that is hotter or colder than the original six pounds will make the [said] ball fall or rise, descending when **115** hot water is added, and rising with the infusion of cold water. So you see how mistaken are those philosophers who would have water to possess some viscosity or cohesion of parts that makes it resistant to division and penetration.

Sagr. On this subject, I have seen very conclusive reasonings in a treatise by our Academician.[46] Yet a strong doubt remained with me, which I do not know how to remove. If no tenacity and coherence exists between the parts of water, then how can great drops, well raised up, sustain themselves without spreading and flattening, as we see especially on cabbage leaves?[47]

Salv. It is true that a man who has the right answer as his own can resolve all objections that are raised against it, but I do not arrogate to myself the power to do that. Still, inability on my part should not detract from the clarity of truth. In the first place, I confess that I don't know how that business of sustaining large and elevated globules of water is accomplished; yet I am certain that it does not derive from any internal tenacity existing among their parts, so the cause of this effect must be situated outside. That it is not internal, I can confirm by another experiment than those previously given.

If there were an internal cause by which the parts of that raised water were sustained when surrounded by air, then the water should be even better sustained when surrounded by a medium in which it has less propensity to sink than it has in air. Such a medium would be any fluid heavier than air; for example, wine. Therefore, some wine being poured around that globule of water, the wine should rise little by little without disturbing the parts of the water, stuck together by their [supposed] internal viscosity. But that is not what happens. Rather, no sooner does the liquor [poured] around approach the globule than, without waiting for this to rise around it, the water comes apart and flattens, staying under the wine [visibly] if that is red.

Therefore the cause of the effect is external, and perhaps it belongs to the surrounding air. Truly, great dissension is observed between air and water, which I have observed in another experiment, and this is that if I fill with water a

---

46. See note 5, above.

47. Galileo noted and discussed several phenomena of surface tension, but, perceiving that they did not depend on any internal properties of water, he ascribed them, as below, to a natural conflict between water and air; cf. *Bodies in Water*, pp. 33–39 (*Opere*, IV, 95–103); *Assayer*, p. 283 (*Opere*, VI, 323).

glass ball that has a small hole, about the size of a straw, and I turn it thus filled mouth downward, then, though water **116** is quite heavy and prone to descend in air, and air is likewise disposed to rise through water, being light, they will not agree the one to fall by coming out through the hole, and the other to rise by entering it, but both remain obstinate and contrary. But if I present to that hole a glass of red wine, which is almost imperceptibly less heavy than water, we promptly see it slowly ascending in rosy streaks through the water, while water with equal slowness descends through the wine, without their mixing, until finally the ball will be filled entirely with wine, and the water will drop quite to the bottom of the glass below.[48]

Now, what should be said here? What is deduced from this but a conflict [*disconvenienza*] between water and air, obscure to me, but perhaps . . .

Simp. I can hardly keep from laughing when I see Salviati's great antipathy for antipathy, since he will not even use the word; yet it is very suitable for solving the problem.[49]

Salv. Well, out of courtesy to Simplicio, let that be the solution of our puzzle; and stopping the digression, let us return to our purpose. We have seen that the difference of speed in moveables of different heaviness is found to be much greater in more resistant mediums. What now? In mercury as the medium, not only does gold go to the bottom more swiftly than lead, but gold alone sinks, and all other metals and stones are moved upward and float in mercury. Yet balls of gold, lead, copper, porphyry, and other heavy materials differ almost insensibly in their inequality of motion through air. Surely a gold ball at the end of a fall through a hundred braccia will not have outrun one of copper by four inches. This seen, I say, I came to the opinion that if one were to remove entirely the resistance of the medium, all materials would descend with equal speed.[50]

48. Despite the solubility of wine in water, very little mixing actually takes place when the experiment is performed as described by Galileo, using an aperture such that water will not flow out against atmospheric pressure.

49. Here Simplicio refers to a specific passage in the *Dialogue*, p. 410 (*Opere*, VII, 436).

50. Cf. note 42, above. The restriction as to specific gravity is now removed. Yet motion in a vacuum is discussed below only in terms of a "probable guess" because actual experiments could not be made by Galileo for want of the air-pump developed soon afterwards.

Simp. That's a fine thing to say, Salviati. I shall never believe that even in the void—if indeed motion could take place there —a lock of wool would be moved as fast as a piece of lead.

Salv. Gently, Simplicio; your difficulty is neither so recondite nor so unforeseeable that you should imagine it not to have occurred to me, and that consequently I have not **117** found the answer to it. Hence for my clarification and your own understanding, hear my reasoning. We are trying to investigate what would happen to moveables very diverse in weight, in a medium quite devoid of resistance, so that the whole difference of speed existing between these moveables would have to be referred to inequality of weight alone. Hence just one space entirely void of air—and of every other body, however thin and yielding—would be suitable for showing us sensibly that which we seek. Since we lack such a space, let us [instead] observe what happens in the thinnest and least resistant mediums, comparing this with what happens in others less thin and more resistant. If we find in fact that moveables of different weight differ less and less in speed as they are situated in more and more yielding mediums; and that finally, despite extreme difference in weight, their diversity of speed in the most tenuous medium of all (though not void) is found to be very small and almost unobservable, then it seems to me that we may believe, by a highly probable guess, that in the void all speeds would be entirely equal.

Let us, then, consider what happens in air. In order to have some form of very light material with a well-defined surface, we shall take an inflated bladder. The air inside this, in air itself as the medium, will weigh little or nothing, for not much can be compressed into it. Hence the [effective] heaviness is merely that of the membrane itself, which will be not one one-thousandth the weight of a quantity of lead the size of the inflated bladder. Now, Simplicio, when both are released from a height of four or six braccia, by how much space do you think the lead will get ahead of the bladder in its fall? Though you would have made it a thousand times as swift, you may be sure that it will not be ahead by a triple or even a double [speed].

Simp. It may be that at the beginning of motion, that is, in the first four or six braccia, things will happen as you say. But in the course of a longer continuation, I believe that the lead would leave the bladder behind not just six parts in twelve of distance, but eight or even ten such parts.

Salv. I, too, believe the same, and I do not doubt that
over very great distances the lead might go a hundred miles   **118**
before the bladder went one mile. But, my good Simplicio,
what you are offering me as an effect contradicting my proposition only confirms it the more. I repeat that my intention
is to explain that the cause of diverse speeds in moveables
of different heaviness is not that different heaviness at all,
but depends on external events, particularly on the resistance
of the medium, in such a way that by taking that away, all
moveables would move at the same degrees of speed.[51] I
deduce this chiefly from what you yourself now admit, and
what is certainly true; that is, that the speeds of [two] moveables very different in weight become more and more different
as the spaces they traverse become greater and greater. This
effect would not follow if the speeds depended on the different
weights; for those being always the same, the ratio between
the spaces traversed would remain always the same. But this
is the ratio we see to be always increasing as motion continues.
A very heavy moveable in a fall of one braccio will not get
ahead of the lightest one by the tenth part of that distance;
but in a fall of twelve braccia it will beat this by one-third;
in a fall of one hundred, by ninety percent; and so on.

Simp. This is all very well; but following in your tracks: If
the difference of weight in moveables of different heaviness
cannot cause the change [with distance] in the ratio of the
speeds, because the heaviness does not change, then neither
can the medium cause any alteration in the ratio of speeds,
since it too is always assumed to stay the same.

Salv. You cleverly bring against what I say an objection
that it is imperative to resolve. I say, then, that a heavy body
has from nature an intrinsic principle of moving toward the
common center of heavy objects (that is, of our terrestrial
globe) with a continually accelerated movement, and always
equally accelerated, so that in equal times there are added
equal new momenta and degrees of speed. This must be
assumed to be verified whenever all accidental and external
impediments are removed. Among these, there is one that
we cannot remove, and that is the impediment of the filled
medium that must be opened and moved laterally by the   **119**

---

51. The plural, "degrees", prepares for the coming discussion of accelerated
motion. For simplicity of treatment, Galileo had dealt with free fall up to
this point in terms of the common and Aristotelian conception of fixed natural
speeds.

falling moveable. The medium, though it be fluid, yielding, and quiet, opposes that transverse motion now with less, and now with greater resistance, according as it must be slowly or swiftly opened to give passage to the moveable, which, as I said, goes by nature continually accelerating, and consequently comes to encounter continually more resistance in the medium.[52] This means [some] retardation and diminution in the acquisition of new degrees of speed, so that ultimately the speed gets to such a point, and the resistance of the medium to such a magnitude, that the two balance each other, prevent further acceleration, and bring the moveable to an equable and uniform motion, in which it always [thereafter] continues to maintain itself. Thus there is an increase of resistance in the medium, not because this changes its essence, but because of change in the speed with which the medium must be opened and moved laterally to yield passage to the falling body that is successively accelerated.

Now since it is seen that there is very great resistance of the air to the small momentum of the bladder, and little to the great weight of the lead, I hold it to be certain that if the air could be entirely removed, greatly accommodating the bladder but aiding the lead very little, their speeds would be equalized. If we then assume the principle that in a medium no resistance exists at all to speed of motion, whether because it is a void or for any other reason, so that the speeds of all moveables would be equal, we can very consistently assign the ratios of speeds of like and of unlike moveables, in the same and in different filled (and therefore resistant) mediums. This we shall do by considering the extent to which the heaviness of the medium detracts from the heaviness of the moveable, which heaviness is the instrument by which the moveable makes its way, driving aside the parts of the medium. No such action occurs in the void [*nel mezzo vacuo*], and therefore no difference [in speed] is derived from different heaviness. And since it is evident that the medium detracts from the heaviness of the body contained in it to the extent of the weight of an equal quantity of its own material, diminishing in that ratio the speeds of the moveables which in a non-resistant medium

52. Here the resistance of the medium over and above its buoyancy effect is first introduced. This is functionally related to the square of the velocity. and not to the simple speed as assumed by Galileo; see note 11 to Fourth Day, below. Years earlier, Galileo had argued the concept of constant terminal speed on a wholly incorrect notion of acceleration; cf. *On Motion*, pp. 104–5 (*Opere* I, 332–33).

would remain equal (as assumed), we shall have our goal.

Assume, then, that lead is ten thousand times as heavy as air, but ebony only one thousand times. From the speeds of these **120** two materials, which would be equal taken absolutely (that is, with all resistance removed), air takes from lead one degree [of speed] in ten thousand, and from ebony one degree in one thousand, or ten in ten thousand. Hence if lead and ebony fall through air from any height, and would have fallen in the same time in the absence of retardation by the air, then the air will take away from the speed of lead one degree in ten thousand, while from ebony it will take away ten degrees. This is to say that dividing the height from which they fall into ten thousand parts, the lead will strike the ground when the ebony remains behind by ten, or rather nine, of the ten thousand parts. This means that a lead ball falling from a tower two hundred braccia high will be found to anticipate an ebony ball by less than four inches.

The ebony weighs one thousand times as much as air, but an inflated bladder weighs only four times as much; so from the instrinsic and natural speed of ebony, air detracts one degree in a thousand; but from that of the bladder, considered absolutely, let it take one degree in four; then the ebony ball falling from the tower will strike the ground when the bladder has passed by only three-quarters of the tower. Lead is twelve times as heavy as water, but ivory only twice; therefore water takes from lead one-twelfth, but from ivory one-half of their equal speeds. Hence when the lead shall have descended eleven braccia in water, the ivory will have dropped only six. And reasoning with this rule, I believe, we shall find that experience fits the computation much better than it fits Aristotle's rule.

Similarly we shall find the ratio between speeds of the same moveable in different fluid mediums, not by comparing the different resistances of the mediums, but by considering the excesses of heaviness of the moveable over the weights of [an equal bulk of] the mediums. Tin is a thousand times as heavy as air, and ten times as heavy as water; therefore, dividing the absolute speed of tin into a thousand degrees, in air, which detracts one one-thousandth, it will move with nine hundred ninety-nine [degrees], but in water with only nine hundred, since water takes away from it one-tenth of its heaviness, and air, one one-thousandth. Assuming a solid that is slightly heavier than water, which might be, for example, a ball of

121   oak that weighs one thousand drachms, while an equal
amount of water weighs nine hundred fifty, and that much air
weighs but two [drachms], it is evident that putting its absolute
speed at one thousand degrees, this will remain nine hundred
ninety-eight in air, while in water [it will be] only fifty,
inasmuch as water takes away nine hundred fifty of the
thousand degrees of weight and leaves only fifty. Hence this
solid would be moved almost twenty times as fast in air as
in water, just as the excess of its weight over that of water
is one-twentieth its own [weight].

Here we should consider that since only those materials that
are of greater specific weight than water can move down in it—
and these are consequently hundreds of times heavier than
air—then when we seek the ratio of their speeds in air and
water, we can assume without notable error that the air takes
nothing much away from the absolute heaviness or the
absolute speed of such materials. The excess of their weights
over the weight of water being easily found, we shall say that
the ratio of their speed through air to that through water is the
same as the ratio of their total weight to its excess over the
weight of water. For example, an ivory ball weighs twenty
ounces, and an equal amount of water weighs seventeen;
therefore the speed of ivory in air is to its speed in water approximately as twenty is to three.[53]

Sagr. I have learned much in this inherently curious matter,
about which I have often troubled my mind without gain.
Nothing is now lacking to put these speculations into practice,
except a method of knowing the weight of air with respect to
that of water, and thereby to other heavy materials.

Simp. What if it is found that air, instead of gravity, has
levity? What must then be said of this reasoning we have been
hearing, which otherwise is so ingenious?

Salv. It would have to be said that it is aerial, light, and
empty. But how can you question that air is heavy, when you
have Aristotle's clear text, affirming that all the elements except
fire have heaviness, even air itself? He adds that a sign of this
is that an inflated leather bottle weighs more than an empty
one.[54]

53. Acceleration is again ignored and steady speed assumed at all
distances, probably because of the great difference in the buoyancy of water
as compared with air; cf. note 51, above.

54. *De caelo* 311a.10–11. The words "except fire" are taken from the
Pieroni MS and were not printed in 1638.

Simp. That a leather bottle, or a football, weighs more when **122** inflated, results, I believe, from heaviness not in the air, but in the many thick vapors that are mixed with air here in our base regions. It is thanks to this, I should say, that heaviness increases in the leather bottle.

Salv. I do not like your saying this, and still less should you make Aristotle say it. For he, speaking of the elements and wishing to persuade me that the element of air is heavy, tries to have me see this by experience; if his proof were to say: "Take a leather bottle and fill it with gross vapors, and observe that its weight increases," I should tell him that it would weigh still more if filled with bran, adding that such experiences prove that bran and gross vapors are heavy, while as to the element of air, I remain in the same doubt as before. Aristotle's [own] experiment, though, is valid; and his proposition is true. But I can't say the same of another argument (taken, however, merely as an indication) by some philosopher whose name I forget, though I know I have read this.[55] The argument is that air is heavy rather than light, because it more readily carries heavy bodies downward than light ones upward.

Sagr. That's great, I swear. So by this argument, air will be much heavier than water, inasmuch as all heavy bodies are more readily carried downward through air than through water, and all light ones more readily [upward?] in water than in air. Indeed,[56] infinitely many materials rise through water that fall through air.

But let [increased] heaviness in the [inflated] bottle exist, Simplicio, whether because of gross vapors or pure air; this in no way bars our purpose of seeking what happens to bodies moved in this vaporous region of ours. Getting back to something else that troubles me more, I should like, for complete instruction in the present matter, not just to rest assured that air is heavy (for I am convinced), but if possible, to know its weight. So if you have anything that will satisfy me on this too, Salviati, I beg you to favor me with it.

55. Girolamo Borri (1512–92), *De motu gravium et levium* (Florence, 1576), p. 231. Borri was one of Galileo's teachers at Pisa, mentioned favourably in his early dialogue on motion; see *Mechanics in Italy*, p. 331 (*Opere*, I, 367). An experiment reported by Borri led Galileo to believe for a time that dense bodies move somewhat more slowly at the beginning of free fall than do less dense ones.

56. Here the Pieroni MS includes an essentially redundant clause omitted in the printed edition: "infinitely many heavy bodies descend in air that ascend in water, and".

Salv. Positive weight exists in air, and not lightness as some have believed; that is perhaps not be found in any material whatever. This is quite conclusively argued by the
123   experience of the inflated football given by Aristotle. For if the quality of absolute and positive lightness existed in air, then when air was multiplied and compressed, the lightness would be increased, and with it the propensity to go upward; but experience shows us the opposite. As to your other question, concerning the method of investigating the weight of air, I have carried that out in the following manner.

I took a large glass flask with a narrow neck to which I applied a leather collar, tied very tightly to the neck of the flask; into this was inserted and firmly tied a football-valve [*animella da pallone*] through which, by means of a syringe, I forced into the flask a great quantity of air; this permits itself to be very greatly condensed, so one can drive in two or three additional volumes [*altri fiaschi*] beyond what is naturally contained. Then, on a very delicate balance, I weighed most precisely that flask with the air compressed inside it, adjusting the balance with fine sand. The valve was then opened, giving exit to the air forcibly held in the flask, which was then put back on the balance and was found to be appreciably lighter. I took away sand from the counter-weight, setting it aside, until the balance came to rest, with the flask and the remaining sand in equilibrium. There can be no doubt that the weight of the sand set aside was that of the air forced into the flask, and afterward released.

Up to this point, the experiment assures me only that the air forcibly held in the vessel weighs as much as the sand saved. I still do not know how much air weighs definitely and unequivocally, with respect to water or other heavy materials, and I cannot know this unless I measure the quantity of air compressed. For this investigation one needs a rule, and I have found two ways in which we can proceed.

One of these is to take another flask, narrow-necked like the first, to the neck of which is tightly tied another collar that will receive the valve of the first, around which it is to be fastened with a tight knot. This second flask must have a [small] hole in the bottom, allowing an iron rod to be inserted, with which we can at will open the valve and give
124   exit to the excess air in the first flask, after it has been weighed; and this second vessel is to be filled with water.

The whole apparatus thus prepared, the valve is opened by means of the rod; the air, coming out impetuously and entering the vessel of water, drives water out through the hole in the bottom. It is obvious that the quantity of water thus expelled is equal to the volume of air coming out of the original vessel; this water is to be saved. Weigh again the [first] flask, now lightened by the [escape of the] compressed air; it is assumed that this flask with the compressed air was already weighed before. The excess sand being removed in the way previously described, it is manifest that this gives the exact weight of as much air in volume as the volume of the water expelled and saved. By weighing that [water], we shall see how many times its weight contains the weight of the sand put aside; and without error we can say that water is that many times heavier than air. This will not be ten times, as Aristotle seems to believe, but about four hundred times, as shown by this experiment.[57]

The other method is quicker and may be carried out with a single flask, namely, the first one, prepared as before. Into this we shall [this time] not put more air beyond what is naturally there, but we shall drive water in without letting any air escape; it must yield to the incoming water and be compressed. Having driven in as much water as possible— and without much force, three-quarters of the capacity of the flask can be put in—we place it on the balance and weigh it very carefully. That done, hold the flask mouth upward and open the valve to free the air, of which exactly as much will escape as the amount of water contained in the flask. The air having escaped, replace the flask on the balance. It will be found to be lighter by the departure of air, and subtracting from the counterweight the excess [as before], from this weight we have the weight of as much air as there is water in the flask.

Simp. The artifices you have invented cannot be called anything except subtle and ingenious; but while they seem to have given entire satisfaction to my mind, they confuse me in another direction. It is undoubtedly true that the elements in their own regions are neither heavy nor light; hence I can't understand of that portion of air that appeared to weigh, say, four drachms of sand, how or where this can

57. Galileo's value is about one-half the correct figure. He first described the experiment in a letter to G.B. Baliani in 1614, where he gave the ratio 460:1 (*Opere*, XII, 354).

125 really be said to have that weight *in air*, where the sand that balanced it indeed does retain its weight. So it seems to me that the experiment should be performed not in the airy element, but in some medium in which air itself can exert its burden [*talento*] of weight, if it truly has any.

Salv. Simplicio's objection is certainly sharp, so it must either be insoluble, or its solution must be equally subtle. It is quite clear that the air which, when compressed, is shown to weigh as much as the sand, being then released into its own element, no longer weighs [anything] there, while the sand still does. Hence in order to make the experiment [properly], we must choose a place and a medium where air, no less than sand, can gravitate. As we said before, from the weight of every material immersed in it, the medium detracts the weight of a volume of the medium equal to the volume immersed in it. Thus air takes from air all its weight, and in order to be performed precisely, the operation must be carried out in the void, where every heavy body exercises its moment without any diminution. Well then, Simplicio; if we were to weigh a quantity of air in the void, would you then rest satisfied and assured of the fact?

Simp. Yes indeed; but this is to ask or wish for the impossible.

Salv. And therefore you should be very much obliged to me when, out of affection for you, I effect the impossible. But I do not want to sell you what I have already given you. In the experiment already adopted, we did weigh air in the void and not in air or any other filled medium. For from the volume immersed in the fluid medium, Simplicio, that medium subtracts weight [only] because it resists being opened, driven aside, and finally lifted up.[58] A sign of this is given to us by its promptness in running immediately back to refill the space that the immersed bulk occupied in it, as soon as it leaves that space. For if it felt nothing of that immersion, it would not oppose it. Now, tell me: When you had, in air, the flask filled with the air naturally contained in it, what division, what driving aside, or in a word, what change did the surrounding external air receive from the

58. There is an implication that if the displaced air were not eventually lifted, its displacement would not resist the moving body at all. Elsewhere, Galileo recognized repeatedly that fluids resist motion of any appreciable speed, quite apart from the buoyancy effect.

additional air that was then forced into the vessel? Did this perhaps enlarge the vessel, so that the ambient [air] had to withdraw a bit to yield it room? Surely not. Hence we may **126** say that the new air is not immersed in the ambient; it occupies no space therein, and is as if it were placed in the void.[59] Indeed, it really is so placed, transfused into the voids that were not filled completely by the original, uncondensed, air.

I really can't see the difference between any two situations of ambit and ambient in which, on the one hand, the ambient does not push against the ambit, and on the other hand the ambit does not push against the ambient. Such are the situations of matter in the void, and of air newly compressed in the flask. The weight, then, that is found in the condensed air is that [weight] which it would have, spread freely in the void. It is indeed true that the weight of the sand that counterpoised it, if weighed in the void rather than in open air, would have been a little more precise: and hence we should say that the air weighed is really somewhat heavier than the sand that balanced it; namely, by as much as an equal bulk of air [to that of the sand] would weigh in the void.

Sagr.[60] A very acute speculation, which contains the solution of a **[126]** problem that seemed to partake of the miraculous. In substance, restricted to a few words, this shows us a way of finding the weight of any body weighed in a void, though we weigh it only in the filled medium of air. The explanation is this. Air detracts, from the absolute weight of every heavy body located in it, as much weight as the weight of a volume of air equal to the volume of the original body. Hence whoever could couple with the given body as much air as its own volume, without enlarging the body, would on weighing it have its absolute weight, that which it would have in the void, since without increasing its volume, there was added that very weight which is subtracted from it by air as the medium. Thus when in the flask already filled by the air naturally contained in it, a quantity of water is introduced without allowing any of the contained air to escape, it is manifest that the air naturally contained is restrained and condensed into a smaller volume, to give place to the water introduced. And it is evident that the volume of air so restricted is equal to the volume of the water introduced. Hence when one weighs in air the flask so prepared, it is evident that the weight of **[127]**

59. This very penetrating remark is promptly related by Galileo to the theory of condensation he had previously propounded on pp. **96** ff.; see below.

60. Sagredo's speech was dictated by Galileo and inserted in his copy of the printed book.

the water is accompanied by [that of] an equal amount of air. Of the [total] weight, part is that of the water together with an equal amount of air, and this is the same weight that the water alone would have in the void. [To find this,] the whole vessel is weighed, and its whole weight is noted down. Then the compressed air is given exit, and everything remaining is reweighed; because of the release of air, this weight will be diminished. Taking the difference of the two weights, we shall have the weight of the compressed air that had been equal in volume to the water. Then taking the weight of the water alone, and adding to this the weight (which we noted separately) of the compressed air, we shall have the weight of the same water alone in the void. Next, to find the weight of the water [in air], empty the water from the vessel, weigh the vessel alone, and subtract this weight from that of the vessel plus water, as weighed before. It is evident that the remainder is the weight of the water alone, in air.

127   *Simp.* It did seem to me that there was still something to be desired in the experiments adduced, but now I am entirely satisfied.

Salv. What I have set forth thus far is new; especially that no difference of weight, however great, plays any part at all in diversifying the speeds of moveables, so that as far as speed depends on weight, all moveables are moved with equal celerity. At first glance, this seems so remote from probability that, if I did not have some way of elucidating it and making it clear as daylight, it would have been better to remain silent than to assert it. So now that it has escaped my lips, I must not neglect any experiment or reason that can corroborate it.

Sagr. Not only this proposition, but many others of yours are so far from the opinions and teachings commonly accepted, that to broadcast them publicly will excite against them a great number of contradictors; for the innate condition of men is to look askance on others working in their field whose studies reveal truth or falsity which they themselves fail to perceive. By calling such men [as you] "innovators of doctrines," a title most unpleasant to the ears of the multitude, they strain to cut those knots they cannot untie, and to demolish with underground mines those edifices which have been built by patient artificers, working with **128** ordinary instruments. But to us, who are far from any such motives, the experiments and reasons adduced up to this point are quite satisfactory.

Salv. The experiment made with two moveables, as different as possible in weight, made to fall from a height in order to observe whether they are of equal speed, labors under certain difficulties. If the height is very great, the medium that must be opened and driven aside by the impetus of the falling body will be of greater prejudice to the small momentum of a very light moveable than to the force of a very heavy one, and over a long distance the light one will remain behind. But in a small height it may be doubtful whether there is really no difference [in speeds], or whether there is a difference but it is unobservable. So I fell to thinking how one might many times repeat descents from small heights, and accumulate many of those minimal differences of time that might intervene between the arrival of the heavy body at the terminus and that of the light one, so that added together in this way they would make up a time not only observable, but easily observable.

In order to make use of motions as slow as possible, in which resistance by the medium does less to alter the effect dependent upon simple heaviness, I also thought of making the moveables descend along an inclined plane not much raised above the horizontal. On this, no less than along the vertical, one may observe what is done by heavy bodies differing in weight. Going further, I wanted to be free of any hindrance that might arise from contact of these moveables with the said tilted plane. Ultimately, I took two balls, one of lead and one of cork, the former being at least a hundred times as heavy as the latter, and I attached them to equal thin strings four or five braccia long, tied high above. Removed from the vertical, these were set going at the same moment, and falling along the circumferences of the circles described by the equal strings that were the radii, they passed the vertical and returned by the same path. Repeating their goings and comings a good hundred times by themselves, they sensibly showed that the heavy one kept time with the **129** light one so well that not in a hundred oscillations, nor in a thousand, does it get ahead in time even by a moment, but the two travel with equal pace. The operation of the medium is also perceived; offering some impediment to the motion, it diminishes the oscillations of the cork much more than those of the lead. But it does not make them more frequent, or less so; indeed, when the arcs passed by the cork were not more than five or six degrees, and those of the lead were fifty or sixty, they were passed over in the same times.[61]

Simp. If that is so, why then will the speed of the lead not be [called] greater than that of the cork, seeing that it travels sixty degrees in the time that the cork hardly passes six?

Salv. And what would you say, Simplicio, if both took the same time in their travels when the cork, removed thirty degrees from the vertical, had to pass an arc of sixty, and the lead, drawn but two degrees from the same point, ran through an arc of four? Would not the cork then be as much the faster? Yet experience shows this to happen. But note that if the lead pendulum is drawn, say, fifty degrees from the vertical and released, it passes beyond the vertical and runs almost another fifty, describing an arc of nearly one hundred degrees. Returning of itself, it describes another slightly smaller arc; and continuing its oscillations, after a great number of these it is finally reduced to rest. Each of those vibrations is made in equal times, as well that of ninety degrees as that of fifty, or twenty, or ten, or of four. Consequently the speed of the moveable is always languishing, since in equal times it passes successively arcs ever smaller and smaller. A similar effect, indeed the same, is produced by the cork that hangs from another thread of equal length, except that this comes to rest in a smaller number of oscillations, as less suited by reason of its lightness to overcome the impediment of the air. Nevertheless, all its vibrations, large and small, are made in times equal among themselves, and also equal to the times of the vibrations of the lead. Whence it is true that if, while the lead passes over an arc of fifty degrees, the cork passes over only ten, then the cork is slower than the lead; but it also happens in reverse that the cork passes along the arc of fifty while the lead passes that of ten, or six; and thus, at different times, the lead will now be faster, and again the cork. But if the same moveables also pass equal arcs in the same equal times, surely one may say that their speeds are equal.

130

61. In an earlier discussion, Galileo had avoided the assertion of exact isochronism for all arcs; cf. *Dialogue*, pp. 230, 450 (*Opere*, VII, 256, 475). The error introduced here seems to have been a deduction from the false assumption that air resistance is proportional to speed, rather than to its square; cf. note 52, above. In the Fourth Day, a fictitious experiment is adduced to support this supposed isochronism (p. **277**, below).

Simp. This reasoning seems to me conclusive, and also it seems it isn't; my mind feels a kind of confusion that arises from the moving of both moveables now quickly, now slowly, and again extremely slowly, so that I can't get it straight in my head whether it is true that their speeds are always equal.

Sagr. I'd like to say a word, Salviati. Tell me, Simplicio, whether you grant that it may be said with absolute truth that the speed of the cork and that of the lead are equal every time that they both start at the same moment from rest and, moving along the same slopes, always pass equal spaces in equal times.

Simp. In this there is no room for doubt; it cannot be contradicted.

Sagr. Now it happens with either pendulum that it passes now sixty degrees, now fifty, now thirty, now ten, now eight, now four, now two, and so on. And when both pass the arc of sixty degrees, they pass this in the same time; in the arc of fifty, both bodies spend the same time; so in the arc of thirty, of ten, and the rest. Thus it is concluded that the speed of the lead in the arc of sixty degrees is equal to the speed of the cork in the same arc of sixty; and that the speeds in the arc of fifty are still equal to each other, and so on in the rest. But nobody says that the speed employed in the arc of sixty [degrees] is equal to that consumed in the arc of fifty, nor this speed to that in the arc of thirty, and so on. The speeds are always less in the smaller arcs, which we deduce by seeing with our own eyes that the same body spends as much time in passing the large arc of sixty degrees as in passing the smaller of fifty or the very small arc of ten; and in sum, that all arcs are passed in equal times. It is therefore true that the lead and the cork do go retarding their motion according to **131** the diminution of the arcs, but their agreement in maintaining equality of speed in every arc that is passed by both of them remains unaltered.

I wanted to say this to learn whether I have correctly understood Salviati's idea, rather than because I believe that Simplicio deserved a clearer explanation than that of Salviati, who here, as in all things, is most lucid. Usually he unravels questions that seem not only obscure, but repugnant to nature and the truth, [and does this] by reasons, observations, or experiences that are well known and familiar to everyone. I have heard from various people that this has given occasion to a certain highly esteemed professor to deprecate his discoveries [*novità*], holding them to be base, as depending on foundations too low and common—as if it were not the most admirable and estimable condition of the demonstrative sciences that they arise and flow from well-known principles, understood and conceded by all.

But let us go on feasting on these light foods, assuming that Simplicio is now willing to assume and grant that the internal heaviness of different moveables has no part at all in diversifying their speeds, so that all, so far as weight is concerned, move with the same speeds. Salviati, tell us how you explain the sensible and obvious inequalities of motion, and reply to Simplicio's objection, which I also confirm, that we see a cannonball fall more swiftly than a lead shot, whereas according to you, the difference of speed will be small. I counter this with some moveables of the same material, of which the larger will fall in less than a pulsebeat, in one medium, through a space that others smaller will not pass in an hour, or four, or twenty. These are stones, and fine sand, to say nothing of that dust that muddies water, a medium in which this does not fall through two braccia in many hours, through which [distance] rocks, and not very big ones at that, fall in a pulsebeat.

Salv. The part played by the medium in more greatly retarding moveables according as they are less in specific gravity has already been explained; this occurs by the subtraction of weight. How a given medium can reduce speed very differently in bodies that differ only in size, and are of the same material and shape, requires for its explanation subtler reasoning than that which suffices to understand how a flat shape in a moveable, or motion of the medium against one, retards its speed.

For the present problem, I reduce the reason to the roughness and porosity found commonly, and for the most part necessarily, at the surface of solid bodies. In motion, those irregularities strike the air or other surrounding medium, an evident sign of which is that we hear bodies hum when they fly rapidly through the air, even when rounded as thoroughly as possible; and they not only hum, but they are heard to whistle and hiss if some notable cavity or protuberance exists in them. It is also seen that every round solid turned on a lathe makes a little breeze. Again, we hear a humming, very high in pitch, made by a top when it spins rapidly on the ground. The pitch of this tone deepens as the spinning languishes bit by bit; this also necessarily argues hindrances by

132

the air of the surface roughnesses, however tiny. It cannot be doubted that in the descent of moveables these [irregularities] rub against the surrounding fluid and bring about retardation of speed, greater as the surface is larger, as is the case with smaller solids in comparison with large ones.

Simp. Wait, please, for here I begin to get confused. Although I understand and grant that friction of the medium with the surface of the moveable slows the motion, and that the slowing is greater (other things being equal) where the surface is larger, I do not understand on what grounds you call the surface of smaller solids greater. Besides, if as you say, a larger surface should bring about greater retardation, then larger solids should be slower, which is not the case. This objection is, however, easily removed by saying that although the larger has the greater surface, it also has greater heaviness; and against this, the impediment of greater surface does not surpass the impediment of a smaller surface as against the smaller heaviness, so that the speed of the larger solid does not become smaller.[62] Hence I see no reason why the equality of speeds should be altered, for to the extent that the motive heaviness is diminished, the retarding property of the surface is diminished equally.

133

Salv. I shall resolve jointly all that you oppose to me. You, Simplicio, assume two equal moveables of the same material and shape, which unquestionably do move equally fast, and then say that if one of these be diminished as much in heaviness as in surface, (but retaining the similarity of shape), the speed will not be reduced in this smaller one.

Simp. It really seems to me that that must follow in your teaching, which has it that greater or less heaviness does not act to accelerate or retard motion.

Salv. This I confirm, and I also grant you your dictum, from which it appears to me that in consequence, when heaviness is diminished more than is surface, some retardation of motion is introduced into any body so reduced; and the more, in proportion as the diminution of weight is greater than the reduction of surface.

Simp. I have no objection to this.

Salv. Know, then, Simplicio, that one cannot diminish the

---

62. What Simplicio here offers is precisely the kind of plausible verbal explanation that had long held back the development of accurately descriptive mathematical physics, influencing also Galileo's early treatise *On Motion*, pp. 106 ff. (*Opere*, I 333 ff.).

surface of a solid exactly as the weight, and still keep the shape similar. For manifestly, in diminishing a heavy solid, the weight is lessened as is the volume; and since in preserving similarity of shape the volume is always diminished more than the surface, the weight will also be reduced more than is the surface. But geometry teaches us that in similar solids, the ratio between volumes is much greater than the ratio of surfaces. For your better understanding of this I shall explain a particular instance.

Imagine for example a die of which the side is two inches long, so that one face will be four square inches, and all six [faces], that is, its whole surface, [will be] twenty-four square inches. Next, imagine that the die is sliced with three cuts into eight smaller dice. The side of each of them will be one inch, and each face one square inch, and its whole surface **134** six square inches, whereas the surface of the uncut die contained twenty-four. Now you see that the surface of the little die is one-quarter the surface of the larger one, this being the ratio of six to twenty-four. But the volume of the same [cut] die is only one-eighth. Thus the volume, and hence the weight, falls off much more quickly than the surface. If you subdivide the little die into eight others, the whole surface of one of these will be one and one-half square inches, which is one-sixteenth of the surface of the original die, while its volume is only one sixty-fourth. See how in just these two divisions, the volumes have diminished four times as much as have the surfaces; and if we continue the subdivision until the original solid is reduced to fine powder, we shall find the weight of the minute atoms to be diminished hundreds and hundreds of times as much as their surfaces.

What I have exemplified for you by cubes happens in all similar solids, the volumes of which are as the three-halves power of their surfaces. You see, therefore, in how much greater ratio the impediment of the surface contact of the moveable with the medium grows in small moveables than in larger ones. And if we add that the roughness of the tiny surfaces in fine powders is perhaps not any less than that of the surfaces of highly polished larger solids, we see how necessary it is that the medium be fluid, and entirely devoid of resistance to its being separated, if it is to give way to the passage of so feeble a force. And note also, Simplicio, that I was by no means mistaken when I said a moment ago that the surface of smaller solids is larger in comparison with that of larger solids.

Simp. I am quite satisfied; and you may both believe me that if I were to begin my studies over again, I should try to follow the advice of Plato and commence from mathematics, which proceeds so carefully, and does not admit as certain anything except what it has conclusively proved.

Sagr. I liked this discussion very well, but before we proceed I want to understand a term that is new to me. You just said that similar solids are to one another as "the three-halves power" of their surfaces. I saw and took in the proposition, **135** with its demonstration in which it is proved that the surfaces of similar solids are in the doubled ratio [i.e., as the squares] of their sides, and the further proof that the volumes are in the tripled ratio [i.e., as the cubes] of the sides. But I cannot recall my ever having heard the ratio of solids to their surfaces named before.

Salv. Well, you are replying for yourself, and answering your own question. For is not that which is the triple of something, of which another thing is double, said to be three-halves of that double? Now, if the surfaces are in doubled ratio of the lines of which the volumes are in tripled ratio, can we not say that the solids are as the three-halves power of the surfaces?[63]

Sagr. I understand; and though some other details concerning the material treated remain for me to ask about, still, if we go on that way from one digression to another, we shall be very late in getting to the questions principally intended, which relate to the various phenomena of resistance by solids to their being broken. So if it suits both of you, we may pick up the original thread from which we started.

Salv. Well said. But the things examined have been so varied that they have robbed us of much time, so that there is little left today to spend on that principal subject, which is full of geometrical demonstrations that must be attentively considered. I think it is better to put this off until tomorrow's meeting, when I can bring along with me some sheets on which I have noted down in order the theorems and problems in which different essentials of that subject are set forth and proved. I should perhaps not call these to mind in the proper order by memory alone.

63. Fractional exponents were not in general use, but this particular relationship of 3:2 was easy enough for Galileo's readers to grasp by reason of the special Euclidean terminology for ratios of squares and cubes; see Glossary for "doubled ratio" and "tripled ratio," which are "powers" in modern language.

Sagr. I willingly accept that counsel, the more so as in finishing today's session I shall have time to hear the explanation of some questions that remain with me on the matters we have just dealt with. One of these is whether we must take the impediment of the medium [alone] as adequate to stop the acceleration of very heavy material in great bulk **136** and spherical shape. I say "spherical" in order to take that which is contained within the smallest surface and is therefore least subject to retardation.

Another [question] concerns the oscillations of pendulums, and it falls into two parts. One is whether all oscillations, large, medium, and small, are truly and precisely made in equal times. The other concerns the ratio of times for bodies hung from unequal threads; the times of their vibrations, I mean.

Salv. The questions are good, and as happens with all truths, I am afraid that in dealing with either of them, other true and curious consequences will be drawn in. I do not know if we shall have time to discuss them all today.

Sagr. If they have the flavor of those already covered, I should be happy to employ on them as many days, let alone as many hours as remain until dark. And I believe that Simplicio will not be wearied by such discussions.

Simp. Certainly not; especially if they deal with physical questions on which no opinions or arguments of other philosophers are to be read in books.

Salv. Then I shall take up the first, and affirm without doubt that there is no sphere so large, nor any material so heavy, that the resistance of the medium, however thin, fails to restrain its acceleration and bring it to uniformity of motion in its continuation. We have a very clear argument of this from experience itself. If any falling moveable were able, by continuing its motion [through a medium], to acquire any degree of speed whatever, then no speed that could be conferred on it by a mover external [to that medium][64] could be so great that the moveable would reject it and be despoiled of it thanks to the impediment of that medium; thus if a

---

64. The limitations added in square brackets seem essential to the sense of the ensuing discussion, in which Galileo did not intend to deny that speed would increase without bound in the void. At the critical moment of entry into water, the speed that has been naturally acquired during fall through air may be (and in Galileo's example is) too great to have been acquired during any fall through water. Yet some speed of entry might be small enough to be continued, or even increased, during the subsequent fall through water.

cannonball that had fallen, say, four braccia through air and acquired ten degrees of speed with which it then entered into water, and the impediment of the water were not able to cancel that impetus in the ball, the impetus would increase, or would at least continue to the bottom, which is not seen to happen. Indeed, the water, though not more than a few braccia deep, impedes and weakens the impetus so that it will make a very light impact on the bed of the river or lake. It **137** is therefore manifest that the speed which the water was able to take from the ball in a short passage would never be permitted [by water] to be acquired even at a depth of a thousand braccia. And why would it permit this to be gained in a thousand, only to be taken away later in a few braccia?

But what next? It is seen that the enormous impetus of the ball shot from the same cannon is so much abated by the interposition of a few braccia of water that with no damage to a ship, it hardly even reaches it to strike it. And air [itself], though very yielding, nevertheless does repress the speed of a falling moveable, however weighty, as we may understand with similar experiments. For if we should fire an arquebus downward from the top of a high tower,[65] the shot would make a smaller dent in the ground than if we had made the shot from a point only four or six braccia above, a clear sign that the impetus with which the ball fired from the top of the tower comes out of the barrel is diminished in descending through air. Therefore no descent, from any height whatever, would be sufficient to make the ball acquire that impetus of which it is deprived by the resistance of the air, no matter how that impetus was conferred on it. Likewise, I believe, the damage done to a wall by a ball shot from a cannon twenty braccia distant would not be done by a ball coming vertically from any immense height. I believe that there is a limit to the acceleration of any natural moveable which leaves from rest, and that the impediment of the medium finally brings this to uniform motion in which the body is thereafter always maintained.

Sagr. These experiments appear to me very much to the purpose, and nothing remains here except that some adversary might entrench himself behind a denial that this can be verified for very great and heavy bulks, [declaring] that a cannonball coming from the orbit of the moon, or just from the highest region of the air, would strike more strongly than

---

65. A wad is assumed to hold the charge in place before firing.

one shot from a cannon.

Salv. No doubt many things could be said in opposition, not all of which can be countered by experiments. In this **138** refutation, however, it appears that something could be brought into consideration; namely, that it is highly probable that the heavy body falling from any height acquires that impetus, on arriving at the ground, which would suffice to drive it up to that height. This is clearly seen in a heavy pendulum which, drawn fifty or sixty degrees from the vertical, gains that speed and force [*virtù*] which precisely suffices to push it to an equal height, except for that little that is taken away by the impediment of the air. Hence to get the cannonball to such a height as would suffice for its acquisition [by fall] of the impetus that is given to it by the powder [*fuoco*] on its emerging from the cannon, it should be enough to shoot it vertically upward with the same cannon and then see, in its falling back, whether it made a blow equal to that of the impact made nearby in emerging [from the cannon]. I believe that it will not be as strong a blow by a long way; hence I think the speed the ball has near the mouth of the cannon to be such that the impediment of the air will never permit this [speed] to occur in natural motion starting from rest at any height whatever.

Next I come to those other questions, pertaining to pendulums, a subject that may seem very dry, especially to philosophers who are forever occupied in the most profound speculations about physics. I do not mean to deprecate these men, inspired by the example of Aristotle himself, in whom I admire above all that it may be said he did not neglect any matter worthy of consideration, or fail to touch on it. Moved by your questions, I shall now tell you something of my thoughts pertaining to music—a most noble subject, on which many great men, including Aristotle himself, have written.[66] He considers many curious problems relating to music; hence if from easy and sensible experiences I too shall draw reasons for marvelous things in the matter of sounds, I may hope that my discussions will be welcome to you.

Sagr. Not just welcome, but highly desired by me at least, as one who is delighted by all musical instruments. Having **139** philosophized much about the consonances, I have always remained puzzled and perplexed by them, inasmuch as one pleases and delights me far more than another, while some

66. Aristotle, *Problemata*, Bk. XIX.

not only fail to delight, but actually offend me. Then there is the old problem of the two strings tuned in unison, of which one moves and audibly resounds to the sound of the other. I am unresolved about this, as I am also unclear about the forms of the consonances, and other particulars.[67]

Salv. We shall see whether these pendulums of ours can bring some satisfaction to all these difficulties. As to the prior question, whether the same pendulum makes all its oscillations—the largest, the average, and the smallest—in truly and exactly equal times, I submit myself to that which I once heard from our Academician. He demonstrated that the moveable which falls along chords subtended by every arc [of a given circle] necessarily passes over them all in equal times, as well that [chord] subtended by one hundred eighty degrees, which is the whole diameter, as those subtended by one hundred, by sixty, by ten, two, one-half, and by a few minutes [of arc]. It is understood that all [these arcs] end at the lowest point [of the circle], touching the horizontal plane.

Now, as to descents along arcs of these chords rising from the horizontal, experience likewise shows us that all those not exceeding ninety degrees, or a quarter-circle, are passed in equal times, shorter, however, than the times of passage along the chords. This effect contains something of the miraculous, since at first glance it seems that the opposite should happen. [The paths] having in common their points of beginning and ending of motion, and the straight line being the shortest that lies between the same ends, it seems reasonable that the motion made along the straight line would have to be completed in the shortest time. This is not the case; the shortest time, and hence the swiftest motion, is that which is made along the arc of which the straight line is the chord.

As to the ratio of times of oscillation of bodies hanging from strings of different lengths, those times are as the square roots of the string lengths; or we should say that the lengths are as the doubled ratios, or squares, of the times. Thus if,

---

67. The "forms" meant were the numerical ratios traditionally associated with musical consonances. The octave, fifth, and fourth were associated with the ratios 2:1, 3:2, and 4:3. Imperfect consonances, not all of which were accepted by all theorists, included the major and minor third and sixth as 5:4, 6:5, 5:3, and 8:5. Galileo's father, Vincenzio Galilei (1520–90), was responsible for many of the experiments and results brought against numerical "forms" by Galileo, especially those he placed in the mouth of Sagredo. Vincenzio had ridiculed mathematical speculations about pure forms, contending that the trained ear is the only proper criterion of musical consonance.

for example, you want the time of oscillation of one pendulum
to be double the time of another, the length of its string must
140   be four times that of the other; or if in the time of one vibration of the first, another is to make three, then the string of
the first will be nine times as long as that of the other. It
follows from this that the lengths of the strings have to one
another the [inverse] ratio of the squares of the numbers of
vibrations made in a given time.[68]

Sagr. Then, if I understood correctly, I can easily know the
length of a string hanging from any great height, even though
the upper end of the attachment is out of my sight, and I see
only the lower end. For if I attach a heavy weight to the
string down here, and set it in oscillation back and forth; and
if a companion counts a number of its vibrations, while at
the same time I likewise count the vibrations made by another
moveable hung to a thread exactly one braccio in length, I
can find the length of the string from the numbers of vibrations of these two pendulums during the same period of time.
For example, let us assume that in the time my friend has
counted twenty vibrations of the long string, I have counted
two hundred forty of my thread, which is one braccio long.
Then after squaring the numbers 20 and 240, giving 400 and
57,600, I shall say that the long string contains 57,600 of those
units [*misure*] of which my thread contains 400; and since my
thread is a single braccio, I divide 57,600 by 400 and get 144,
so 144 braccia is the length of the string.

Salv. Nor will you be in error by a span, especially if you
take a very large number of vibrations.

Sagr. You often give me occasion to admire the richness
of nature and her great liberality, when from such common
things, or I might even say such base ones, you draw new
and curious knowledge that is often far beyond my imagining.
A thousand times I have given attention to oscillations, in
particular those of lamps in some churches hanging from very
long cords, inadvertently set in motion by someone, but the
most that I ever got from such observations was the improbability of the opinion of many, who would have it that
motions of this kind are maintained and continued by the
medium, that is, the air. It would seem to me that the air
must have exquisite judgment and little else to do, consuming
141   hours and hours in pushing back and forth a hanging weight

68. See note 74, below; Galileo's omission here of the word "inverse"
was noted by Viviani.

with such regularity. And now I learn that a given moveable, hung from a cord one hundred braccia long, and drawn from its lowest point now ninety degrees and again but one degree, or half a degree, would consume as much time in passing this smallest arc as that maximum arc. I certainly do not believe that I would ever have discovered this, which still seems to me to have in it something of the impossible. Now I wait to hear how these same simple minutiae provide me with reasons that can set my mind at least partly at rest concerning musical problems.

Salv. First of all, it is necessary to note that each pendulum has its own time of vibration, so limited and fixed in advance that it is impossible to move it in any other period than its own unique and natural one. Take in hand any string you like, to which a weight is attached, and try the best you can to increase or diminish the frequency of its vibrations; this will be a mere waste of effort. On the other hand, we confer motion on any pendulum, though heavy and at rest, by merely blowing on it. This motion may be made quite large if we repeat our puffs; yet it will take place only in accord with the time appropriate to its oscillations. If at the first puff we shall have removed it half an inch from the vertical, by adding the second when, returned toward us, it would commence its second vibration, we confer a new motion on it; and thus successively with more puffs given at the right time (not when the pendulum is going toward us, for thus we should impede the motion and not assist it), and continuing with many impulses, we shall confer on it impetus such that much greater force than a breath would be needed to stop it.

Sagr. As a boy, I observed that with impulses given at the right time, one man alone could ring a very large bell, and in trying to stop it later, several men would take hold of the rope and all of them would be lifted up into the air; nor could many men together [immediately] arrest the impetus that one man alone had conferred on it by regular pulls.

Salv. Your example explains my meaning no less acutely than my prefatory remarks fit in with the answer to that remarkable problem of the zither- or harpsichord-string that **142** moves and even resounds, and not only with one in unison and concord, but also with its octave and fifth. The cord struck begins and continues its vibrations during the whole time that its sound is heard; these vibrations make the air near it vibrate and shake; the tremors and waves [*increspa-*

menti] extend through a wide space and strike on all the strings of the same instrument as well as on those of any others nearby. A string tuned in unison with the one struck, being disposed to make its vibrations in the same times, commences at the first impulse to be moved a little; the second, third, twentieth, and many more [impulses] being added, all in exact periodic times, it finally receives the same tremor as that originally struck, and its vibrations are seen to go widening until they are as spacious as those of the mover.

This wave action [*ondeggiamento*] that expands through the air moves and sets in vibration not only other strings, but any other body disposed to tremble and vibrate in the same time as the vibrating string. If you attach to the base of the instrument various bits of bristle or other flexible material, it will be seen that when the harpsichord is played, this little body or that one trembles according as that string shall be struck whose vibrations are made in time with it. The others are not moved at the sound of this string, nor does the one in question tremble to the sound of a different string. If a thick viola-string is strongly bowed near a cup of thin and delicate glass, and the tone of the string is in unison with that of the goblet, the latter will shake and sensibly resound. The fuller waving of the medium close to the resonant body is easily seen by making the goblet sound when it contains water, by rubbing the ball of the finger on its edge. The contained water is seen to become wavy in a regular order, and the effect is still better seen by holding the base of the goblet on the base of some much larger vessel in which there is water almost up to the brim of the goblet. This being again made to resound by friction of the finger, this regular waving in the water will be seen to spread with great speed to a good distance around the goblet. Sounding in this way a very large vessel almost full of water, I have often seen waves formed in the water with extreme regularity; and sometimes it happens that the tone of the goblet jumps one octave higher, at which moment I have seen each of the waves divided in two; an event that very clearly proves the form of the octave to be the double [ratio].

Sagr. The same has happened for me more than once, to my delight, and to my profit also, since I had long been perplexed about the [ideal] forms of the consonances. It seemed to me that the reason ordinarily adduced for it by authors that have written on music up to the present was insufficiently conclusive. They say that the diapason, or octave, is contained in the double [ratio]; the diapente, which we call the [perfect] fifth, by the sesquialter [three-to-two] ratio; and so on. Their reason is that when a string is stretched on the monochord, and sounded first entire and then in the half, by placing the bridge at the center, one hears the octave; when the bridge is placed at one-third the whole [length of the] string, and the whole string is sounded against two-thirds of it, the fifth is heard; hence, they say, the octave is embraced between two and one, and the fifth between three and two. This reasoning, I say, did not seem conclusive to me in assigning by law the double and the sesquialterate as the natural forms of octave and fifth, my idea being as follows.

There are three ways in which the pitch of a string can be raised. One is to shorten it; another is to stretch, or let us say pull, it more; the third is to make it thinner. If we keep the same tension and thickness of string, and want to hear the octave, we must shorten the string by half; that is, strike it and then its half. Keeping the same length and thickness, however, if we wish to make it rise an octave by pulling it harder, it will not suffice to pull twice as hard, but one needs four times [the tension]; thus if at first it was pulled by a one-pound weight, it will be necessary to attach four [pounds] to raise it one octave. Finally, if we keep the same length and tension, and want a string that will give the octave by being thinner, we must retain only one-fourth the thickness of the deeper string.

What I say of the octave—that is, that its form [when] derived from the tension or the thickness of string is in squared ratio of that which we have from its length—is to be understood of all the musical intervals. For what a length **144** in three-halves ratio gives us, as when we sound the whole and then two-thirds, may be derived from tension or from thinness, but this requires the square of the three-halves ratio, taking that of nine to four. Thus if the lower string is stretched by four pounds of weight, not six but nine must be attached to the higher string; and as to thickness, to get the fifth, the lower string must be to the higher one in the ratio of nine to four.[69] These being true experiments, I saw no reason why

69. This relationship had been discovered experimentally by Vincenzio Galilei, who published it in 1589. It may be the first physical law to have been discovered by systematic experiment for the purpose of overthrowing a previously accepted mathematical rule.

wise philosophers should have established the form of the octave as the double [ratio] any more than as the quadruple, or that of the fifth as three-halves rather than nine-fourths.

But inasmuch as it is quite impossible to count the vibrations of a sounding string, since it makes so many of them, I should have remained always in doubt whether it was true that the higher string of the octave makes double the number of vibrations in the same time as the lower string, had it not been that the waves persist as long as you like in the sounding and vibrating goblet. This showed me sensibly that at the very moment at which the tone is sometimes heard to jump an octave, smaller waves are seen to be generated that precisely bisect those that were there before.

Salv. A beautiful observation, making possible the distinction one by one of the waves from the vibration of the resounding body. It is these which, diffused through the air, go to make that titillation of the eardrum which in the mind becomes sound. But since such observations, and the sight of waves in water, last only as long as friction with the finger is continued, and even during that time the waves are not permanent, but are continually generated and dissolved, would it not be a fine thing if one might arrange that they remain with great exactness for a long time, months or years, to make it easier to measure and number them at leisure?

Sagr. That really would be an invention!

Salv. The invention was by accident, and my observation amounted only to making capital of this and esteeming it as a new proof of a noble theory, though a very humble achievement in itself.

145   Scraping a brass plate with an iron chisel to remove some spots from it, I heard the plate emit a rather strong and clear note once or twice in many strokes as I moved the chisel rapidly over it. Looking at the plate, I saw a long row of thin lines, parallel to one another and at exactly equal distances apart. Scraping again, many times, I noticed that it was only when a stroke made this noise that the chisel left marks on the plate, and when it went without the shrill tone, there was not the faintest trace of such lines. As I repeated the trick again and again, stroking now with greater and again with less speed, the sound was of higher or lower pitch; and I observed that the marks made during the shriller tone were closer together, and those made during the lower tone less so. Sometimes also, according as the stroke itself was made faster at the end than the beginning, the sound was heard to rise in pitch and the lines were seen to increase in frequency, though always marked with extreme neatness and absolutely parallel.

During the sibilant strokes, moreover, I felt the iron tremble in my hand, from which a kind of tenseness ran through me; in the iron is felt, to be brief, precisely what we feel in speaking *sotto voce* and then raising this to a loud voice. For the breath being sent in a whisper to form sound, we feel hardly any movement in the throat and mouth in comparison with the great tremor in larynx and jaws when we speak, especially in a low and powerful tone. Sometimes I have also noted, among the strings of a harpsichord, two [vibrating in] unison with two sounds made by scraping in the way described. These two, different in pitch, were separated by a perfect fifth; and measuring the intervals between the lines for each of the two strokes, it was seen that the distance containing forty-five spaces in one, contained thirty in the other, which indeed is the form [of ratio] attributed to the diapente.

Before going further, I want to call your attention here to the fact that of the three ways in which pitch may be raised, that which you assigned to thinness of string should more **146** properly be attributed to the weight. For the change [in pitch] due to thickness answers [to the squared ratio] when the strings are of the same material, so that one gut string must be four times as thick as another gut string to sound the octave; or one brass string four times the thickness of another brass string. But if I should want to form the octave between a brass string and one of gut, it would be done not by thickening [the lower] one four times, but by making it four times as heavy. As to thickness, the metal string would not be four times as thick at all, but four times as heavy, and in some cases this would even be thinner than the corresponding gut an octave higher in pitch. So it comes about that stringing one harpsichord with gold strings, and another with brass strings of the same length, tension, and thickness, the first tuning comes out about a fifth lower, since gold is about twice as heavy.[70] Here note that the heaviness of the moveable is more resistant to speed than is its thickness, contrary to what one might at first suppose, since it seems reasonable

---

70. Gold being twice as dense, the effect is as the square root of two, or about 2.8 to 2, which is close to the musical interval of the fifth; that is, 3:2.

that speed should be more retarded by the resistance of the medium to being separated by a thick but light moveable than by a heavy and thin one; yet in this instance, the contrary happens.[71]

Going back to our original purpose, I say that the length of strings is not the direct and immediate reason behind the forms of musical intervals, nor is their tension, nor their thickness, but rather, the ratio of the numbers of vibrations and impacts of air waves that go to strike our eardrum, which likewise vibrates according to the same measure of times. This point established, we may perhaps assign a very congruous reason why it comes about that among sounds differing in pitch, some pairs are received in our sensorium with great delight, others with less, and some strike us with great irritation; we may thus arrive at the reason behind perfect consonances, and imperfect, and dissonances. The irritation from the latter is born, I believe, of the discordant pulsations of two different tones that strike on our eardrums all out of proportion; and very harsh indeed will be the dissonances whose times of vibration are incommensurables. One such

147 [noise] will occur when two strings are sounded together, of which one is to the other as the side of a square is to its diagonal, a dissonance such as that of the tritone or minor third. Those pairs of sounds will be consonant, and heard with pleasure, which strike the eardrum with good order; this requires first that the impacts made within the same period are commensurable in number, so that the cartilage of the eardrum need not be in a perpetual torment of bending in two different ways to accept and obey ever-discordant beatings. Hence the first and most welcome consonance is the octave, in which for every impact that the lower string delivers to the eardrum, the higher gives two, and both go to strike unitedly in alternate vibrations of the high string, so that one-half of the total number of impacts agree in beating together. In unison, the blows of strings are always joined together and are therefore as [those] of a single string, and do not form a consonance [properly speaking]. The fifth also gives pleasure, inasmuch as for every two pulsations of the low string, the high string gives three, whence it follows that counting the vibrations of the high string, one-third of all

71. The Pieroni MS did not contain this paragraph, which went considerably beyond the findings of Galileo's father.

[pulses] agree in beating together, with two solitary [pulses] interposed between each pair of concords; in the [perfect] fourth, three [solitary beats] intervene. In the whole tone, or sesquioctave, only one in every nine pulsations comes to strike in agreement with that of the lower string; all the rest are discordant, and being received with irritation on the eardrum are judged dissonant by our hearing.

Simp. I should like to hear this argument explained more clearly.

Salv. Let this line *AB* be the amplitude [*spazio e la dilatazione*] of one vibration of the lower string, and let line *CD* be that of the higher string which gives the octave with the first. Divide *AB* in the center at *E*, and let the strings begin to move at the points *A* and *C*. It is manifest that when the higher vibration has come to the point *D*, the lower has got only to the center, *E*, which, not being an end of its motion, causes no impact, though a stroke will be made at *D*. The vibration *D* returning then to *C*, the other passes from *E* to *B*, wherefore the two impacts at *B* and *C* beat unitedly on the eardrum. Returning and repeating similar vibrations thereafter, it is concluded that alternately, in one but not the other of the vibrations *C* and *D*, union of impacts will occur with *A* and *B*. But the pulsations at the ends [*A* and *B*] always have as companions either *C* or *D*, and always the same one. This is obvious, because assuming that *A* and *C* beat together, while *A* goes to *B*, *C* goes to *D* and returns to *C*, so that *C* beats with *B*; and in the time that *B* returns to *A*, *C* passes through *D* and returns to *C*, so that the blows *A* and *C* are made together.

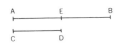

148

Next, let the two vibrations *AB* and *CD* be those that produce the fifth, of which the times are in the ratio of three to two; divide *AB* of the lower string into three equal parts at *E* and *O*. Supposing the vibrations to commence at the same moment from points *A* and *C*, it is evident that at the stroke to be made at *D*, the vibration *AB* will have got only to *O*, so that the eardrum receives the impact of *D* alone; in the return of *D* to *C*, the other vibration passes from *O* to *B* and gets back to *O*, making a pulsation at *B*; this, however, is solitary and countertimed, a fact yet to be considered. For having assumed the first pulsations to be made at the same moment at points *A* and *C*, the second, which was only that at point *D*, is made after as much time as the transit *CD*, or *AO*. But the next, which is made at *B*, is spaced from the other

solitary [pulse] only as much as the time of *OB*, which is half [the previous time]. Continuing now the return from *O* to *A*, while *C* goes to *D*, there come to be made two united pulsations at *A* and *D*. Then there follow other periods similar to the above; that is, with the interposition of two pulsations of the high string unaccompanied and solitary, and one of the low string, also solitary and interposed between the two solitary [pulsations] of the high string.

Thus if we imagine time divided into moments, that is, into minimum equal tiny parts, and assume that in the first two [moments] from concordant pulsations made at *A* and *C*, we go to *O* and *D*, and at *D* there is a stroke, then in the third and fourth moments there is a return from *D* to *C* with a stroke at *C*,[72] and from *O* there is a passage through *B* and a return to *O*, with a stroke at *B*, and finally in the fifth and sixth moments there is passage from *O* and *C* to *A* and *D*, with strokes at both. Then we shall have on the eardrum pulsations distributed in such order that, assuming the pulsations of the two strings at the same instant, two moments later the eardrum will receive a solitary impact; at the third **149** moment, another solitary one; at the fourth, another solitary one; and then two moments later, that is, at the sixth moment, two [pulses] joined together; and this finishes the period, or so to speak, the anomaly, which period is thereafter many times repeated.[73]

Sagr. No longer can I remain silent; I must exclaim over the great pleasure I take in hearing such a complete explanation of phenomena which have so long held me in the dark and blinded. Now I understand why unison does not differ at all from one single tone; I see why the octave is the principal consonance, but so like unison that, like unison, it is taken and mixed with other consonances. It resembles unison because where all the pulsations of strings in unison always strike together, those of the lower string in the octave are always accompanied by those of the upper string, but one [pulsation] of the latter is interposed alone and at equal intervals and (so to speak) without any foolery, so that the

---

72. The words "with a stroke at *C*" do not occur in the Pieroni MS.

73. Galileo's theory of consonances differs essentially both from its predecessors and from the modern view. It mistakenly assumes dependence on phase relationships, and fails to escape the implication that string-lengths of, say, 3001:2000 should sound harsh together; in fact, such a chord would be indistinguishable from the perfect fifth, 3:2.

resulting consonance is rather too bland, and lacks fire.

The fifth, however, is characterized by its displaced beats; that is, by the interposition of two solitary beats of the upper string and one solitary beat of the lower string between each pair of united pulsations. These three [solitary beats] are moreover separated by an interval of time equal to one-half of that between each pair of united beats and a solitary beat of the upper string. This produces a tickling and teasing of the cartilage of the eardrum so that the sweetness is tempered by a sprinkling of sharpness, giving the impression of being simultaneously sweetly kissed, and bitten.

Salv. Seeing that you like these novelties so well, I must show you how the eye, too, and not just the hearing, can be amused by seeing the same play that the ear hears.

Hang lead balls, or similar heavy bodies, from three threads of different lengths, so that in the time that the longest makes two oscillations, the shortest makes four and the other makes three. This will happen when the longest contains sixteen spans, or other units, of which the middle [length] contains nine,[74] and the smallest, four. Removing all these from the vertical and then releasing them, an interesting interlacing of the threads will be seen, with varied meetings such that at every fourth oscillation of the longest, all three arrive unitedly at the same terminus; and from this they depart, repeating again the same period. The mixture of oscillations is such that when made by [tuned] strings, it renders to the hearing an octave with the intermediate fifth. And if with similar arrange-   **150** ments we modify the lengths of other strings so that their vibrations answer to those of other musical intervals which are consonances, other interlacings will be seen in which, at determinate times and after definite numbers of vibrations, all the strings (let them be three or four) agree in coming at the same moment to the terminus of their oscillations, and begin from there another like period. But if the vibrations of two or more strings are either incommensurables, so that they never return to concord at the end of a definite number of oscillations, or if, not being incommensurables, they return [only] after a long time and a large number of oscillations, then vision is confused by the disorderly order so irregularly interlaced, as the ear is annoyed by untempered pulses of air

74. The appropriate lengths are not 4, 9, and 16 as given in the text, but $\frac{64}{16}$, $\frac{64}{9}$, and $\frac{64}{4}$; that is, 4, $7\frac{1}{9}$, and 16; cf. note 68, above. The correction of 9 to $7\frac{1}{9}$ was later noted by Viviani.

tremors that go without order or law to strike the eardrum.

But gentlemen, where have we allowed ourselves to be carried through so many hours by various problems and unforeseen discussions? It is evening, and we have said little or nothing about the matters proposed; rather, we have gone astray in such a way that I can hardly remember the original introduction and that small start that we made by way of hypothesis and principle for future demonstrations.

Sagr. It will be best, then, to put an end for today to our discussions, giving time for our minds to compose themselves tranquilly at night, so that we may return tomorrow (if you are pleased to favor us) to the discussions desired and in the main agreed upon.

Salv. I shall not fail to be here at the same hour as today, to serve and please you.

The First Day Ends

Sagr. Simplicio and I have been awaiting your arrival, and in the meantime we have been reviewing in memory the last consideration, to be taken as a principle and assumption for the conclusions that you intended to demonstrate to us. This concerned that resistance which all bodies have to fracture, and depends on that cement that holds their parts attached and conjoined so that they do not yield and separate without a powerful pull. There was then a search for the cause of that coherence, which is extremely strong in some solids; and the chief cause proposed was that of the void. This was then the occasion of many digressions that kept us occupied all day, without our getting near to the principal subject originally chosen. This, as I said, was the investigation of the resistances of solids to being broken.

Salv. I remember it all quite well. And taking up the original thread, then, whatever may be the resistance of solid bodies to parting under a violent pull, its presence in them is beyond any doubt. This resistance is very great against a force that pulls them in a straight line, but is observed to be much less when the force is across them. Thus we see that a steel or glass rod, for example, supports a weight of a thousand pounds lengthwise, but when fixed horizontally in a wall, it is broken by attaching only fifty [pounds] to it. We must speak of this second resistance, seeking the proportions in which it is found in prisms **152** and cylinders of the same material, whether similar or dissimilar in shape, length, and thickness. In such speculations I take as a known principle one which is demonstrated in mechanics about the properties of the rod which we call the lever: that in using a lever, the force is to the resistance in the inverse ratio of the distances from the fulcrum to the force and to the resistance.

Simp. This was demonstrated by Aristotle in his [*Questions of*] *Mechanics* before anyone else.[1]

1. The pseudo-Aristotelian treatise noted that points on a rotating bar travel in the same time through distances proportional to their distances from the center of rotation, and argued that this measures the ease of motion, whence this motion is inverse to the heaviness of the respective weights; see Loeb edition, pp. 343–47; 375–77. From this stress on time and ease of motion, Galileo took his principle of virtual velocities, for which he credited the ancient treatise in his *Bodies in Water*, p. 71 (*Opere*, IV, 69).

Salv. I admit that we should concede priority to him, but in rigor of proof I think we must put Archimedes a long way in front of him. Upon a single proposition proved in his [*Plane*] *Equilibrium* there depends the reasoning not only for the lever, but for most other mechanical instruments.[2]

Sagr. Inasmuch as that principle is to be assumed as the foundation of everything that you mean to demonstrate to us, it would be much to the purpose for you to give us also its proof, if that matter is not too prolix. Thus you will be giving us entire and complete instruction.

Salv. If this must be done, it will be better that I introduce you by another approach, somewhat different from that of Archimedes, to our whole field of future speculations. Without assuming anything except that equal weights placed in a balance of equal arms are in equilibrium (a principle likewise assumed by Archimedes),[3] I shall next prove to you that it is equally true that unequal weights rest in equilibrium on a steelyard when they are suspended at distances unequal in the inverse ratio of these weights; and not only that, but that it is also the same thing to hang equal weights at equal distances as it is to put unequal weights at distances having the inverse ratio of the weights.

Now for a clear demonstration of what I say, I draw a solid prism or cylinder *AB*, hung by its ends from the line *HI* by two threads, *HA* and *IB*. It is evident that if I suspend all this by the

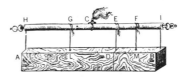

string *C*, placed at the middle of the balance *HI*, the prism *AB* will remain in equilibrium by the principle assumed, one-half

2. Archimedes, *On Plane Equilibrum*, Bk. I, Props. 6, 7; see T. L. Heath, *The Works of Archimedes*, (Cambridge, 1897), pp. 192–94. What Galileo calls a single proposition made up two for Archimedes, who proved separately the cases of commensurable and incommensurable distances.

3. Two other Archimedean assumptions are omitted by Galileo; namely, that of equal weights at unequal distances, the more distant one will descend, and that weights in equilibrum will be disturbed by any addition or removal of weight on either side. He brings in later the further Archimedean assumption that equilibrium is preserved when proportional weights or distances are substituted. Galileo's proof, which is much easier to follow than that of Archimedes, appeared first in a treatise written for his students; see *On Mechanics*, pp. 153–55 (*Opere*, II, 161–63).

of its weight being on one side, and one-half on the other [side] of the point of suspension *C*. Next, suppose the prism to be **153** divided in two unequal parts by a plane through the line at *D*; let *DA* be the greater part, and *DB* the smaller. And in order that when this cut is made, the parts of the prism will remain in place, in the same arrangement with respect to the line *HI*, let us assist with a thread *ED*, tied at the point *E*, which thread shall sustain both parts of the prism, *AD* and *DB*. There is no question that since there has been no change of place on the part of the prism with respect to the balance *HI*, it will remain in the previous state of equilibrium. But the part of the prism which is now suspended from its two ends by the threads *AH* and *DE* also remains in the same arrangement if hung by a single thread *GL* placed at its center; and likewise the other part, *DB*, will not change position if suspended from the middle and sustained by the thread *FM*.[4] Hence, removing threads *HA, ED*, and *IB*, and leaving only the two [threads] *GL* and *FM*, the previous equilibrium will prevail, the suspension from point *C* always remaining.

Now here we turn to consider that we have two heavy bodies, *AD* and *DB*, hanging from the ends *G* and *F* of a steelyard *GF*, in equilibrium around the point *C* in such a way that the distance of suspension of the heavy body *AD* from point *C* is the line *CG*, while the other part, *CF*, is the distance from which the body *DB* is hung. So it remains only to demonstrate that these distances have the same ratio between them as those same weights, but taken inversely. That is, the distance *GC* is to *CF* as the prism *DB* is to the prism *DA*. This we prove as follows.

Line *GE* being one-half of the line *EH*, and *EF* one-half of *EI*, all *GF* is one-half of the whole *HI*, and is therefore equal to *CI*. Subtracting the common part *CF*, the remainder *GC* will be equal to the remainder *FI*, that is, to *FE*. Taking *CE* common to both, *GE* is equal to *CF*; hence as *GE* is to *EF*, so *FC* is to *CG*. But as *GE* is to *EF*, so is one double to the other; that is, so *HE* to *EI*, or the prism *AD* to the prism *DB*. Therefore, by equidistance of ratios and inversion, as distance *GC* is to distance *CF*, so is weight *BD* to weight *DA*; which is what I wished to prove to you.

---

4. It is assumed that so long as no motion takes place as a result of manipulations within the system, no motion of the system as a whole will occur.

154   This understood, I believe you will have no difficulty in
admitting that the two prisms *AD* and *DB* are in equilibrium
around point *C*, because half the entire solid *AB* is to the right
of the suspension *C*, and the other half is to its left, so that they
represent two equal weights disposed and extended over two
equal distances. And the two prisms *AD* and *DB*, reduced to

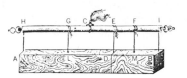

cubes or balls or any other shapes whatever, provided only
that they keep the same suspensions, *G* and *F*, will continue
to be in equilibrium around point *C*.[5] I believe no one can
doubt this, for it is quite evident that shape does not change
a weight so long as the same quantity of material is preserved.
From this we may draw the general conclusion that two
weights, whatever they may be, are in equilibrium at distances
inversely corresponding to their heavinesses.

This principle is therefore established. Before we go on
further, I must next point out that these forces, resistances,
moments, shapes, and so on may be considered in the abstract
and separated from matter; or alternatively, in the concrete
and conjoined with matter. In the latter way, the phenomena
that conform to the diagrams considered as immaterial receive
some modifications when we add to them material, and hence
heaviness. For example, let us take a lever, which shall be *BA*

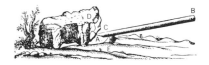

here, placed on the fulcrum *E* and used to raise the heavy
stone *D*. It is obvious, from the principle just proved, that the
force applied at the end *B* will suffice to match the resistance
of the heavy body *D*, if its moment has the same ratio to the
moment of *D* that distance *AC* has to distance *CB*; or rather,
this is true without taking into consideration other moments

5. This involves the further assumption that the center of gravity may be
taken as representing the whole body; cf. *On Mechanics*, p. 152 (*Opere*,
II, 160).

than those of the simple force at *B* and the resistance at *D*, as if the lever were immaterial and without heaviness. But if we also take into account the heaviness of the lever-instrument itself, which will be sometimes of wood and sometimes of iron, it is manifest that when the weight of the lever is added to the force at *B*, the ratio will be altered and will have to be stated **155** in other terms. Hence, before we proceed, it is necessary for us to agree in making a distinction between these two manners of considering things, saying that the instrument is "taken absolutely" when we mean it to be taken in the abstract, separated from the heaviness of its actual material; but joining both material and heaviness to the simple and absolute figures, we shall call the figures joined with matter "moment, or compound force."[6]

Sagr. Here I must break my resolve to give no occasion for digression, for I cannot apply myself attentively to what is to come until a certain doubt, just born in me, is removed. It seems to me that you make comparison between the force applied at *B* and the total heaviness of the stone *D*, though I think that a part of this, and perhaps most of it, is supported on the horizontal plane, so that . . .

Salv. I quite understand. Say no more; yet please notice that I have not yet spoken of the total heaviness of the stone, but only of the moment that it has and exercises on the point *A*, the very end of the lever *BA*. This [moment] is always less than the entire weight of the [supported] stone, and it varies according to the shape of the stone, and whether it is to be lifted more, or less.

Sagr. All right, as to this; but another desire now awakens in me, which is that for a complete understanding, I be shown the way (if there is any) by which we can determine the part of the total weight that is sustained on the plane beneath, and the part that weighs on the bar at its extermity *A*.

Salv. Since I can give you satisfaction in a few words, I shall not fail to serve you. Therefore, drawing a little diagram, take

---

6. Thus far, the device or instrument is the lever, but later it becomes the beam whose strength is to be analyzed. Either is said to be "taken absolutely" when its own weight is neglected; when that is taken into account, the text refers to the "moment" of the lever or beam, or to its "compound force." The moment, or compound force, belongs to the lever as such, separately from the weights and forces applied to it, and its effect depends upon the point of support in relation to that of application. It represents, so to speak, the "net leverage" of a heavy beam acting against itself; cf. Prop. VI, below.

the weight with center of gravity *A*, supported on the horizontal plane at the end *B*; at the other end, let it be sustained by the

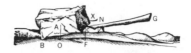

lever *CG* with fulcrum *N* by a power [*potenza*] placed at *G*. From the center *A* and the end *C*, drop perpendiculars *AO* and *CF* to the horizontal. I say that the moment of the whole weight has to the moment of power at *G*, the ratio compounded from the ratio of the distance *GN* to the distance *NC*, and [the ratio of the distance] *FB* to *BO*.

Find [a line *X*] such that as line *FB* is to *BO*, *NC* is to *X*. Now, the whole weight *A* being sustained by the two powers placed at *B* and *C*, the power *B* is to *C* as the distance *FO* is to *OB*; and by composition [of each ratio], the two powers *B* and *C* together, that is, the whole moment of all the weight *A*, is to the power at *C*, as line *FB* is to *BO*; that is, as *NC* is to *X*. But the moment of the power at *C* is to the moment of the power at *G* as distance *GN* is to *NC*; therefore, by perturbed equidistance of ratios, the total weight *A* is to the moment of the power at *G* as *GN* is to *X*. But the ratio of *GN* to *X* is compounded from the ratio of *GN* to *NC* and that of *NC* to *X*, which is [that of] *FB* to *BO*. Hence the weight *A* has to the power that sustains it at *G*, the ratio compounded from [that of] *GN* to *NC* and that of *FB* to *BO*; which is what was to be demonstrated.

Now, getting back to our first purpose, it will not be difficult to understand the reason whence it comes about that:

PROPOSITION I

A solid prism or cylinder of glass, steel, wood, or other material capable of fracture, which suspended lengthwise will sustain a very heavy weight attached to it, will sometimes be broken across (as said earlier) by a very much smaller weight, according as its length exceeds its thickness.

Let us imagine the solid prism *ABCD* fixed into a wall at the part *AB*; and at the other end is understood to be the force of the weight *E* (assuming always that the wall is vertical and the prism or cylinder is fixed into the wall at right angles). It is evident that if it must break, it will break at the place *B*, where

the niche in the wall serves as support, *BC* being the arm of the lever on which the force is applied. The thickness *BA* of the solid is the other arm of this lever, wherein resides the resistance, which consists of the attachment that must exist between the part of the solid outside the wall and the part that is inside. Now, by what has been said above, the moment of the force applied at *C* has, to the moment of the resistance which exists in the thickness of the prism (that is, in the attachment of the base *BA* with its contiguous part), the same ratio that the length *CB* has to one-half of *BA*. Hence the absolute resistance to fracture in the prism *BD*, (being that **157** which it makes against being pulled [apart] lengthwise, for then the motion of the mover is equal to that of the moved) has, to resistance against breakage by means of the lever *BC*, the same ratio as that of the length *BC* to one-half of *AB*, in the prism; or, in the cylinder, to the radius of its base. And let this be our first proposition.[7]

Note that what I say is to be understood without consideration of the weight of the solid *BD* itself, which solid has been taken as weighing nothing. When we come to take into account

7. Propositions are numbered in the original only in the margins, and the later ones are not numbered; here they will be given numbers in brackets for convenience of reference. The first proposition (which underlies the rest) is a postulate or assumption rather than a theorem and is followed by an explanation rather than a proof. In making this assumption, Galileo deliberately neglected various properties that differentiate the actual materials named as examples. Note that adhesion or coherence is treated as if spread uniformly over the area at *AB*, and that points *A* and *C* are regarded as rigidly connected. Galileo's is the first known attempt to formulate a mathematical theory of strength of materials; in it, as in his treatment of motion (Third Day), he concerns himself only with ratios, whence the factors left out of account (particularly elasticity) do not invalidate his results expressed only as proportions applicable to a given type of material.

its own heaviness, adding this to the weight *E*, we must add to the weight *E* one-half the weight of the solid *BD*. Thus the weight of *BD* being, say, two pounds, and the weight *E* ten pounds, the weight *E* must be taken as if it were eleven.

Simp. And why not as if it were twelve?

Salv. The weight *E*, my good Simplicio, hangs from the end *C* and presses on the lever *BC* with its full moment of ten pounds; and if *BD* alone were hung there, it would weigh down with its full moment of two pounds. But as you see, that solid is uniformly distributed along the entire length *BC*, whence the parts near the extremity *B* press down less than do those farther away. In short, balancing the near parts with the far, the weight of the whole prism comes to operate at its center of gravity, which corresponds to the center of the lever *BC*. But a weight hanging from the extremity *C* has double the moment it would have if hung from the middle; and hence one-half the weight of the prism must be added to the weight *E* when we treat the moment of both as located at the end *C*.

158

Simp. I understand; and if I am not mistaken, the power of both weights *BD* and *E*, thus placed, would have the same moment as if all the weight of *BD* and double the weight *E* were hung from the center of the lever *BC*.

Salv. Precisely so, and this must be kept in mind. And now we can immediately understand:

PROPOSITION II

How, and in what ratio, a rod, or rather a prism of greater breadth than thickness, more greatly resists breaking when loaded across its breadth than across its thickness.

For an understanding of this, imagine a ruler *AD* whose breadth is *AC* and whose thickness, much less, is *CB*. It is

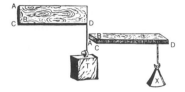

asked why, when we wish to break it on edge as in the first figure, it will resist the great weight *T*; but placed flat, as in the second figure, it will not [even] resist *X*, which is less than *T*. This is clear when we understand the fulcrum in one [the latter] case to be under the line *BC*, and in the other case under *CA*, the distances of the force being equal to the length *BD* in both cases. For in the first case the distance of the resistance from the fulcrum, which is one-half the line *CA*, is greater than in the other case, where this is one-half of *BC*, whence it is necessary that the force of the weight *T* be greater than *X* by the same amount that one-half the breadth *CA* is greater than one-half the thickness *BC*, *CA* serving in the former, and *CB* in the latter, as counterlever to overcome the same resistance, which is in the [same] quantity of fibers of the whole base *AB*. It is thus concluded that the same ruler or prism, broader than it is thick, more greatly resists being broken when on edge than when flat, according to the ratio of its breadth to its thickness.

Next, it is appropriate for us to commence this investigation:

PROPOSITION III

The ratio in which the moment of heaviness of a horizontal prism or cylinder increases, in relation to its own resistance to being broken by elongation, I find to be in squared proportion to the lengthening.

159

For this demonstration, consider the prism or cylinder *AD* fixed solidly into the wall at the end *A*, parallel to the horizon, and understand this to be lengthened to *E* by adding the part *BE*. It is evident that the lengthening of the lever *AB* out to *C* increases the moment of the downward [*premente*] force against resistance to fracture and detachment made at *A*. [This increase], taken absolutely, is in the ratio of *CA* to *BA* [alone]; but besides this, the weight of the solid *BE* added to the weight of the solid *AB* increases the downward moment of

heaviness according to the ratio of the prism *AE* to the prism *AB*, which ratio is the same as that of the length *AC* to *AB*. Hence, combining the two increases of length and of heaviness, it is manifest that the moment compounded from both is in the squared ratio of either one. It is therefore concluded that the moments of the forces of prisms (or cylinders) of equal thickness but unequal length are to each other in the squared ratio of their lengths; that is, are as the squares of the lengths.

We shall now show, in the second place, the ratio according to which resistance to being broken increases in prisms and

160   cylinders of the same length, when they are increased in thickness. Here I say that:

PROPOSITION IV

In prisms and cylinders of equal length but unequal thickness, resistance to fracture increases as the cubed ratios of the thicknesses or the diameters [respectively] of their bases.

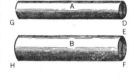

Let the two cylinders *A* and *B* have equal lengths *DG* and *FH*, and unequal bases, these being circles of diameters *CD* and *EF*. I say that the resistance to fracture of cylinder *B* is to the resistance of cylinder *A* as the cube of the ratio of diameter *FE* to diameter *DC*. For first, consider the absolute and simple resistance that resides in the bases (that is, in the [areas of] circles *EF* and *DC*), when these are to be broken by exerting a force that stretches them lengthwise. There is no doubt that the resistance of cylinder *B* is [in that case] greater than that of cylinder *A* by as much as circle *EF* is greater than [circle] *CD*, because so many the more are the fibers, filaments, or holding elements that keep the parts of such solids together.

Now let us consider that in exerting force crossways, we employ two levers. The arms (or distances at which the forces are applied) are the lines *DG* and *FH*; the fulcrums are at points *D* and *F*; and the other arms (or distances at which the resistances are situated) are the radii of circles *DC* and *EF*, for the filaments [being] spread throughout the surfaces of these circles, it is as if all were concentrated at their centers. In such levers, I say, consider the resistance at the center of the base *EF* against the force at *H*; this is as much greater than the resistance of the base *CD* against the force applied at *G* as the radius [of] *FE* is greater than the radius [of] *DC*; and the forces at *G* and *H* act on the equal levers *DG* and *FH*. Therefore the resistance to fracture in cylinder *B* exceeds the resistance of cylinder *A* according to both ratios—that of the circles *EF* and *DC*, and [that] of their radii (or diameters). Now the ratio of the circles is the square of that of the diameters; but the ratio of resistances, being compounded from these [two ratios], is the triplicate ratio of the same diameters; which is **161** what was to be proved. And since cubes are in the triplicate ratio of their sides, we may similarly conclude that the resistances of cylinders of equal length are to one another as the cubes of their diameters.

From what has been demonstrated, we can also conclude:

COROLLARY

The resistances of prisms and cylinders of equal length are as the three-halves power of the [ratio of volumes of the] said cylinders.[8]

This is manifest, since prisms (or cylinders) of equal altitude have to one another the same ratio as their bases, which is the square of the [ratio of] sides (or diameters) of those bases. But the resistances, as demonstrated above, are as the cubes of the same sides (or diameters); therefore the ratio of resistances is the three-halves [power] of the ratio of the solids themselves, and consequently of the weights of those solids [of like material and equal length].

Simp. Before we go on, it is necessary for me to be relieved of a certain difficulty. Thus far, I have heard nothing said in consideration of a certain other kind of resistance which appears to me to decrease in solids as they increase in length, and [to weaken them] not only transversely but also longitudinally. Thus we see a very long rope to be much less able to hold a great weight than if shorter; and I believe that a short wooden or iron rod can support much more weight than a very long one when loaded lengthwise (not [just] crosswise), and also taking into account its own weight, which is greater in the longer.

Salv. I think that you, together with many other people, are mistaken on this point, Simplicio, at least if I have correctly grasped your idea. You mean that a rope, say forty braccia in length, cannot sustain as much weight as one or two braccia of the same rope.

8. The exponential terminology is used here in place of Galileo's ratio terminology; see Introduction and note 63 to First Day, above.

Simp. That is what I meant, and at present it appears to me a highly probable statement.

Salv. And I take it to be not just improbable, but false; and I believe that I can easily remove the error. So let us assume this rope *AB*, fastened above at one end, *A*, and at the other end let there be the weight *C*, by the force of which this rope is to break. Now assign for me, Simplicio, the exact place at which the break occurs.

162   *Simp.* Let it break at point *D*.

Salv. I ask you the cause of breaking at *D*.

Simp. The cause of this is that the rope at that point has not the strength to bear, for instance, one hundred pounds of weight, which is the weight of the part *DB* together with [that of] the stone *C*.

Salv. Then whenever the rope is strained at point *D* by the same 100 pounds of weight, it will break there.

Simp. So I believe.

Salv. But now tell me: if the same weight is attached not to the end of the rope, *B*, but close to point *D*, say at *E*; or the rope being fastened not at *A*, but closer to and above the same point *D*, say at *F*; then tell me whether the point *D* will not feel the same weight of 100 pounds.

Simp. It will indeed, provided that the length of rope *EB* accompanies the stone *C*.

Salv. If, then, the rope, pulled by the same hundred pounds of weight, will break at the point *D* by your own admission, and if *FE* is but a small part of the length *AB*, how can you say that the long rope is weaker than the short? Be pleased therefore to have been delivered from an error, in which you had plenty of company, even among men who are otherwise very well informed, and let us proceed.

Having demonstrated that prisms and cylinders of constant thickness increase in moment beyond their own resistances as the squares of their lengths, and likewise that those of equal length but differing in thickness increase their resistances in the ratio of the cubes of the sides (or diameters) of their bases, let us go on to investigate what happens to such solids when they differ in both length and thickness. In these, I note that:

PROPOSITION V

Prisms and cylinders differing in length and thickness have their resistances to fracture in the ratio compounded
163   from the ratio of the cubes of the diameters of their bases

and from the inverse ratio of their lengths.[9]

Let *ABC* and *DEF* be two such cylinders; I say that the resistance of cylinder *AC* has to the resistance of cylinder *DF* the ratio compounded from the ratio of the cube of diameter *AB* to the cube of diameter *DE*, and the ratio of length *EF* to length *BC*.

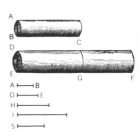

Make *EG* equal to *BC*. Let *H* be the third and *I* the fourth proportional to lines *AB* and *DE*, and let *I* be to *S* as *EF* is to *BC*. Since the resistance of cylinder *AC* is to the resistance of cylinder *DG* as the cube of *AB* is to the cube of *DE*, it will be as the line *AB* is to the line *I*; and since the resistance of cylinder *DG* is to the resistance of cylinder *DF* as the length *FE* is to *EG*, it will be as the line *I* is to the line *S*. Therefore, by equidistance of ratios, as the resistance of cylinder *AC* is to the resistance of cylinder *DF*, so is line *AB* to line *S*. But line *AB* has to line *S* the ratio compounded from [the ratios of] *AB* to *I* and *I* to *S*; therefore the resistance of cylinder *AC* has to the resistance of cylinder *DF* the ratio compounded from [the ratio of] *AB* to *I* (that is, the cube of *AB* to the cube of *DE*) and from the ratio of line *I* to line *S* (that is, the length *EF* to the length *BC*). And that is what was to be demonstrated.

Having demonstrated this proposition, we are to consider what happens among [geometrically] similar cylinders and prisms. We shall prove that:

PROPOSITION VI

The compound moments[10] of [two geometrically] similar cylinders or prisms, resulting from their own weights and [from their own] lengths serving as levers, have to one another the ratio that is the three-halves power of the ratio of the resistances of their bases.

To demonstrate this, let us draw two similar cylinders *AB* and *CD*; I say that the moment of cylinder *AB* in overcoming the resistance of its base *B* has, to the moment of *CD* in

9. This is probably the first expression of s strictly physical property in terms of two independent variables. Archimedes had used the compounding of ratios in a similar way, but only for mathematical relationships. Cf. Heath, *Archimedes*, p. clixxix.

10. See note 6, above, regarding "compound moments."

164   overcoming the resistance of its [base] *D*, that ratio which is
the three-halves power of the ratio which the resistance of the
base *B* has to the resistance of the base *D*.

The moments of solids *AB* and *CD*, [in acting] to overcome
the resistances of their bases *B* and *D*, are compounded from
their [respective] weights and net leverages [*forze delle lor
leve*].[11] The net leverage of *AB* is equal to the net leverage of
CD, because length *AB* has the same ratio to the radius
of base *B* that length *CD* has to the radius of base *D*, by

[geometrical] similarity of the cylinders. It follows that the
total moment of cylinder *AB* [with respect to its resistance]
is to the total moment of *CD* [with respect to its resistance]
as the weight alone of cylinder *AB* is to the weight alone of
cylinder *CD*; that is, as cylinder *AB* itself is to cylinder *CD*.
But these [volumes] are in cubed ratio of the diameters of the
bases *B* and *D*; and the resistances of the bases being to one
another as the [areas of the] bases themselves, these resistances
are in squared ratio of those same diameters. Therefore the
moments of the cylinders are as the three-halves power of the
[ratio of the] resistances of their bases.

Simp. This proposition strikes me as not only new but
surprising, and at first glance very remote from the judgment I
had conjecturally formed. For since the shapes are similar in
all other respects, I should have thought it certain that their
moments against their own resistances would also be in the
same ratio.

Sagr. This demonstrates the proposition which, as I said
at the beginning of our discussions, seemed then to reveal
itself to me through shadows.

Salv. What is now happening to Simplicio happened also to
me for some time. I believed the resistances of [geometrically]
similar solids to be similar,[12] until a certain observation,

---

11. In this translation, the phrase "net leverages" (note 6, above) is introduced to distinguish Galileo's own phrase, "forces of their levers," from
the modern implications of that phrase. What is meant is not the ratio of
mechanical advantages of the cylinders used as levers, but the ratio of the
leverages of two similar solids against their own weights when each is
supported at one end.

12. That is, to be proportional to their volumes. Similarity of materials

itself not very definite or correct, suggested to me that among similar solids there is not to be found an equal tenor of robustness, and that the larger are less fitted to suffer violent shocks. Thus large men are injured more by falling than are small **165** boys; and as we said at the beginning, a great beam or a column is seen to go to pieces where a stick or a small marble cylinder falling from the same height does not. It was this observation that put my mind to the investigation of that truly remarkable property which I am about to demonstrate; and indeed, among the infinite [possible] shapes of [geometrically] similar solids, not even two have the same ratio of moments with respect to their own resistances.

Simp. Now I recall something or other that was proposed by Aristotle in his *Mechanical questions*, where he tries to give a reason for the fact that the longer pieces of wood are, the weaker they are and the more they bend, even though the shorter [pieces] are quite thin, and the long ones very thick. If I recall correctly, he reduces this to the simple lever.[13]

Salv. Quite true, and since his solution seems to leave some reason for doubt, Monsignor di Guevara, who has greatly ennobled and illuminated that treatise with his learned commentaries, adds other very acute speculations to resolve all difficulties.[14] But he too remains perplexed on one point: whether, by increasing in constant ratio the lengths and thicknesses of such solid shapes, one may retain the same level of robustness in their resistance to fracture as well as bending. After long thought about this, I found what I am about to put before you, in proper order, concerning this point. And first I shall demonstrate that:

PROPOSITION VII

Among [geometrically] similar prisms or cylinders having weight [*gravi*], there is a single and unique case of the critical [*ultimato*] state between breaking and remaining whole when [the solid is] pulled down [*gravato*] by its own weight, such that if greater, unable to resist its own weight, it will break; and if smaller, it resists with some

---

is assumed throughout (note 7, above), and the word "geometrically" has been added in brackets because similar figures are meant here. With regard to Sagredo's remark, above, cf. pp. **51–52**.

13. *Questions of mechanics*, 27 (Loeb ed., p. 401).

14. The book meant is identified in note 11 to First Day, above.

force whatever is done to break it.[15]

166

Let the material [*grave*] prism *AB* be brought to the greatest length of its holding together, so that lengthened a trifle [*minimo*], it breaks. I say that *AB* is unique in reduction to this neutral [*ancipite*] state among all [geometrically] similar [prisms], which are infinitely many, so that every one greater, pressed by its own weight, breaks, and every smaller does not, but will resist up to some additional load a new force beyond that of its own weight.

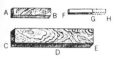

Let the prism *CE* be similar to, but greater than, *AB*; I say that it cannot hold together but will break, overcome by its own heaviness. Take in it the part *CD*, as long as *AB*; since the resistance of *CD* is to that of *AB* as the cube of the thickness of *CD* is to the cube of the thickness of *AB* (that is, as the prism *CE* is to the prism *AB*, these being similar), the weight of *CE* is the greatest that can be sustained, spread over the length of prism *CD*. But the length *CE* is greater [than *CD*], whence the prism *CE* will break.[16] Now let *FG* be smaller [than *AB*]; it will be likewise demonstrated, by putting *FH* equal to *BA*, that the resistance of *FG* to that of *AB* would be as the prism *FG* to the prism *AB* if the distance *AB* (that is, *FH*) were equal to *FG*. But it is greater; therefore the moment of prism *FG* if placed at *G* does not suffice to break the prism *FG*.

Sagr. A very clear and concise demonstration, which proves the truth and the necessity of a proposition which at first glance seemed far from probable. It will therefore be necessary, in order to achieve that neutral state between holding and breaking, to alter greatly the ratio between length and thickness of the greater prism [*CE*] by thickening or shortening it. The investigation of that state, I think, might require equal ingenuity.

Salv. Even more, and more labor too; I know, for I spent no small time in finding it. But now I wish to share it with you.

PROPOSITION VIII

Given a cylinder or prism of the maximum length that is not broken by its own weight, and given also a greater length, to find the thickness of some cylinder or prism

---

15. This represents one of the first physical problems to be treated in terms of maxima and minima, concepts that were well known in pure geometry but were of limited application before the invention of the calculus.

16. By reason of the greater moment when the same total weight is spread out over a greater length.

which, at this given length, is the unique and maximum that resists its own weight.

Let the cylinder *BC* be of the greatest [length] resisting its own weight, and let *DE* be a greater length than *AC*; it is required to find the thickness of the cylinder of length *DE* that will be the greatest to resist *its* own weight. Take *I*, the third proportional between lengths *DE* and *AC*; and as *DE* is to *I*, let the diameter *FD* be to *BA*, and construct the cylinder *FE*. I say that this is the unique and maximum, among all those similar [to *FE*], that can resist its own weight. Let *M* be the third and *O* the fourth proportional to lines *DE* and *I*, and make *FG* equal to *AC*. Since the diameter *FD* is to the diameter *AB* as line *DE* is to *I*, and *O* is the fourth proportional to *DE* and *I*, the cube of *FD* will be to the cube of *BA* as *DE* is to *O*. But as the cube of *FD* is to the cube of *BA*, so is the resistance of cylinder *DG* to the resistance of cylinder *BC*. Therefore the resistance of cylinder *DG* is to that of cylinder *BC* as line *DE* is to *O*. And since the moment of cylinder *BC* is equal to its resistance, if we show that the moment of cylinder *FE* is to the moment of cylinder *BC* as the resistance *DF* is to the resistance *BA* (that is, as the cube of *FD* is to the cube of *BA*, or as line *DE* is to *O*), we shall have our goal, that the moment of cylinder *FE* is equal to the resistance situated at *FD*.

The moment of cylinder *FE* is to the moment of cylinder *DG* as the square of *DE* is to the square of *AC*; that is, as line *DE* is to *I*. But the moment of cylinder *DG* to the moment of cylinder *BC* is as the square of *DF* to the square of *BA*, which is as the square of *DE* to the square of *I*, which is as the square of *I* to the square of *M*, or as *I* is to *O*. Therefore, by equidistance of ratios, as the moment of cylinder *FE* is to the moment of cylinder *BC*, so is line *DE* to *O*, which is as the cube of *DF* is to the cube of *BA*, which is as the resistance of the base *DF* is to the resistance of the base *BA*; and that is what was sought.

Sagr. This is a long proof, Salviati, and very difficult to keep in mind by hearing it only once. Hence I should like you to be so kind as to repeat the demonstration.

Salv. I shall obey your request, but perhaps it would be better to give you a quicker and more concise proof. This will require a somewhat different diagram.[17]

---

17. The ensuing demonstration is a simplified form of one that had been sent to Galileo at Siena in 1633 by Andrea Arrighetti (1592–1672); see (*Opere*, XV, 279–81). Galileo replied that he wished to include it in this book (ibid., pp. 283–84).

168   *Sagr.* So much greater will be the favor, but I should be obliged if you would also give me the above proof in writing, so that I may study it at leisure.

Salv. I shall be happy to oblige you. Now let us take the cylinder *A*, of which the base diameter is the line *DC*, and let *A* be the maximum [cylinder of the given material] that can sustain itself; we wish to find a larger [cylinder] than this which is again the maximum and unique one that sustains itself. Let *E* be [a cylinder geometrically] similar to *A* but of the assigned length, and let the diameter of its base be *KL*. Let *MN*, the third proportional of the two lines *DC* and *KL*, be the diameter of cylinder *X*, equal in length to *E*; I say that *X* is that which we seek. The resistance *DC* is to the resistance *KL* as the square of *DC* is to the square of *KL*, which is as the square of *KL* is to the square of *MN*, which is as cylinder *E* is to cylinder *X*, which is as the moment of *E* is to the moment of *X*. The resistance *KL* is to the resistance *MN* as the cube of *KL* is to the cube of *MN*, or as the cube of *DC* is to the cube of *KL*, or as cylinder *A* is to cylinder *E*, or as the moment of *A* is to the moment of *E*. Hence, by perturbed equidistance of ratios, as the resistance *DC* is to *MN*, so is the moment *A* to the moment *X*; whence prism *X* has the same relation of moment and resistance as does prism *A*.

I wish now to make the problem still more general, so that the proposition will be this:

[PROPOSITION IX]

Given the cylinder *AC*, of any moment whatever against its own resistance, and given any length *DE*, to find the thickness of the cylinder of length *DE*, such that its moment against its resistance shall have the same ratio as that of the moment of cylinder *AC* against its [resistance].[18]

Returning to the earlier diagram and taking once more nearly the same steps, let us say: Since the moment of cylinder *FE* has to the moment of its part *DG* the same ratio that the square of *ED* has to the square of *FG*, which is that of line *DE* to *I*; and since the moment of cylinder *FG* is to the moment of cylinder *AC* as the square of *FD* is to the square of *AB*, or as the square of *DE* is to the

---

18. The earlier diagram required here was not repeated in the original edition.

square of *I*, or as the square of *I* is to the square of *M*, or **169**
as the line *I* is to *O*; then, by equidistance of ratios, the
moment of cylinder *FE* has, to the moment of cylinder *AC*,
the same ratio as that of line *DE* to *O*, or the cube of *DE*
to the cube of *I*, or the cube of *FD* to the cube of *AB*;
that is, [the ratio] of the resistance of the base *FD* to the
resistance of the base *AB*; which is what was to be done.[19]

You now see how, from the things demonstrated thus far,
there clearly follows the impossibility (not only for art, but
for nature herself) of increasing machines to immense size.
Thus it is impossible to build enormous ships, palaces, or
temples, for which oars, masts, beamwork, iron chains, and
in sum all parts shall hold together; nor could nature make
trees of immeasurable size, because their branches would
eventually fail of their own weight; and likewise it would
be impossible to fashion skeletons for men, horses, or other
animals which could exist and carry out their functions
proportionably when such animals were increased to immense
height—unless the bones were made of much harder and
more resistant material than the usual, or were deformed
by disproportionate thickening, so that the shape and
appearance of the animal would become monstrously gross.
Perhaps this was noticed by our very alert poet when, in
describing a huge giant, he said:

> His height is quite beyond comparison,
> So immeasurably gross is he all over.[20]

To give one short example of what I mean, I once drew
the shape of a bone, lengthened only three times, and then
thickened in such proportion that it could function in its
large animal relatively as the smaller bone serves the smaller

19. This proposition is an appropriate conclusion to the first set of problems on strengths of uniform solid beams, since it corrects the intuitive
answer offered near the beginning of the First Day by Sagredo (p. **50**) and
affirmed by Salviati to be a common misapprehension (p. **51**).

20. Ariosto, *Orlando Furioso*, xvii, 30. Galileo's wording contains two
minor departures from the standard modern text translated here.

animal; here are the pictures. You see how disproportionate
the shape becomes in the enlarged bone. From this it is
manifest that if one wished to maintain in an enormous
giant those proportions of members that exist in an ordinary
man, it would be necessary either to find much harder and
170   more resistant material to form his bones, or else to allow

his robustness to be proportionately weaker than in men of
average stature; otherwise, growing to unreasonable height,
he would be seen crushed by his own weight and fallen.
On the other hand it follows that when bodies are diminished,
their strengths do not diminish in like ratio; rather, in very
small bodies the strength grows in greater ratio, and I believe
that a little dog might carry on his back two or three dogs
of the same size, whereas I doubt if a horse could carry
even one horse of his own size.

Simp. But the immense bulks that we encounter among
fishes give me grave reason to doubt whether this is so.
From what I hear, a whale is as large as ten elephants; yet
whales hold together.

Salv. Your doubt, Simplicio, enables me to deduce something that I did not mention before, a condition capable of
making giants and other vast animals hold together and
move around as well as smaller ones. That would follow if,
but not only if, strength were added to the bones and other
parts whose function it is to sustain their own weight and
that which rests on them. But leaving the skeleton in the
same proportions, these structures would hold together just
as well, or even better, if one were to diminish in the same
ratio the heaviness of the material of the bones themselves,
and that of the flesh or other [material] that must be
supported on the bones. It is this latter artifice that nature
uses in the structure of fish, making the bones and flesh
not merely somewhat lighter, but without any heaviness
whatever.

Simp. I perceive the direction of your reasoning, Salviati.
You mean that the habitat of fishes being the element of

water, which by its bodily nature, or as some will have it, its heaviness, reduces the weight of bodies that are submerged in it; and for that reason the material of fishes, weighing nothing, can be sustained without overloading their bones. But this does not suffice. Even if most of the substance of fishes does not weigh down, there is no doubt that the material of their bones does do so. Who would deny that a whale's rib, large as a beam, weighs a great deal, and would sink to the bottom in water? Hence those bones must **171** be unable to sustain so vast a bulk.

Salv. Your objection is clever. Before I reply to your question, tell me: have you observed fish remaining motionless at their pleasure under water, neither descending to the bottom nor rising to the top, [yet] without applying any force by swimming?

Simp. This is very easily observed.

Salv. Well, the ability of fish to stay motionless in water is a convincing argument that the composition of their corporeal bulk is equal to water in specific gravity. So if some parts heavier than water are found in them, there must necessarily be an equivalent amount less heavy in order for equilibrium to hold. So if the bones are heavier [than water], it must be that the flesh, or some other material present, is lighter, and that these offset with their lightness the weight of the bones. Thus, what happens in aquatic animals is the opposite of the case with terrestrial animals; namely, that in the latter, it is the task of the skeleton to sustain its own weight and that of the flesh, while in the former, the flesh supports its own weight and that of the bones. And there the marvel ceases that there can be very vast animals in the water, but not on the earth, that is to say, in the air.

Simp. This satisfies me; and I note further that these animals which we call "terrestrial" might more reasonably be called "aerial," since they truly live in air, are surrounded by air, and it is air that they breathe.[21]

Sagr. Simplicio's reasoning pleases me, both as to the question and its solution. Furthermore, I understand quite easily that one of these enormous fishes, drawn up on land, would perhaps be unable to support itself very long; the

21. Evangelista Torricelli (1608–47) used the memorable phrase, "We live at the bottom of a sea of air" in his explanation of atmospheric pressure in 1644.

attachments of its bones becoming weakened, its vast bulk would flatten out.

Salv. For the present I am inclined to believe this, nor am I far from thinking that the same might happen to yonder huge ship which, floating in the sea, does not come apart under the weight and load of its many goods and furnishings, but which would perhaps burst its seams on land, surrounded by air. But let us get on with our subject, and show that:

172

[PROPOSITION X]

Given a prism or cylinder and its [own] weight, and [given] the maximum weight it sustains [at one end], we can find the maximum length beyond which the prism itself, if prolonged, would break of its own weight.

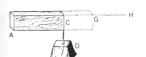

Given the prism *AC* with its own weight, let the given weight *D* be the maximum that can be sustained at its end *C*; the maximum length must be found, beyond which the prism cannot be extended without breaking. Extend *CA* to *HA* [in the same ratio] as [that of] the weight of prism *AC* to the combination of the weight *AC* with double the weight of *D*. Let *AG* be the mean proportional between these [*CA* and *HA*]; I say that *AG* is the length sought. Inasmuch as the downward [*gravante*] moment of weight *D* at *C* is equal to the moment of a weight double that of *D* but placed at the middle of *AC*, which is the center of moment of prism *AC*, the moment of the resistance at *A* of prism *AC* is equivalent to the downward tendency of double the weight *D* [together] with the weight of *AC*, attached at the middle of *AC*. What is sought is that the moment of the said [combined] weights (that is, of double *D* plus *AC*), so situated, shall be to the moment of *AC* as *HA* is to *AC*. Between these, the mean proportional is *AG*; therefore the moment of double *D* plus the moment of *AC* is to moment *AC* as the square of *GA* is to the square of *AC*. But the downward moment of prism *GA* is to the moment of *AC* as the square of *GA* is to the square of *AC*. Hence the length *AG* is the maximum sought, that is, the length to which prism *AC* would sustain itself, but beyond which it would break.

Thus far there have been considered the moments and resistances of solid prisms and cylinders of which one extremity is assumed to be fixed, and only at the other end is the force of a pressing weight applied, this [weight] being

considered alone, or in conjunction with the heaviness of **173**
the solid [prism] itself, or again, only the heaviness of that
solid [being considered]. Now I wish some discussion of the
same prisms and cylinders, but when they are sustained at
both ends, or are supported on a single point taken between
their extremities.

I say first that the cylinder pressed [*gravato*] by its own
weight [alone] and brought to that maximum length beyond
which it can no longer sustain itself, either on a single
support exactly at its middle, or supported by two at its
extremities, can be twice as long as when fixed in a wall or
sustained at one end only. This is sufficiently manifest in
itself, for if we take the half *AB* of the cylinder *ABC* as the

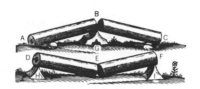

greatest length capable of sustaining itself when fixed at the
end *B*, just so will it sustain itself when placed on the support
G and counterbalanced by the other half, *BC*. And similarly,
if the length of the cylinder *DEF* is such that only half of
it can sustain itself when fixed at the end *D*, and only the
other [half] *EF* when fixed at end *F*, it is manifest that
putting the supports *H* and *I* under the ends *D* and *F*, any
moment of force or weight that is added at *E* will make a
break there.

Deeper speculation is required when, abstracting their own
heaviness from such solids, it is proposed to investigate
whether that force or weight which would suffice, when
applied at the middle of a cylinder sustained at both extremities, to break this, could have the same effect when
applied at any other place, closer to one end than the other.
For example, if we want to break a staff by taking its ends
in hand and pressing the knee at its center, will the force
[just] sufficient to break it in that way suffice also when the
knee is placed not at the center, but closer to one of the
ends?

Sagr. I think that this problem was touched on by Aristotle
in his *Questions of Mechanics*.[22]

22. *Questions of Mechanics*, 14 (Loeb ed., p. 369).

174   *Salv.* Aristotle's question was not precisely the same. All he sought was to give a reason why less effort is required to break the stick by holding the hands at its ends, away from the knee, than by holding them closer together; and he gave a general reason, reducing the cause to the greater length of the levers [applied] when one's arms are separated to grasp the ends. Our question adds something further; we inquire whether the same force serves at all places, with the knee at the center or elsewhere, but keeping the hands always at the ends [of the stick].

Sagr. At first glance it would seem that it does, since these two levers in a certain way preserve the same [total] moment, inasmuch as to the extent that one is shortened, the other is lengthened.

Salv. Just see how ready at hand mistakes can be, and with what caution and circumspection one must proceed in order not to run into them. What you say, and what seems at first to have so much probability, is in a word so false that whether the knee (which is the fulcrum of both levers) is placed at the center or not makes so great a difference that the force required to cause fracture at the center, when applied at some other place, will sometimes remain inadequate even if multiplied four, or ten, or a hundred times, or a thousand.

Let us consider this generally, and then we may come to the specific determination of the ratio in which the forces that cause fracture vary from one point to another. First we shall draw this timber *AB*, to be broken at the middle over the support *C*, and then the same [timber], but designated *DE*, to be broken over the support *F*, some distance from the middle. The distances *AC* and *CB* being equal, it is manifest first that the applied force will be divided equally between the ends *B* and *A*. Second, as the distance *DF* becomes less than the distance *AC*, the moment of the force applied at *D* becomes less than the moment [of the force] at *A*, applied at distance *CA*. The former diminishes in the ratio of line *DF* to *AC*; hence this [force at *D*] must be increased in order to equal or overcome the resistance at *F*. But distance *DF* can diminish *in infinitum* in relation to

175   distance *AC*; hence it is necessary to increase *in infinitum* the force applied at *D* in order to match the resistance at *F* [as *F* recedes toward *D*].

On the other hand, as distance *FE* increases beyond *CB*,

one must diminish the force at E to match the resistance at F;[23] but the distance FE cannot increase *in infinitum* in relation to CB as the support F is withdrawn toward end D; in fact, it cannot even double. Therefore the force required at E to match the resistance at F will always be [less than, but] more than one-half, the force at B. Thus you understand the necessity of infinitely increasing the combined moments of the forces at E and D, in order to equal or overcome the resistance located at F, as the support F approaches the extremity D.

Sagr. What shall we say, Simplicio? Must we not confess that the power of geometry is the most potent instrument of all to sharpen the mind and dispose it to reason perfectly, and to speculate? Didn't Plato have good reason to want his pupils to be first well grounded in mathematics? I understood quite well the action [*facoltà*] of the lever, and how by increasing or reducing its length, the moment of its force and of the resistance grew or diminished; yet for all that, I was mistaken in the solution of the present problem, and not a little, but infinitely.

Simp. Truly I begin to understand that although logic is a very excellent instrument to govern our reasoning, it does not compare with the sharpness of geometry in awakening the mind to discovery [*invenzione*].

Sagr. It seems to me that logic teaches how to know whether or not reasonings and demonstrations already discovered are conclusive, but I do not believe that it teaches how to find conclusive reasonings and demonstrations.

But it will be better for Salviati to show us the ratio of increase of the moments of the forces required to overcome resistance in the same timber, with regard to its different places of breaking.

Salv. The ratio which you seek has the following form: **176**

[PROPOSITION XI]

If two places are taken in the length of a cylinder at which the cylinder is to be broken, then the resistances at those two places have to each other the inverse ratio [of areas] of rectangles whose sides are the distances of those two places [from the two ends.]

23. In order that breaking shall occur and not a mere pulling of one hand by the other.

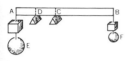

Let forces *A* and *B* be the least [forces required] for breakage at *C*, and likewise let *E* and *F* be the least for breakage at *D*; I say that forces *A* and *B* have to forces *E* and *F* the same ratio as the rectangle *AD–DB* has to the rectangle *AC–CB*. Forces *A* and *B* have to forces *E* and *F* the ratio compounded from the [ratio of the sum of] forces *A* and *B* to force *B*; that of *B* to *F*; and that of *F* to [the sum of] *F* and *E*. But as forces *A* and *B* are to force *B*, so is the length *BA* to *AC*; and as the force *B* is to *F*, so is line *DB* to *BC*; and as the force *F* is to forces *F* and *E*, so is line *DA* to *AB*. Therefore forces *A* and *B* have to forces *E* and *F* the ratio compounded from the three; that is, from the said[24] *BA* to *AC*, *DB* to *BC*, and *DA* to *AB*. But from the two [ratios] *DA* to *AB* and *AB* to *AC* is compounded the ratio of *DA* to *AC*; hence forces A and B have to forces *E* and *F* the ratio compounded from [those of] *DA* to *AC* and *DB* to *BC*. But rectangle *AD–DB* has to rectangle *AC–CB* the ratio compounded from [those of] *DA* to *AC* and *DB* to *BC*; therefore the forces *A* and *B* stand to *E* and *F* as rectangle *AD–DB* to rectangle *AC–CB*. This is to say that the resistance to breakage at *C* has to resistance to breaking at *D* the same ratio that rectangle *AD–DB* has to rectangle *AC–CB*; which was to be proved.[25]

177   In consequence of this theorem, we can solve another very curious problem, which is:

[PROPOSITION XII]

Given the maximum weight supported at the middle of a cylinder (or prism), where its resistance is least, and given a weight greater than this, to find the point in the cylinder at which the given greater weight is supported as a maximum weight.

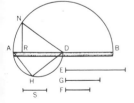

Let the given weight, greater than the maximum supported at the middle of cylinder *AB*, be in the same ratio to that maximum as line *E* is to line *F*; it is required to find the point in the cylinder at which the given weight is sustained as the maximum. Let *G* be the mean proportional between *E* and *F*, and as *E* is to *G*, make *AD* to *S*; *S* will be less than *AD*. Let *AD* be the diameter of the semicircle *AHD*, in which

24. The original text reads *retta*, but the context shows that *detta* was meant.
25. A different, shorter proof appears in the Pieroni MS.

take *AH* equal to *S*. Draw *HD*, and cut off *DR* equal to *HD*; I say that *R* is the point sought, at which the given weight, greater than the maximum supported by the middle of cylinder *D*, will be supported as the maximum.

On *BA* construct the semicircle *ANB* and erect the perpendicular *RN*; join *ND*. Since the squares of *NR* and *RD* are equal to the square of *ND*, that is, to the square of *AD*, which is [equal to the squares of] *AH* and *HD*, and [the square of] *HD* is equal to the square of *DR*, then the square of *NR*, or the rectangle *AR–RB*, will be equal to the square of *AH*, which is the square of *S*. But the square of *S* is to the square of *AD* as *F* is to *E*; that is, as the maximum weight supported at *D* is to the given greater weight. Hence this greater [weight] will be supported at *R* as the maximum that can be sustained there; which is what was sought.

Sagr. I understand perfectly. And I am considering that since prism *AB* is always stronger and more resistant to pressure at points farther and farther from the middle, then from very large and heavy beams a considerable part might be removed toward the ends, with notable lightening of weight. This would be of no small advantage and utility in the rafters of great halls. It would be a fine thing to know **178** the shape that must be given to a solid in order that it would be equally resistant at all points, and no more easily broken by a given weight pressing on it in the middle than at any other place.[26]

Salv. I was about to tell you something very noteworthy and wonderful to this purpose; here is a diagram, the better to explain this. Here, *DB* is a prism in which the resistance to fracture by a force pressing on end *B* is, as previously demonstrated, less at the end *AD* than is the resistance at *CI*, by as much as length *CB* is less than *BA*.[27] Next, consider the same prism sawed through diagonally along the line *FB*, so that the opposite faces form two triangles, one of which, *FAB*, is facing us. This solid has a nature contrary to that of the prism, since it less resists being broken over the point

26. The problem is to find a beam supported at both ends that would bear some constant given weight as a maximum at any point. Salviati's next remark suggests that he will discuss a related but different problem; cf. note 31, below.

27. What had been proved was not this, but that the weights required for breaking prisms supported at *A* and *C* were inverse to the lengths *AB* and *BC*, as Viviani noted in his copy of the book.

C than over *A*, by a force applied at *B*, in proportion as *CB* is less than *BA*. This is easily proved.

Consider the section *CNO* parallel to *AFD*; in triangle *FAB*, line *FA* has to *CN* the same ratio that line *AB* has to *BC*. Now understand points *A* and *C* to be the fulcrums of two similar levers whose arms are *BA* and *AF*, *BC* and *CN*. The moment of the force applied at *B* [acting] through distance *BA* against the resistance situated at distance *AF* will be that which the same force [applied] at *B* has, acting through distance *BC* against the same resistance situated at distance *CN*. But the resistance to be overcome by the force applied at *B*, at the fulcrum *C* situated at distance *CN*, is as much less than the resistance at *A* as rectangle *CO* is less than rectangle *AD*; that is, as much as line *CN* is less than *AF*, or *CB* than *BA*. Hence the resistance of part *OCB* to being broken [off] at *C* is as much less than the resistance of all *DAB* to being broken at *A* as the length *CB* is less than *AB*. Thus we have taken away from the beam or prism *DB* a part, in fact one-half, by cutting it diagonally, leaving the wedge or triangular prism *FBA*; and these two solids are of

179   contrary condition, the former being more resistant the more it is shortened [in the direction of B], and the latter losing robustness as it is shortened. Now, this being the case, it seems quite reasonable and even necessary that a cut can be made after which, the superfluous part being removed, there remains a solid of such shape that it is equally resistant in all its parts.

Simp. Indeed, it is necessary that where we pass from the greater to the less, we also meet with the equal.

Sagr. But the point now is to find how to guide the saw so as to make this cut.

Simp. It seems to me that this should be an easy task. By cutting the prism diagonally and taking away half, the shape that remains has its nature contrary to that of the entire prism, in such a way that wherever the latter gained strength, the former lost as much. So I believe that we should take the middle path; that is, by taking only one-half of the half, or one-quarter part of the whole, the remaining figures will neither gain nor lose robustness at any of those places at which the other two figures had equal losses and gains.

Salv. You have not hit the target, Simplicio. As I shall show you, that which can be sawed from the prism and removed without weakening it is in truth not one-quarter, but one-third. Now, as Sagredo has mentioned, it remains to find the line by which the saw should travel, which line I shall prove to be parabolic. But first it is necessary to demonstrate a certain lemma, which is this:

[LEMMA]

If there are two balances or levers, divided by their supports in such a way that the two distances at which the forces [to be compared] are applied shall be to each other in the squared ratio of the distances of the resistances, and if those resistances are to each other as their distances, then the sustaining powers [at the two points] will be equal.

Let *AB* and *CD* be two levers, divided by their fulcrums *E* and *F* in such a way that distance *EB* has to *FD* the squared ratio of distance *EA* to *FC*, and assume, at *A* and *C*, **180** resistances in the ratio of *EA* and *FC*.[28] I say that equal powers at *B* and *D* will sustain the resistances *A* and *C*. Take *EG* as the mean proportional between *EB* and *FD*; then as *BE* to *EG*, so *GE* will be to *FD*, and *AE* to *CF*, which was taken to be the [ratio of the] resistance *A* to the resistance *C*. Since *AE* is to *CF* as *EG* is to *FD*, by permuting, *GE* will be to *EA* as *DF* is to *FC*. Therefore, since the two levers, *DC* and *GA*, are proportionately divided at points *F* and *E*, the power which, when applied at *D*, balances resistance *C*, when moved to *G* will balance the same resistance *C* moved to *A*. But by the assumption, resistance *A* has to resistance *C* the same ratio that *AE* has to *CF*, or *BE* to *EG*. Therefore the power [at] *G*, or we may say [at] *D*, when placed at *B*, will sustain the resistance situated at *A*; which was to be proved.

This understood, let the parabolic line *FNB*, whose apex is *B*, be drawn on the face *FB* of prism *DB*, and let the prism be sawed along this line, leaving the solid that lies between the base *AD*, the rectangular plane *AG*, the straight line *BG*, and the surface *BGDF*, which has the curvature of the parabolic line *FNB*. I say that this solid [taken absolutely] is equally resistant throughout.[29]

Take the plane section *CO*, parallel to *AD*, and think of

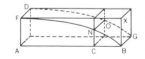

---

28. The words from "and assume . . ." to the end of this sentence appear in the Pieroni MS but not in printed editions prior to the *Opere*.

29. In the sense of breaking indifferently at *F* or *N* when the weight is applied at *B*.

two levers divided by and placed on the fulcrums *A* and *C*, so that the arms of one lever are *BA* and *AF*, and of the other, *BC* and *CN*. Since in the parabola *FBA*, *AB* is to *BC* as the square of *FA* is to the square of *CN*, it is manifest that arm *BA* of one lever has to arm *BC* of the other, the squared ratio of arm *AF* to arm *CN*. And since the resistance to be balanced by lever *BA* has to the resistance to be balanced by lever *BC* the same ratio that rectangle *DA* has to rectangle *OC*, which is the same as that of line *AF* to *NC* (the other two arms of the levers), it is evident from the above lemma that the same force which, applied by the line *BG*, balances resistance *DA*, will also balance resistance *CO*. The same is demonstrated when the solid is cut at any other place; and therefore this parabolic solid [taken absolutely] is equally resistant throughout.[30]

181

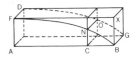

That one-third is removed by cutting the prism along the parabolic line *FNB* becomes apparent, because the semi-parabola *FNBA* and the rectangle *FB* are the bases of two solids lying between two parallel planes, that is, between rectangles *FB* and *DG*, whence they have the same ratio as that of their bases. But rectangle *FB* is three-halves of semi-parabola *FNBA*; therefore, the prism being cut along the parabolic line, one-third is taken away. From this it is seen that beamwork can be built, without any reduction in strength, while diminishing weight by over thirty-three per cent. In large ships, especially to support the decks, this may be quite useful, since lightness is extremely important in such structures.[31]

Sagr. Its uses are so numerous that it would be a long task, or impossible, to record them all. But leaving that aside, I had rather understand that the lightening made is in the ratio you have assigned. I understand quite well that the cut along the diagonal removes one-half the weight; I

30. The words "... to a force applied at *BG*" are required at the very least, and even then this would hold only for a weightless beam—or as Galileo would say, a beam "taken absolutely." This oversight became the topic of further study and controversy by Francois Blondel (1626–86), Viviani, and Alessandro Marchetti (1633–1714). Yet it should be noted that the entire discussion by Galileo opened (p. **173**) as if intended "absolutely"; that is, by abstracting heaviness.

31. Obviously Galileo did not perceive the essential difference between his problem and that of beams supported at both ends, for which the shape sought turned out to be elliptical rather than parabolic; cf. notes 26 and 30, above.

can believe that this parabolical cut carries away one-third of the prism on Salviati's word, always truthful, but on this I should be thankful for science rather than faith.

Salv. Then you would like to have the proof that the excess of the prism over what we may here call the parabolic solid is one-third of the whole prism. I know that I once demonstrated this, and I shall now try whether I can put the proof together again. I recall that for this I made use of a certain lemma of Archimedes in his book *On Spiral Lines*, and this is that if any given number of lines equally exceed one another, the excess being equal to the shortest of them; and given an equal number of lines each equal to the longest, then the [sum of the] squares of all the latter is less than triple the [sum of the] squares of the former lines, while it is more than triple the same after deducting **182** the square of the longest line.[32]

Assuming this lemma, let the parabolic line *AB* be inscribed in the rectangle *ACBP*. We must prove that the mixed triangle[33] *BAP*, whose sides are *BP* and *PA*, and whose base is the parabolic line *BA*, is one-third of the whole rectangle *CP*. If it is not, it will be either more than one-third, or less. Let it be less, if possible, and let it be short by the space *X*. Divide the rectangle continually into equal parts by lines parallel to the sides *BP* and *CA*; eventually we shall arrive at parts less than space *X*. Let one such part be rectangle *OB*, and through the points at which the other parallels cut the parabolic line, pass lines parallel to *AP*.

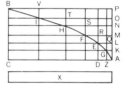

Now, by "circumscribed about our mixed triangle," I shall mean the figure composed of rectangles *BO, IN, HM, FL, EK,* and *GA,* which [broken-line] figure will be less than one-third of the rectangle *CP*, since the excess of this figure over the mixed triangle is much less than rectangle *BO*, and that in turn is less than space *X* [by construction].

Sagr. A moment, please, for I do not see why the excess of this circumscribed figure, over and above the mixed triangle, is much less than rectangle *BO*.

Salv. Rectangle *BO* is equal, is it not, to the sum of all these little rectangles through which our parabolic line

32. Archimedes, *On Spiral Lines*, Prop. 10 (Heath, *Archimedes*, p. 162, with proof on pp. 107–9, the same theorem having been used as a lemma to Prop. 2, *On Conoids and Spheroids*). Galileo's next demonstration illustrates the Archimedean method of exhaustion.
33. The three-sided figure of which one side is a curved line.

passes? I am speaking of *BI, IH, HF, FE, EG*, and *GA*, of each of which only a part lies outside the mixed triangle. And wasn't rectangle *BO* assumed to be also less than space *X*? Therefore if, for the adversary, the [mixed] triangle plus *X* equaled one-third of the rectangle *CP*, then the circumscribed figure, which adds to the [mixed] triangle less than space *X*, will still have to be less than one-third of rectangle *CP*. But this cannot be, since [as will be shown] it is greater than one-third; hence it is not true that our mixed triangle is less than one-third of the rectangle.

183 *Sagr.* I understand the answer to my question, but now you must prove to us that the circumscribed figure is more than one-third of rectangle *CP*, in which I believe we shall have a great deal more trouble.

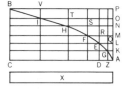

Salv. Oh, there is no great difficulty about it. In the parabola, the square of line *DE* has to the square of *ZG* the same ratio that line *DA* has to *AZ*, which is the ratio of rectangle *KE* to rectangle *AG*, the altitudes *AK* and *KL* being equal. Therefore the ratio of the square *ED* to the square *ZG* (that is, of square *LA* to square *AK*) is also that of rectangle *KE* to rectangle *KZ*. And in just the same way the other rectangles, *LF, MH, NI*, and *OB*, are proved to be to one another as the squares of lines *MA, NA, OA*, and *PA*.

Next, consider that the circumscribed figure is composed of spaces that are to one another as the squares of lines that exceed one another by differences equal to the shortest, and that the rectangle *CP* is composed of that same number of spaces, each of which is equal to the longest, namely, all the rectangles equal to *OB*. Then, by the lemma from Archimedes, the circumscribed figure is more than one-third of rectangle *CP*.[34] But it was also less, which is impossible. And thus the mixed triangle is not less than one-third of rectangle *CP*.

Likewise I say that it is not more. For if it is more than one-third of rectangle *CP*, make space *X* [now] equal to the excess of the [mixed] triangle over one-third of rectangle *CP*. Then, having made the division and subdivision of the rectangle into ever-equal rectangles, we shall again arrive at one such that it is less than space *X*. This done, and rectangle

34. Strictly speaking, the Archimedean theorem applies to sums of squares only, but a corollary to it states that any geometrically similar figures of any kind may be substituted for squares.

BO being less than *X*, describe the figure as above, and we shall have inscribed within the mixed triangle a figure, composed of the rectangles *VO, TN, SM, RL*, and *QK*, which will not be less than one-third of rectangle *CP*. For the mixed triangle exceeds this inscribed figure by much less than it surpasses one-third of rectangle *CP*, inasmuch as the excess of the [mixed] triangle over and above one-third of rectangle *CP* is equal to space *X*, which is [again] less than rectangle *BO*, and this latter is still less than the excess of the [mixed] triangle over the inscribed figure. For *BO* is equal to all the rectangles *AG, GE, EF, FH, HI*, and *IB*, and the excesses of the [mixed] triangle over the inscribed **184** figure are less than one-half of these. And since the [mixed] triangle exceeds one-third of the rectangle *CP* by much more (that is, by space *X*) than it exceeds the inscribed figure, this figure will still be greater than one-third of rectangle *CP*. But by the assumed lemma, it is less, since the rectangle *CP*, as aggregate of all the long rectangles, has the same ratio to the component rectangles of the inscribed figure, that the aggregate of all the squares of lines equal to the longest has to the squares of all the lines that equally exceed [one another, after] deducting the square of the longest. And thus, as happens with the squares [of the lines], the whole aggregate of long [rectangles], which is rectangle *CP*, is more than triple the aggregate of the [rectangles] exceeding one another, omitting the longest, that compose the inscribed figure. Therefore the mixed triangle is neither greater nor less than one-third of rectangle *CP*, and they are accordingly equal.

Sagr. A beautiful and ingenious demonstration, so much the more so in that it gives us the quadrature of the parabola, showing this to be four-thirds of the triangle inscribed in it.[35] This proves something that Archimedes demonstrated by two different trains of many propositions, both of them admirable, and which was also demonstrated more recently by Luca Valerio, a second Archimedes according to our age[36], whose demonstration is given in the book he wrote

35. The true triangle *ABC*, which is not drawn in Galileo's diagram; this is not to be confused with the "mixed" triangle (note 33, above) used in the proof. See Archimedes, *Quadrature of the Parabola*, Props. 17, 24 (Heath, *Archimedes*, pp. 246, 251–52).

36. Cf. note 20 to First Day. Valerio's proof is found in Prop. IX of his *Quadratura parabolae* (Rome, 1606).

on the center of gravity in solids.

Salv. A book truly not to be placed below anything written by the most famous geometers of the present or all past centuries. When it was seen by our Academician, it caused him to desist from pursuing the discoveries that he had been writing about the same subject, since he saw the whole thing so happily revealed and demonstrated by Signor Valerio.[37]

185 *Sagr.* I was told of all these events by the Academician himself, and also tried to get him to let me see the demonstrations he had already found when he met with Signor Valerio's book, but I did not succeed in seeing them.

Salv. I have a copy and will show it to you, for it will please you to see the difference in the methods by which these two authors move through the investigation of the same conclusions and their demonstrations. Some of the conclusions have different explanations, though in fact equally true.

Sagr. I shall be very happy to see them; when you return to our customary meetings, do me the favor of bringing them along. Meanwhile, since this [matter] of the resistance of a solid removed from a prism by a parabolic cut is an operation no less elegant than useful in many mechanical works, it would be a good thing for artisans to have some easy and speedy rule for drawing the parabolic line on the surface of the prism.

Salv. There are many ways of drawing such lines, of which two are speedier than the rest; I shall tell these to you. One is really marvelous, for by this method, in less time than someone else can draw finely with a compass on paper four or six circles of different sizes, I can draw thirty or forty parabolic lines no less fine, exact, and neat than the circumferences of those circles. I use an exquisitely round bronze ball, no larger than a nut; this is rolled [*tirata*] on a metal mirror held not vertically but somewhat tilted, so that the ball in motion runs over it and presses it lightly. In moving, it leaves a parabolic line, very thin, and smoothly traced. This [parabola] will be wider or narrower, according

37. At Valerio's request, Galileo had withheld publication of his early work on the same subject, here included as an appendix, when that was planned in 1613, because Valerio was at work on a revised edition of the work cited in the text. Galileo's posthumous tribute to Valerio is thus more than generous, particularly in view of Valerio's opposition to his Copernican campaign at Rome in 1616.

as the ball is rolled higher or lower. From this, we have a clear and sensible experience that the motion of projectiles is made along parabolic lines, an effect first observed by our friend, who also gives a demonstration of it.[38] We shall all see this in his book on motion at the next [*primo*] meeting. To describe parabolas in this way, the ball must be somewhat warmed and moistened by manipulating it in the hand, so that the traces it will leave shall be more apparent on the mirror.

186

The other way to draw on the prism the line we seek is to fix two nails in a wall in a horizontal line, separated by double the width of the rectangle in which we wish to draw the semiparabola. From these two nails hang a fine chain, of such length that its curve [*sacca*] will extend over the length of the prism. This chain curves in a parabolic shape, so that if we mark points on the wall along the path of the chain, we shall have drawn a full parabola.[39] By means of a perpendicular hung from the center between the two nails, this will be divided into equal parts. There is then no difficulty about transferring such a line onto the opposite faces of the prism; an average craftsman will know how to do this. Or one may use the geometrical lines marked on our friend's [*proportional*] compass to mark out the points of the same line on the face of the prism directly, without any other stratagem.[40]

Thus far we have demonstrated many conclusions relating to the theory of resistances of solids to fracture, having first opened the door to this science by supposing known their longitudinal resistance. It is thus possible to go on ahead, discovering more and more conclusions and their demonstrations, which are inexhaustible [*infinite*] in nature. But now, as the final end of today's discussions, I want to add the theory of resistances of hollow [*vacui*] solids. Art, and nature even more, makes use of these in thousands of operations in which robustness is increased without adding weight, as

38. The parabola underlay Galileo's first mathematical treatise composed in 1587. The same two methods of tracing parabolas are also described in the undated notebook now preserved at Paris, left by Galileo's patron, Guidobaldo del Monte (1545–1607).

39. The curve formed by a hanging chain is a catenary, not a parabola, but closely approximates one under the conditions given in the Fourth Day (p. **310**).

40. Galileo's "geometric and military compass," devised about 1597, included a scale of squares facilitating the drawing of parabolas.

is seen in the bones of birds and in many stalks [*canne*] that are light and very resistant to bending and breaking. For a straw sustains an ear much heavier than the whole stem, but if this were made of the same quantity of material **187** compacted, it would be much less resistant to bending and breaking. Hence art has observed, and experience has confirmed, that a hollow rod or a tube of wood or metal is much firmer than it would be if it were of the same weight and length, but solid, and consequently thinner; and thus art has found how to make lances hollow when it is desired to have them strong and light. We shall, therefore, show that:

[PROPOSITION XIII]

The resistances of two cylinders of equal weight and length, one of which is hollow and the other solid, are to each other as the diameters.

Let *AE* be the tube or hollow cylinder, and *IN* the solid cylinder, equal in weight and equally long; I say that the

resistance of the tube *AE* to fracture has to the resistance of the solid cylinder *IN* the same ratio that the diameter *AB* has to the diameter *IL*. This is manifest, for the tube [*AE*] and the cylinder *IN* being equal [in volume and material] and equally long, the circle *IL* that is the base of the cylinder will be equal to the doughnut [*ciambella*][41] *AB* that is the base of the tube *AE* (I call the surface that remains when a smaller circle is taken from a larger one concentric to it a "doughnut"); whence their absolute resistances will be equal. In breaking the cylinder *IN* across, we use the length *LN* as a lever with its fulcrum at point *L*, and the radius (or diameter) *LI* as its counterlever. But in the tube, the arm of the lever *BE* is equal to *LN*, while the counterlever beyond the fulcrum *B* is the radius (or diameter) *AB*. Hence it is clear that the resistance of the tube will exceed that of the solid cylinder in the ratio of the diameter *AB* over the

41. *Ciambella* is the name of a flat pastry having a central hole. See p. **74** for similar uses by Galileo of homely expressions as technical terms for geometric forms.

diameter *IL*; which is what we sought. The robustness of **188**
the tube therefore gains over the robustness of the solid
cylinder in proportion to the diameters, provided always
that both are of the same material, weight, and length.

It will be good next to investigate what happens in other
cases, in general, between all equally long tubes and solid
cylinders unequal in weight, and more or less widely hollowed
out. And first we shall deomonstrate how:

[PROPOSITION XIV]

Given a hollow tube, to find a filled cylinder equal to
it [in resistance to fracture].

The operation is very easy. Let line *AB* be the diameter
of the tube, and *CD* the diameter of its hollow. In the larger
circle draw line *AE* equal to the diameter *CD*, and join *E*
and *B*. Since in the semicircle *AEB, E* is a right angle, the
circle whose diameter is *AB* will be equal to the two circles
of diameters *AE* and *EB*. But *AE* is the diameter of the hollow
of the tube; therefore the circle whose diameter is *EB* will
be equal to the doughnut *ACBD*. Hence the solid cylinder
whose base is the circle of diameter *EB* will be equal [in
area, and hence resistance] to the tube, the two being of
equal length.

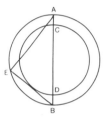

This proved, we shall quickly:

[PROPOSITION XV]

Find the ratio of the resistances of any tube and any
cylinder whatever, of equal lengths.

Let there be the tube *ABE* and the cylinder *RSM*, of equal
length; we must find the ratio between their resistances. By

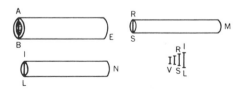

the preceding [proposition], find the cylinder *ILN* equal to
the tube and of the same length. Let line *V* be the fourth
proportional of lines *IL* and *RS*, the diameters of the bases
of cylinders *IN* and *RM*. I say that the resistance of the tube
AE is to that of the cylinder *RM* as line *AB* is to *V*. For the **189**
tube *AE* being equal to and of equal length with the cylinder

IN, the resistance of the tube will be to the resistance of the cylinder as line *AB* is to *IL*; but the resistance of cylinder *IN* is to the resistance of cylinder *RM* as the cube of *IL*

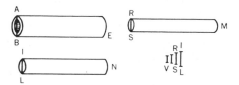

is to the cube of *RS*; that is, as line *IL* is to *V*. Therefore, by equidistance of ratios, the resistance of tube *AE* has to the resistance of cylinder *RM* the same ratio that line *AB* has to *V*; which is what was sought.

The Second Day Ends[42]

---

42. Division of the book at this point was made by the publishers. Lack of the usual conversational conclusion here, and of a conversational opening for the Third Day, suggests that Galileo intended to add material here but failed to get it to Leyden in time. See note 30 to Fourth Day, below.

Third Day

[*Salviati* (reading from Galileo's Latin treatise):]

On Local Motion

We bring forward [promovemus] *a brand new science concerning a very old subject.*

There is perhaps nothing in nature older than MOTION, about which volumes neither few nor small have been written by philosophers; yet I find many essentials [symptomata] *of it that are worth knowing which have not even been remarked, let alone demonstrated. Certain commonplaces have been noted, as for example that in natural motion, heavy falling things continually accelerate; but the proportion according to which this acceleration takes place has not yet been set forth. Indeed no one, so far as I know, has demonstrated that the spaces run through in equal times by a moveable descending from rest maintain among themselves the same rule* [rationem] *as do the odd numbers following upon unity.[1] It has been observed that missiles or projectiles trace out a line somehow curved, but no one has brought out that this is a parabola. That it is, and other things neither few nor less worthy* [than this] *of being known, will be demonstrated by me, and (what is in my opinion more worthwhile) there will be opened a gateway and a road to a large and excellent science of which these labors of ours shall be the elements,* [a science] *into which minds more piercing than mine shall penetrate to recesses still deeper.*

We shall divide this treatise into three parts. In the first part we consider that which relates to equable or uniform

---

1. Although important rules of uniformly accelerated motion had been given by medieval writers, those to which Galileo alludes were not among them. Neither the progression of spaces traversed according to the odd numbers, nor the relation of total distances to the squares of times, had been related to free fall. Those relations had been found by Galileo in 1604, and were utilized in his *Dialogue*, pp. 221–23, 227–29 (*Opere*, VII, 248–50, 253–56). His neglect to mention this may have been due to the fact that the *Dialogue* was a prohibited book. That Galileo did not mention here Cavalieri's application of those rules in 1632 is understandable, as is his omission in the next sentence of Cavalieri's derivation of the parabolic trajectory: cf. note 30 to First Day.

motion; in the second, we write of motion naturally accelerated; and in the third, of violent motion, or of projectiles.

On Equable Motion

191

Concerning equable or uniform motion, we require a single definition which I offer in this form:

DEFINITION

Equal or uniform motion I understand to be that of which the parts run through by the moveable in any equal times whatever are equal to one another.

   NOTE: To the old definition,[2] which simply calls motion "equable" when equal spaces are completed [transiguntur] in equal times, it seems good to add the qualifier "any whatever," that is, in all equal times; for it may happen that a moveable passes through equal spaces in some equal times although the spaces completed in smaller parts of those same times, themselves equal, are not equal.

From the definition there hang four axioms, as follows:

AXIOM I

During the same equable motion, the space completed in a longer time is greater than the space completed in shorter time.

AXIOM II

The time in which a greater space is traversed in the same equable motion is longer than the time in which a smaller space is traversed.

AXIOM III [3]

The space traversed with greater speed is greater than the

2. The "old definition" is presumably that of Aristotle, *Physica* 237b. 27–30: "In all cases where a thing is in motion with uniform velocity . . . if we take a part of the motion which shall be commensurable with the whole, the whole motion is completed in as many equal periods of time as there are parts of the motion." Galileo's definition removes the restriction to commensurables. Archimedes did not define uniform motion in any surviving work, but his book *On Spiral Lines*, Prop. 1, implied Galileo's addition, which had also been correctly given by Richard Swineshead in the fourteenth century.

3. This and the next axiom, unlike the first two, are not restricted to uniform motion; like Props. II and III below, which depend on them, they are of perfectly general applicability.

space traversed in the same time with lesser speed.

AXIOM IV

The speed with which more space is traversed in the same time is greater than the speed with which less space is traversed.

PROPOSITION I. THEOREM I[4]

If a moveable equably carried [latum] with the same speed passes through two spaces, the times of motion will be to one another as the spaces passed through.

Let the moveable equably carried with the same speed pass through two spaces, AB *and* BC; *and let the time of motion through* AB *be* DE, *while the time of motion through* BC *is* EF; *I say that space* AB *is to space* BC *as time* DE *is to time* EF.

Extend the spaces toward G *and* H, *and the times toward* I *and* K. *In* AG *take any number of spaces [each] equal to* AB, *and in* DI *likewise as many times [each] equal to* DE. *Further, let there be taken in* CH *any multitude of spaces [each] equal to* CB, *and in* FK *that multitude of times [each] equal to* EF. *Space* BG *and time* EI *will now be equimultiples of space* BA *and time* ED *[respectively], according to whatever multiplication was taken. Similarly, space* HB *and time* KE *will be equimultiples of space* CB *and time* EF *in such multiplication. And since* DE *is the time of movement through* AB, *the whole of* EI *will be the time of the whole [space]* BG, *since this motion is assumed equable, and in* EI *there are as many equal times* DE *as there are equal spaces* BA *in* BG; *and similarly it is concluded that* KE *is the time of movement through* HB. *But since the motion is assumed equable, if the space* GB *is equal to* BH, *the time* IE *will be equal to time* EK, *while if* GB *is greater than* BH, *so will* IE *be greater than* EK; *and if less, less. Thus there are four magnitudes,* AB *first,*

---

4. For ease of reference in this translation, the proposition numbers are placed before the theorem or problem numbers, reversing the original order. Theorem I is the converse of Archimedes, *On Spiral Lines*, Prop. 1, whose proof was likewise based on the Eudoxian definition of "same ratio" (Euclid, *Elements* V, Def. 5).

BC *second,* DE *third, and* EF *fourth; and of the first and third
(that is, of space* AB *and time* DE*), equimultiples are taken
according to any multiplication,* [*i.e.*] *the time* IE *and the
space* GB*; and it has been demonstrated that these either both
equal, both fall short of, or both exceed the time* EK *and the
space* BH*, which are equimultiples of the second and fourth.
Therefore the first has to the second (that is, space* AB *has
to space* BC*) the same ratio as the third to the fourth (that
is, time* DE *to time* EF*); which was to be demonstrated.*

PROPOSITION II. THEOREM II

*If a moveable passes through two spaces in equal times,
these spaces will be to one another as the speeds. And if
the spaces are as the speeds, the times will be equal.*[5]

Taking the previous diagram, let there be two spaces, AB
and BC*, completed in equal times, space* AB *with speed* DE
and space BC *with speed* EF*; I say that space* AB *is to space*
BC *as speed* DE *is to speed* EF*. Again, as above, taking
equimultiples both of spaces and of speeds according to any
multiplication—that is,* GB *and* IE [*equimultiples*] *of* AB *and*
DE*, and likewise* HB *and* KE [*equimultiples*] *of* BC *and* EF
—it is concluded in the same way as above that multiples GB
and IE *either both fall short of, or equal, or exceed equimultiples*
BH *and* EK*. Therefore the proposition is manifest.*

PROPOSITION III. THEOREM III

*Of movements through the same space at unequal speeds,
the times and speeds are inversely proportional.*

Let there be unequal speeds, A *greater and* B *lesser, and
let there be motion through the same space* CD *according to
each* [*speed*]*; I say that the time in which speed* A *goes through*
[*permeat*] CD *is, to the time in which speed* B *goes through
the same space, as speed* B *is to speed* A*.*[6] *For let* CD *be to*

194

<hr>

5. The statements of this and the next theorem contain no restriction to
equable or uniform motion, nor do their proofs involve restricted axioms;
cf. note 3, above. Galileo used Prop. II in certain proofs relating to accelerated
motion, for example, on p. **222**.

6. The wording here which makes a speed go through or traverse a distance
is curious. Elsewhere Galileo sometimes speaks of speeds as being spent,
consumed, or used up (as we say of time) in the traversing of a space by a
moveable.

CE *as* A *is to* B*; then from the preceding, the time in which speed* A *traverses* [conficit] CD *is the same as the time in which* B *traverses* CE*; but the time in which speed* B *traverses* CE *is, to the time in which the same* [B] *traverses* CD*, as* CE *is to* CD*. Therefore the time in which speed* A *traverses* CD *is, to the time in which speed* B *traverses the same* CD*, as* CE *is to* CD*, or as speed* B *is to speed* A*; which was proposed.*

PROPOSITION IV.   THEOREM IV

> *If two moveables are carried in equable motion but at unequal speeds, the spaces run through by them in unequal times have the ratio compounded from the ratio of speeds and from the ratio of times.*[7]

Let two moveables, E *and* F*, be moved in equable motion, and let the ratio of the speed of moveable* E *be to the speed of moveable* F *as* A *is to* B*, while the ratio of the time in which* E *is moved, to the time in which* F *is moved, is as* C *is to* D*; I say that the space run through by* E *at speed* A *in time* C *has, to the space run through by* F *at speed* B *in time* D*, the ratio compounded from the ratio of speed* A *to speed* B *and from the ratio of time* C *to time* D*.*

Let G *be the space run through by* E *at speed* A *in time* C*, and let* G *be to* I *as speed* A *is to speed* B*, and let* I *be to* L *as time* C *is to time* D*. It follows that* I *is the space through which* F *is moved in the same time as that in which* E *is moved through* G*, since spaces* G *and* I *are as speeds* A *and* B*. Since* I *is to* L *as time* C *is to time* D*, and* I *is the space that is traversed by moveable* F *in time* C*, then* L *will be the space traversed by* F *in time* D *with speed* B*. Hence the ratio of* G *to* L *is compounded from the ratios of* G *to* I *and of* I *to* L*; that is, from the ratios of speed* A *to speed* B *and of time* C *to time* D*; therefore the proposition holds.*

PROPOSITION V.   THEOREM V   **195**

> *If two moveables are carried in equable motion but with unequal speeds, and unequal spaces are run through, then the ratio of the times will be compounded from the ratio of spaces and from the the inverse ratio of speeds.*

Let there be two moveables A *and* B*, and let the speed of* A *be to the speed of* B *as* V *is to* T*; and let the spaces*

---

7. The Archimedean concept of compound ratios is essential here, as elsewhere in Galileo's applications of mathematics to physics; cf. Introduction, Glossary, note 9 to Second Day, and pp. **210, 220**, below.

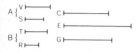

run through be as S *is to* R; *I say that the ratio of the time in which* A *is moved, to the time in which* B *is moved, is compounded from the ratio of speed* T *to speed* V *and from the ratio of space* S *to space* R.

Let C *be the time of motion* A, *and as speed* T *is to speed* V, *so let time* C *be to time* E. *Since* C *is the time in which* A, *at speed* V, *traverses space* S, *and since time* C *is to time* E *as speed* T *of moveable* B *is to speed* V, *time* E *will be that in which moveable* B *traverses the same space* S. *Now make time* E *to time* G *as space* S *is to space* R. *Clearly,* G *is the time in which* B *will traverse space* R. *And since the ratio of* C *to* G *is compounded from the ratios* C *to* E *and* E *to* G, *the ratio of* C *to* E *is the same as the inverse ratio of the speeds of moveables* A *and* B; *that is, [the same] as the ratio of* T *to* V. *But the ratio of* E *to* G *is the same as the ratio of spaces* S *and* R; *therefore the proposition holds.*

196

PROPOSITION VI.   THEOREM VI

If two moveables are carried in equable motion, the ratio of their speeds will be compounded from the ratio of spaces run through and from the inverse ratio of times.

Let two moveables, A *and* B, *be carried in equable motion, and let the spaces run through by them be in the ratio of* V *to* T, *while the times are as* S *is to* R; *I say that the speed of moveable* A *has to the speed of moveable* B *the ratio compounded from the ratios of space* V *to space* T *and of time* R *to time* S.

Let speed C *be that with which moveable* A *traverses space* V *in time* S, *and let speed* C *have to another [speed],* E, *the ratio that space* V *has to space* T. *Then* E *will be the speed with which moveable* B *traverses space* T *in the same time,* S. *But if speed* E *is made to another [speed],* G, *as time* R *is to time* S, *then speed* G *will be that with which moveable* B *traverses space* T *in time* R. *Thus we have speed* C, *with which moveable* A *traverses space* V *in time* S, *and speed* G, *with which moveable* B *traverses space* T *in time* R; *and the ratio of* C *to* G *is compounded from the ratios* C *to* E *and* E *to* G. *But the ratio* C *is assumed to be the same as the ratio of space* V *to space* T, *while [ratio]* E *to* G *is the same as ratio* R *to* S; *therefore the proposition holds.*

Salv. What we have just seen is all that our Author has written of equable motion. We therefore pass on to a new and more subtle contemplation, concerning naturally accelerated motion, which is that which is universally carried out by heavy falling moveables. Here is his title and his introduction:

On Naturally Accelerated Motion[8]   **197**

Those things that happen which relate to equable motion have been considered in the preceding book; next, accelerated motion is to be treated of.

And first, it is appropriate to seek out and clarify the definition that best agrees with that [accelerated motion] which nature employs. Not that there is anything wrong with inventing at pleasure some kind of motion and theorizing about its consequent properties, in the way that some men have derived spiral and conchoidal lines from certain motions, though nature makes no use of these [paths]; and by pretending these, men have laudably demonstrated their essentials from assumptions [ex suppositione.] But since nature does employ a certain kind of acceleration for descending heavy things, we decided to look into their properties so that we might be sure that the definition of accelerated motion which we are about to adduce agrees with the essence of naturally accelerated motion. And at length, after continual agitation of mind, we are confident that this has been found, chiefly for the very powerful reason that the essentials successively demonstrated by us correspond to, and are seen to be in agreement with, that which physical experiments [naturalia experimenta] show forth to the senses.[9] Further, it is as though we have been led by the hand to the investigation of naturally accelerated motion by consideration of the custom and procedure of nature herself in all her other works, in the performance of which she habitually employs the first, simplest, and easiest means. And indeed, no one of judgment believes that swimming or flying can be accomplished in a simpler or easier way than that which fish and birds employ by natural instinct.

8. It is significant that this title refers to natural rather than to uniform acceleration. Galileo's central topic is free fall, and he defines uniformity on the basis of natural phenomena. This reverses the medieval procedure, in which a purely mathematical analysis of accelerated motion was carried out, often illustrated by ingenious examples but never based on reference to free fall.

9. Compare the statement of Heinrich Hertz cited in the Introduction.

Thus when I consider that a stone, falling from rest at some height, successively acquires new increments of speed, why should I not believe that those additions are made by the simplest and most evident rule?[10] *For if we look into this attentively, we can discover no simpler addition and increase than that which is added on always in the same way. We easily understand that the closest affinity holds between time and motion, and thus equable and uniform motion is defined through uniformities of times and spaces; and indeed, we call movement equable when in equal times equal spaces are traversed. And*

198   *by this same equality of parts of time, we can perceive the increase of swiftness to be made simply, conceiving mentally that this motion is uniformly and continually accelerated in the same way whenever, in any equal times, equal additions of swiftness are added on.*

Thus, taking any equal particles of time whatever, from the first instant in which the moveable departs from rest and descent is begun, the degree of swiftness acquired in the first and second little parts of time [together] is double the degree that the moveable acquired in the first little part [of time]; and the degree that it gets in three little parts of time is triple; and in four, quadruple that same degree [acquired] in the first particle of time. So, for clearer understanding, if the moveable were to continue its motion at the degree of momentum of speed acquired in the first little part of time, and were to extend its motion successively and equably with that degree, this movement would be twice as slow as [that] at the degree of speed obtained in two little parts of time. And thus it is seen that we shall not depart far from the correct rule if we assume that intensification of speed is made according to the extension of time; from which the definition of the motion of which we are going to treat may be put thus:

[DEFINITION]

I say that that motion is equably or uniformly accelerated which, abandoning rest, adds on to itself equal momenta of swiftness in equal times.

Sagr. Just as it would be unreasonable for me to oppose this, or any other definition whatever assigned by any author, all

10. The ordinary and traditional view was that the simplest rule was to take the ever-changing speeds as proportional to distances traversed from rest. An essential mathematical disparity between that rule and Galileo's is shown in the discussion on pp. **203–4**, below.

[definitions] being arbitrary, so I may, without offence, doubt whether this definition, conceived and assumed in the abstract, is adapted to, suitable for, and verified in the kind of accelerated motion that heavy bodies in fact employ in falling naturally. And since it seems that the Author promises us that what he has defined is the natural motion of heavy bodies, I should like to hear you remove certain doubts that disturb my mind, so that I can then apply myself with better attention to the propositions that are expected, and their demonstrations.

Salv. It will be good for you and Simplicio to propound the difficulties, which I imagine will be the same ones that occurred to me when I first saw this treatise, and that our Author himself put to rest for me in our discussions, or that I removed for myself by thinking them out.

Sagr. I picture to myself a heavy body falling. It leaves from rest; that is, from the deprivation of any speed whatever, and enters into motion in which it goes accelerating according to **199** the ratio of increase of time from its first instant of motion. It will have obtained, for example, eight degrees of speed in eight pulse-beats, of which at the fourth beat it will have gained four; at the second [beat], two; and at the first, one. Now, time being infinitely divisible, what follows from this? The speed being always diminished in this ratio, there will be no degree of speed, however small (or we might say, "no degree of slowness, however great"), such that the moveable will not be found to have this [at some time] after its departure from infinite slowness, that is, from rest. Thus if the degree of speed that it had at four beats of time were such that, maintaining this uniformly, it would run two miles in one hour, while with the degree of speed that it had at the second beat it would have made one mile an hour, it must be said that in instants of time closer and closer to the first [instant] of its moving from rest, it would be found to be so slow that, continuing to move with this slowness, it would not pass a mile in an hour, nor in a day, nor in a year, nor in a thousand [years], and it would not pass even one span in some still longer time. Such events I find very hard to accommodate in my imagination, when our senses show us that a heavy body in falling arrives immediately at a very great speed.

Salv. This is one of the difficulties that gave me pause at the outset; but not long afterward I removed it, and its removal was effected by the same experience that presently sustains it for you.

You say that it appears to you that experience shows the heavy body, having hardly left from rest, entering into a very considerable speed; and I say that this same experience makes it clear to us that the first impetuses of the falling body, however heavy it may be, are very slow indeed. Place a heavy body on some yielding material, and leave it until it has pressed as much as it can with its mere weight. It is obvious that if you now raise it one or two braccia, and then let it fall on the same material, it will make a new pressure on impact, greater than it made by its weight alone. This effect will be caused by the falling moveable in conjunction with the speed gained in fall, and will be greater and greater according as the height is greater from which the impact is made; that is, according as the speed of the striking body is greater. The amount of speed of a falling body, then, we can estimate without error from the quality and quantity of its impact.

200   But tell me, gentlemen: if you let a sledge fall on a pole from a height of four braccia, and it drives this, say, four inches into the ground, and will drive it much less from a height of two braccia, and still less from a height of one, and less yet from a span only; if finally it is raised but a single inch, how much more will it accomplish than if it were placed on top [of the pole] without striking it at all? Certainly very little. And its effect would be quite imperceptible if it were lifted only the thickness of a leaf. Now, since the effect of impact is governed by the speed of a given percussant, who can doubt that its motion is very slow and minimal when its action is imperceptible? You now see how great is the force of truth, when the same experience that seemed to prove one thing at first glance assures us of the contrary when it is better considered.

But without restricting ourselves to this experience, though no doubt it is quite conclusive, it seems to me not difficult to penetrate this truth by simple reasoning. We have a heavy stone, held in the air at rest. It is freed from support and set at liberty; being heavier than air, it goes falling downward, not with uniform motion, but slowly at first and continually accelerated thereafter. Now, since speed may be increased or diminished *in infinitum*, what argument can persuade me that this moveable, departing from infinite slowness (which is rest), enters immediately into a speed of ten degrees rather than into one of four, or into the latter before a speed of two, or one, or one-half, or one one-hundredth? Or, in short, into all the lesser [degrees] *in infinitum*?

Please hear me out. I believe you would not hesitate to grant me that the acquisition of degrees of speed by the stone falling from the state of rest may occur in the same order as the diminution and loss of those same degrees when, driven by impelling force, the stone is hurled upward to the same height. But if that is so, I do not see how it can be supposed that in the diminution of speed in the ascending stone, consuming the whole speed, the stone can arrive at rest before passing through every degree of slowness.

Simp. But if the degrees of greater and greater tardity are infinite, it will never consume them all, and this rising heavy body will never come to rest, but will move forever while always slowing down—something that is not seen to happen.

Salv. This would be so, Simplicio, if the moveable were to hold itself for any time in each degree; but it merely passes there, without remaining beyond an instant. And since in any finite time [*tempo quanto*], however small, there are infinitely many instants, there are enough to correspond to the infinitely many degrees of diminished speed. It is obvious that this rising heavy body does not persist for any finite time in any one degree of speed, for if any finite time is assigned, and if the moveable had the same degree of speed at the first instant of that time and also at the last, then it could likewise be driven upward with this latter degree [of speed] through as much space [again], just as it was carried from the first [instant] to the second; and at the same rate it would pass from the second to a third, and finally, it would continue its uniform motion *in infinitum*.

201

Sagr. From this reasoning, it seems to me that a very appropriate answer can be deduced for the question agitated among philosophers as to the possible cause of acceleration of the natural motion of heavy bodies. For let us consider that in the heavy body hurled upwards, the force [*virtù*] impressed upon it by the thrower is continually diminishing, and that this is the force that drives it upward as long as this remains greater than the contrary force of its heaviness; then when these two [forces] reach equilibrium, the moveable stops rising and passes through a state of rest. Here the impressed impetus is [still] not annihilated, but merely that excess has been consumed that it previously had over the heaviness of the moveable, by which [excess] it prevailed over this [heaviness] and drove [the body] upward. The diminutions of this alien impetus then continuing, and in consequence the advantage passing over to the side of the heaviness, descent commences, though slowly because of the opposition of the impressed force, a good part of which still remains in the moveable. And since this continues to diminish, and comes to be overpowered in ever-greater ratio by the heaviness, the continual acceleration of the motion arises therefrom.[11]

Simp. The idea is clever, but more subtle than sound; for if it were valid, it would explain only those natural motions which had been preceded by violent motion, in which some part of the external impetus still remained alive. But where there is no such residue, and the moveable leaves from longstanding rest, the whole argument loses its force.

Sagr. I believe you are mistaken, and that the distinction of cases made by you is superfluous, or rather, is idle. For tell me: can the thrower impress on the projectile sometimes much force, and sometimes little, so that it may be driven upward a hundred braccia, or twenty, or four, or only one?

202   *Simp.* No doubt he can.

Sagr. No less will the force impressed be able to overcome the resistance of heaviness by so little that it would not raise [the body] more than an inch. And finally, the force of projection may be so small as just to equal the resistance of the heaviness, so that the moveable is not thrown upward, but merely sustained. Thus, when you support a rock in your hand, what else are you doing but impressing on it just as much of that upward impelling force as equals the power of its heaviness to draw it downward? And do you not continue this force of yours, keeping it impressed through the whole time that you support [the rock] in your hand? Does the force perhaps diminish during the length of time that you support the rock? Now, as to this sustaining that prevents the fall of the rock, what difference does it make whether it comes from your hand, or a table, or a rope tied to it? None whatever. You must conclude, then, Simplicio, that it makes no difference at all whether the fall of the rock is preceded by a long rest, or a short one, or one only momentary, and that the rock always starts with just as much of the force contrary to its heaviness as was needed to hold it at rest.

Salv. The present does not seem to me to be an opportune time to enter into the investigation of the cause of the accel-

---

11. What Sagredo presents here was Galileo's own first approach to the question of natural acceleration by seeking its cause; cf. *On Motion*, pp. 89–91 (*Opere*, I, 319–20) and note 12, below.

eration of natural motion, concerning which various philosophers have produced various opinions, some of them reducing
this to approach to the center; others to the presence of
successively less parts of the medium [remaining] to be divided;
and others to a certain extrusion by the surrounding medium
which, in rejoining itself behind the moveable, goes pressing
and continually pushing it out. Such fantasies, and others like
them, would have to be examined and resolved, with little
gain. For the present, it suffices our Author that we understand
him to want us to investigate and demonstrate some attributes
[*passiones*] of a motion so accelerated (whatever be the cause
of its acceleration) that the momenta of its speed go increasing,
after its departure from rest, in that simple ratio with which the
continuation of time increases, which is the same as to say
that in equal times, equal additions of speed are made. And if
it shall be found that the events that then shall have been
demonstrated are verified in the motion of naturally falling and
accelerated heavy bodies,[12] we may deem that the definition
assumed includes that motion of heavy things, and that it is
true that their acceleration goes increasing as the time and the   **203**
duration of motion increases.

Sagr. By what I now picture to myself in my mind, it appears
to me that this could perhaps be defined with greater clarity,
without varying the concept, [as follows]: Uniformly accelerated motion is that in which the speed goes increasing
according to the increase of space traversed. Thus for example, the degree of speed acquired by the moveable in the
descent of four braccia would be double that which it had after
falling through the space of two, and this would be the double
of that resulting in the space of the first braccio. For there
seems to me to be no doubt that the heavy body coming from a
height of six braccia has, and strikes with, double the impetus
that it would have from falling three braccia, and triple that
which it would have from two, and six times that had in the
space of one.[13]

Salv. It is very comforting to have had such a companion in

12. Note the similarity to the statement of Hertz cited in the Introduction. Rejection of causal inquiries was Galileo's most revolutionary
proposal in physics, inasmuch as the traditional goal of that science was the
determination of causes.

13. It is true that impact is proportional to the height of fall, but this
does not apply to the speed acquired, as Sagredo assumes; cf. note 17,
below. Galileo had made this assumption in 1604, in effect using it to define
"velocity" physically (*Opere*, X, 115; VIII, 373).

error, and I can tell you that your reasoning has in it so much of the plausible and probable, that our Author himself did not deny to me, when I proposed it to him, that he had labored for some time under the same fallacy. But what made me marvel then was to see revealed, in a few simple words, to be not only false but impossible, two propositions which are so plausible that I have propounded them to many people, and have not found one who did not freely concede them to me.

Simp. Truly, I should be one of those who concede them. That the falling heavy body *vires acquirat eundo* [acquires force in going],[14] the speed increasing in the ratio of the space, while the momentum of the same percussent is double when it comes from double height, appear to me as propositions to be granted without repugnance or controversy.

Salv. And yet they are as false and impossible as [it is] that motion should be made instantaneously, and here is a very clear proof of it. When speeds have the same ratio as the spaces passed or to be passed, those spaces come to be passed in equal times;[15] if therefore the speeds with which the falling body passed the space of four braccia were the doubles of the speeds[16] with which it passed the first two braccia, as one space is double the other space, then the times of those passages are equal; but for the same moveable to pass the four braccia and the two in the same time cannot take place except in instanta-neous motion. But we see that the falling heavy body makes its motion in time, and passes the two braccia in less [time] than the four; therefore it is false that its speed increases as the space.

The other proposition is shown to be false with the same clarity. For that which strikes being the same body, the difference and momenta of the impacts must be determined only by the difference of the speeds;[17] if therefore the per-

204

14. Virgil, *Aeneid* iv.175, where the reference is to rumor.

15. Cf. Prop. II and notes 3, 5, above. The ensuing argument may be an application of this rule to instantaneous velocities, whereas it had pre-viously been proved only for finite motions.

16. The plurals are essential to Galileo's concept, which is that of es-tablishing a one-to-one correspondence between all possible speeds in the whole motion and all possible speeds in the first half of it. For speeds proportional to distances, this leads to a contradiction of experience, though for speeds proportional to time it does not.

17. If "determined by" means "proportional to," this inference is incorrect, since impact is proportional not to velocity but to its square; cf. note 13, above. Fall through doubled height does in fact double the impact, but this results from speed increased as the square root of two, and not from doubled speed. Galileo appears here to believe the apparent doubling of impact

cussent coming from a double height delivers a blow of double momentum, it must strike with double speed; but double speed passes the double space in the same time, and we see the time of descent to be longer from the greater height.[18]

Sagr. Too evident and too easy is this [reasoning] with which you make hidden conclusions manifest. This great facility renders the conclusions less prized than when they were under seeming contradiction. I think that people generally will little esteem ideas gained with so little trouble, in comparison with those over which long and unresolvable altercations are waged.

Salv. Things would not be so bad if men who show with great brevity and clarity the fallacies of propositions that have commonly been held to be true by people in general received only such bearable injury as scorn in place of thanks. What is truly unpleasant and annoying is a certain other attitude that some people habitually take. Claiming, in the same studies, at least parity with anyone that exists, these men see that the conclusions they have been putting forth as true are later exposed by someone else, and shown to be false by short and easy reasoning. I shall not call their reaction envy, which then usually transforms itself into rage and hatred against those who reveal such fallacies, but I do say that they are goaded by a desire to maintain inveterate errors rather than to permit newly discovered truths to be accepted. This desire sometimes induces them to write in contradiction to those truths of which they themselves are only too aware in their own hearts, merely to keep down the reputations of other men in the estimation of the common herd of little understanding. I have heard from our Academician not a few such false conclusions, accepted as true and [yet] easy to refute; and I have kept a record of some of these.

Sagr. And you must not keep them from us, but must share them with us some time, even if we need a special session for the purpose. But now, taking up our thread again, it seems to me that we have at this point fixed the definition of uniformly accelerated motion, of which we shall treat in the ensuing discussion; and it is this:

205

---

to be illusory; cf. note 18, below.

18. The logical conclusion here is that the blow delivered is not one of doubled momentum, since it cannot be of doubled speed (denial of consequent). The argument is so elliptical as to suggest a confusion of terminal speed with overall speed, which in the context is improbable. More lakely, Galileo expected the reader to review the preceding argument in full.

[DEFINITION]

We shall call that motion equably or uniformly accelerated which, abandoning rest, adds on to itself equal momenta of swiftness in equal times.

Salv. This definition established, the Author requires and takes as true one single assumption; that is:

[POSTULATE]

I assume that the degrees of speed acquired by the same moveable over different inclinations of planes are equal whenever the heights of those planes are equal.[19]

He calls the "height" of an inclined plane that vertical from the upper end of the plane which falls on the horizontal line extended through the lower end of the said inclined plane. For an understanding of this, take line AB parallel to the horizon, upon which are the two inclined planes CA and CD; the vertical CB, falling to the horizontal BA, is called by the Author the height [or altitude, or elevation] of planes CA and CD. Here he assumes that the degrees of speed of the same moveable, descending along the inclined planes CA and CD to points A and D, are equal, because their height is the same CB; and the like is also to be understood of the degree of speed that the same body falling from the point C would have at B.

Sagr. This assumption truly seems to me to be so probable as to be granted without argument, supposing always that all accidental and external impediments are removed, and that the planes are quite solid and smooth, and that the moveable is of perfectly round shape, so that both plane and moveable alike have no roughness. With all obstacles and impediments removed, my good sense [*il lume naturale*] tells me without difficulty that a heavy and perfectly round ball, descending along the lines CA, CD, and CB, would arrive at the terminal points A, D, and B with equal impetus.

Salv. You reason from good probability. But apart from mere plausibility, I wish to increase the probability so much by an experiment that it will fall little short of equality with

206 necessary demonstration. Imagine this page to be a vertical wall, and that from a nail driven into it, a lead ball of one or two ounces hangs vertically, suspended by a fine thread

19. An attempted demonstration of this postulate was added at Galileo's request to editions after 1638, and was placed immediately before Prop. III, below, for reasons explained in note 26, below.

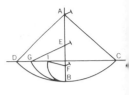

two or three braccia in length, *AB*. Draw on the wall a horizontal line *DC*, cutting at right angles the vertical *AB*, which hangs a couple of inches out from the wall; then, moving the thread *AB* with its ball to *AC*, set the ball free. It will be seen first to descend, describing the arc *CB*, and then to pass the point *B*, running along the arc *BD* and rising almost up to the parallel marked *CD*, falling short of this by a very small interval and being prevented from arriving there exactly by the impediment of the air and the thread.[20] From this we can truthfully conclude that the impetus acquired by the ball at point *B* in descent through arc *CB* was sufficient to drive it back up again to the same height through a similar arc *BD*. Having made and repeated this experiment several times, let us fix in the wall along the vertical *AB*, as at *E* or *F*, a nail extending out several inches, so that the thread *AC*, moving as before to carry the ball *C* through the arc *CB*, is stopped when it comes to *B* by this nail, *E*, and is constrained to travel along the circumference *BG*, described about the center *E*. We shall see from this that the same impetus can be made that, when reached at *B* before, drove this same moveable through the arc *BD* to the height of horizontal *CD*, but now, gentlemen, you will be pleased to see that the ball is conducted to the horizontal at point *G*. And the same thing happens if the nail is placed lower down, as at *F*, whence the ball will describe the arc *BI*, ending its rise always precisely at the same line, *CD*. If the interfering nail is so low that the thread advancing under it could not get up to the height *CD*, as would happen when the nail was closer to point *B* than to the intersection of *AB* with the horizontal *CD*, then the thread will ride on the nail and wind itself around it.

207

This experiment leaves no room for doubt as to the truth of our assumption, for the two arcs *CB* and *DB* being equal and similarly situated, the acquisition of momentum made by descent through the arc *CB* is the same as that made by descent through the arc *DB*; but the momentum acquired at *B* through arc *CB* is able to drive the same moveable back up through arc *BD*, whence also the momentum acquired in the descent *DB* is equal to that which drives the same moveable through the same arc from *B* to *D*. So that in general, every momentum acquired by descent through an arc equals one which can make

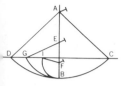

the same moveable rise through that same arc; and all the momenta that make it rise through all the arcs *BD, BG*, and *BI* are equal, because they are created by the same momentum acquired through the descent *CB*, as experiment shows. Hence all the momenta acquired through descents along arcs *DB, GB*, and *IB* are equal.

Sagr. The argument appears to me conclusive, and the experiment is so well adapted to verify the postulate that it may very well be worthy of being conceded as if it had been proved.

Salv. I do not want any of us to assume more than need be, Sagredo; especially because we are going to make use of this assumption chiefly in motions made along straight surfaces, and not curved ones, in which acceleration proceeds by degrees very different from those that we assume it to take when it proceeds in straight lines.[21] The experiment adduced thus shows us that descent through arc *CB* confers such momentum on the moveable as to reconduct it to the same height along any of the arcs *BD, BG*, or *BI*. But we cannot show on this evidence that the same would happen when [even] a most perfect sphere is to descend along straight planes inclined according to the tilt of the chords of those arcs. Indeed, we may believe that since straight planes would form angles at point *B*, a ball that had descended along the incline through the chord *CB* would encounter obstruction from planes ascending according to chords *BD, BG*, or *BI*; and in striking against those, it would lose some of its impetus, so that in rising it could not get back to the height of line *CD*. But if the obstacle that prejudices this experiment were removed, it seems to me that the mind understands that the

208   impetus, which in fact takes [its] strength from the amount of the drop, would be able to carry the moveable back up to the same height.

Hence let us take this for the present as a postulate, of which the absolute truth will be later established for us by our seeing that other conclusions, built on this hypothesis, do indeed correspond with and exactly conform to experience.[22]

21. Galileo had known this as early as 1602 from the fact that a body descends more swiftly along conjugate chords of a circular arc than along its chord, though the latter path is the shorter (*Opere*, X, 100). Cf. the scholium to Prop. XXXVI, below, and note 48.

22. The reference here is not to the attempted demonstration preceding Theorem III, below, which was composed later (notes 19, above, 26, 27, below). The certainty referred to here derives from observed agreements

This postulate alone having been assumed by the Author, he passes on to the propositions, proving them demonstratively; and the first is this:

PROPOSITION I. THEOREM I

The time in which a certain space is traversed by a moveable in uniformly accelerated movement from rest is equal to the time in which the same space would be traversed by the same moveable carried in uniform motion whose degree of speed is one-half the maximum and final degree of speed of the previous, uniformly accelerated, motion.[23]

Let line AB *represent the time in which the space* CD *is traversed by a moveable in uniformly accelerated movement from rest at* C. *Let* EB, *drawn in any way upon* AB, *represent the maximum and final degree of speed increased in the instants of the time* AB. *All the lines reaching* AE *from single points of the line* AB *and drawn parallel to* BE *will represent the increasing degrees of speed after the instant* A. *Next, I bisect* BE *at* F, *and I draw* FG *and* AG *parallel to* BA *and* BF; *the parallelogram* AGFB *will* [*thus*] *be constructed, equal to the triangle* AEB, *its side* GF *bisecting* AE *at* I.

Now if the parallels in triangle AEB *are extended as far as* IG, *we shall have the aggregate of all parallels contained in the the quadrilateral equal to the aggregate of those included in triangle* AEB, *for those in triangle* IEF *are matched by those contained in triangle* GIA, *while those which are in the trapezium* AIFB *are common. Since each instant and all instants of time* AB *correspond to each point and all points of line* AB, *from which points the parallels drawn and included within triangle* AEB *represent increasing degrees of the increased speed, while the parallels contained within the parallelogram represent in the same way just as many degrees of speed not increased but*

---

in accordance with the Hertzian principle cited in the Introduction; cf. also notes 8, 9, 12, above, and note 25, below.

23. Characteristic of Galileo's concern with actual events (note 8, above) is his utilization of one-half the terminal speed, which could be measured by observing horizontally deflected bodies. Medieval writers assumed an ideal mean-speed to measure every uniformly accelerated motion directly Galileo's proof matched elements in two infinite aggregates for each instant and all instants, conceiving that in uniform motion there is not one single speed but infinitely many, all equal, and corresponding to the infinitely many speeds, all different, in accelerated motion.

209 *equable, it appears that there are just as many momenta of speed consumed in the accelerated motion according to the increasing parallels of triangle AEB, as in the equable motion according to the parallels of the parallelogram GB. For the deficit of momenta in the first half of the accelerated motion (the momenta represented by the parallels in triangle AGI falling short) is made up by the momenta represented by the parallels of triangle IEF.*

It is therefore evident that equal spaces will be run through in the same time by two moveables, of which one is moved with a motion uniformly accelerated from rest, and the other with equable motion having a momentum one-half the momentum of the maximum speed of the accelerated motion; which was [the proposition] intended.

PROPOSITION II. THEOREM II

If a moveable descends from rest in uniformly accelerated motion, the spaces run through in any times whatever are to each other as the duplicate ratio of their times; that is, are as the squares of those times.

Let the flow of time from some first instant A be represented by the line AB, in which let there be taken any two times, AD and AE. Let HI be the line in which the uniformly accelerated moveable descends from point H as the first beginning of motion; let space HL be run through in the first time AD, and HM be the space through which it descends in time AE. I say that space MH is to space HL in the duplicate ratio of time EA to time AD. Or let us say that spaces MH and HL have the same ratio as do the squares of EA and AD.

Draw line AC at any angle with AB. From points D and E draw the parallels DO and EP, of which DO will represent the maximum degree of speed acquired at instant D of time AD, and PE the maximum degree of speed acquired at instant E of time AE. Since it was demonstrated above that as to spaces run through, those are equal to one another of which one is traversed by a moveable in uniformly accelerated motion from rest, and the other is traversed in the same time by a moveable carried in equable motion whose speed is one-half the maximum acquired in the accelerated motion, it follows that spaces MH and LH **210** *are the same that would be traversed in times EA and DA in equable motions whose speeds are as the halves of PE and OD. Therefore if it is shown that these spaces MH and LH are in the duplicate ratio of the times EA and DA, what is intended will be proved.*

*Now in Proposition IV of Book I ["On Uniform Motion,"
above] it was demonstrated that the spaces run through by
moveables carried in equable motion have to one another the
ratio compounded from the ratio of speeds and from the ratio of
times. Here, indeed, the ratio of speeds is the same as the ratio
of times, since the ratio of one-half* PE *to one-half* OD, *or of*
PE *to* OD, *is that of* AE *to* AD. *Hence the ratio of spaces run
through is the duplicate ratio of the times; which was to be
demonstrated.*

*It also follows from this that this same ratio of spaces is the
duplicate ratio of the maximum degrees of speed; that is, of lines*
PE *and* OD, *since* PE *is to* OD *as* EA *is to* DA.

*can a theorem
for uniform
motion be
applied to
accelerated
motion?*

COROLLARY I

*From this it is manifest that if there are any number of
equal times taken successively from the first instant or
beginning of motion, say* AD, DE, EF, *and* FG, *in which
spaces* HL, LM, MN, *and* NI *are traversed, then these
spaces will be to one another as are the odd numbers from
unity, that is, as 1, 3, 5, 7; but this is the rule [ratio] for
excesses of squares of lines equally exceeding one another
[and] whose [common] excess is equal to the least of the
same lines, or, let us say, of the squares successively from
unity. Thus when the degrees of speed are increased in equal
times according to the simple series of natural numbers, the
spaces run through in the same times undergo increases
according with the series of odd numbers from unity.*

Sagr. Please suspend the reading for a bit, while I develop a
fancy that has come to my mind about a certain conception. To
explain this, and for my own as well as for your clearer understanding. I'll draw a little diagram. I imagine by this line *AI*
the progress of time after the first instant at *A*; and going from
A at any angle you wish, I draw the straight line *AF*. And
joining points *I* and *F*, I divide the time *AI* at the middle in
C, and I draw *CB* parallel to *IF*, taking *CB* to be the maximum
degree of the speed which, commencing from rest at *A*, grows
according to the increase of the parallels to *BC* extended in
triangle *ABC*; which is the same as to increase [according]
as the time increases.

I assume without argument, from the discussion up to this
point, that the space passed by the moveable falling with its
speed increased in the said way is equal to the space that would

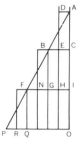

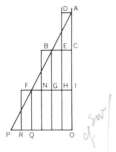

be passed by the same moveable if it were moved during the
211 same time *AC* in uniform motion whose degree of speed was
equal to *EC*, one-half of *BC*. I now go on to imagine the moveable [to have] descended with accelerated motion and to be
found at instant *C* to have the degree of speed *BC*. It is manifest
that if it continued to be moved with the same degree of speed
BC, without accelerating further, then in the ensuing time *CI*
it would pass a space double that which it passed in the equal
time *AC* with degree of uniform speed *EC*, one-half the degree
BC.[24] But since the moveable descends with speed always
uniformly increased in all equal times, it will add to the
degree *CB*, in the ensuing time *CI*, those same momenta of
speed growing according to the parallels of triangle *BFG*,
equal to triangle *ABC*; so that to the degree of speed *GI* there
being added one-half the degree *FG*, the maximum of those
[speeds] acquired in the accelerated motion governed by the
parallels of triangle *BFG*, we shall have the degree of speed *IN*,
with which it would be moved with uniform motion during
time *CI*. That degree *IN* is triple the degree *EC* convinces [us]
that the space passed in the second time *CI* must be triple that
[which was] passed in the first time *CA*.

And if we assume added to *AI* a further equal part of time
IO, and enlarge the triangle out to *APO*, then it is manifest
that if the motion continued through the whole time *IO* with
the degree of speed *IF* acquired in the accelerated motion
during time *AI*, this degree *IF* being quadruple *EC*, the space
passed in time *IO* would be quadruple that passed in the first
equal time *AC*. Continuing the growth of uniform acceleration
in triangle *FPQ*, simlar to that of triangle *ABC* which, reduced
to equable motion, adds the degree equal to *EC*, and adding
QR equal to *EC*, we shall have the entire equable speed
exercised over time *IO* quintuple the equable [speed] of the
first time *AC*; and hence the space passed [will be] quintuple
that [which was] passed in the first time *AC*.

Thus you see also, in this simple calculation, that the spaces
passed in equal times by a moveable which, parting from rest,
212 acquires speed in agreement with the growth of time, are to
one another as the odd numbers from unity, 1, 3, 5; and taking
jointly the spaces passed, that which is passed in double the
time is four times that passed in the half [i.e., in the given time],

24. This "double-distance" rule was in fact not found by Galileo until
after his odd-number rule, so that here the order of presentation follows his
order of discovery. Cf. scholium to Prop. XXIII, below, and Prop. XXV.

and that passed in triple the time is nine times [as great.] And in short, the spaces passed are in the duplicate ratio of the times; that is, are as the squares of those times.

Simp. Really I have taken more pleasure from this simple and clear reasoning of Sagredo's than from the (for me) more obscure demonstration of the Author, so that I am better able to see why the matter must proceed in this way, once the definition of uniformly accelerated motion has been postulated and accepted. But I am still doubtful whether this is the acceleration employed by nature in the motion of her falling heavy bodies. Hence, for my understanding and for that of other people like me, I think that it would be suitable at this place [for you] to adduce some experiment from those (of which you have said that there are many) that agree in various cases with the demonstrated conclusions.

Salv. Like a true scientist, you make a very reasonable demand, for this is usual and necessary in those sciences which apply mathematical demonstrations to physical conclusions, as may be seen among writers on optics, astronomers, mechanics, musicians, and others who confirm their principles with sensory experiences that are the foundations of all the resulting structure. I do not want to have it appear a waste of time [*superfluo*] on our part, [as] if we had reasoned at excessive length about this first and chief foundation upon which rests an immense framework of infinitely many conclusions—of which we have only a tiny part put down in this book by the Author, who will have gone far to open the entrance and portal that has until now been closed to speculative minds. Therefore as to the experiments: the Author has not failed to make them, and in order to be assured that the acceleration of heavy bodies falling naturally does follow the ratio expounded above, I have often made the test [*prova*] in the following manner, and in his company.

In a wooden beam or rafter about twelve braccia long, half a braccio wide, and three inches thick, a channel was rabbeted in along the narrowest dimension, a little over an inch wide and made very straight; so that this would be clean and smooth, there was glued within it a piece of vellum, as much smoothed and cleaned as possible. In this there was made to descend a very hard bronze ball, well rounded and polished, the beam having been tilted by elevating one end of **213** it above the horizontal plane from one to two braccia, at will. As I said, the ball was allowed to descend along [*per*] the said groove, and we noted (in the manner I shall presently tell you) the time that it consumed in running all the way, repeating the same process many times, in order to be quite sure as to the amount of time, in which we never found a difference of even the tenth part of a pulse-beat.[25]

This operation being precisely established, we made the same ball descend only one-quarter the length of this channel, and the time of its descent being measured, this was found always to be precisely one-half the other. Next making the experiment for other lengths, examining now the time for the whole length [in comparison] with the time of one-half, or with that of two-thirds, or of three-quarters, and finally with any other divison, by experiments repeated a full hundred times, the spaces were always found to be to one another as the squares of the times. And this [held] for all inclinations of the plane; that is, of the channel in which the ball was made to descend, where we observed also that the times of descent for diverse inclinations maintained among themselves accurately that ratio that we shall find later assigned and demonstrated by our Author.

As to the measure of time, we had a large pail filled with water and fastened from above, which had a slender tube affixed to its bottom, through which a narrow thread of water ran; this was received in a little beaker during the entire time that the ball descended along the channel or parts of it. The little amounts of water collected in this way were weighed from time to time on a delicate balance, the differences and ratios of the weights giving us the differences and ratios of the times, and with such precision that, as I have said, these operations repeated time and again never differed by any notable amount.

Simp. It would have given me great satisfaction to have been present at these experiments. But being certain of your diligence in making them and your fidelity in relating them, I am content to assume them as most certain and true.

Salv. Then we may resume our reading, and proceed.

COROLLARY II

It is deduced, second, that if at the beginning of motion

25. Actual results obtained by procedures similar to Galileo's vindicate his claim as to their reliability. His manuscript records of another type of inclined plane experiment show him to have obtained results within one percent of modern theoretical values.

there are taken any two spaces whatever, run through in any [two] times, the times will be to each other as either of these two spaces is to the mean proportional space between the two given spaces.

From the beginning of motion, S, take two spaces, ST and SV, of which the mean proportional shall be SX; the time of fall through ST will be to the time of fall through SV as ST is to SX; or let us say that the time through SV is to the time through ST as VS is to SX. Since it has been demonstrated that the spaces run through are in the duplicate ratio of the times (or what is the same thing, are as the squares of the times), the ratio of space VS to space ST is the doubled ratio of VS to SX, or is the same as that of the squares of VS and SX. It follows that the ratio of times of motion through SV and ST are as the spaces, or the lines, VS and SX.

SCHOLIUM

What we have demonstrated for movements run through along verticals is to be understood also to apply to planes, however inclined; for these, it is indeed assumed that the degree of increased speed [accelerationis] *grows in the same ratio; that is, according to the increase of time, or let us say according to the series of natural numbers from unity.*[26]

Salv. Here, Sagredo, I want permission to defer the present reading **[214]** for a time, though perhaps I shall bore Simplicio, in order that I may explain further what has been said and proved up to this point. At the same time it occurs to me that, by telling you of some mechanical conclusions reached long ago by our Academician, I can add new confirmation of the truth of that principle which has already been examined by us with probable reasonings and by experiments. More important, this will be geometrically proved after the prior demonstration of a single lemma that is elementary in the study of impetuses.

Sagr. When you promise such gains, there is no amount of time I should not willingly spend in trying to confirm and completely establish these sciences of motion. For my part, I not only grant permission to you to satisfy us on this matter, but I even beg you to allay **[215]** as swiftly as possible the curiosity you have aroused in me. I think Simplicio feels the same way about this.

26. The ensuing section was added later (note 19, above) without disturbing the original text order. Dictated by the blind Galileo about October 1638 and revised in November 1639, it was put into dialogue form by Viviani and inserted in the 1655 edition. It was placed at this point rather than with the earlier statement of the postulate because it requires prior demonstration of Prop. II, in which the postulate was not used.

Simp. How can I say otherwise?

Salv. Then, since you give me leave, consider it in the first place as a well-known effect that the momenta or speeds of the same moveable are different on diverse inclined planes, and that the greatest [speed] is along the vertical. The speed diminishes along other inclines according as they depart more from the vertical and are more obliquely tilted. Whence the impetus, power [*talento*], energy, or let us say momentum of descent, comes to be reduced in the underlying plane on which the moveable is supported and descends.

The better to explain this, let the line *AB* be assumed to be erected vertically on the horizontal *AC*, and then let it be tilted at different inclinations with respect to the horizontal, as at *AD*, *AE*, *AF*, etc. I say that the impetus of the heavy body for descending is maximal and total along the vertical *BA*, is less than that along *DA*, still less along *EA*, successively diminishes along the more inclined *FA*, and is finally completely extinguished on the horizontal *CA*, where the moveable is found to be indifferent to motion and to rest, and has in itself no inclination to move in any direction, nor yet any resistance to being moved. Thus it is impossible that a heavy body (or combination thereof) should naturally move upward, departing from the common center toward which all heavy bodies mutually converge [*conspirano*]; and hence it is impossible that these be moved spontaneously except with that motion by which their own center of gravity approaches the said common center.[27] Whence, on the horizontal, which here means a surface [everywhere] equidistant from the said [common] center, and therefore quite devoid of tilt, the impetus or momentum of the moveable will be null.

[216]   This change of impetus assumed, I must next explain something that our Academician, in an old treatise on mechanics written at Padua for the use of his pupils,[28] demonstrated at length and conclusively in connection with his treatment of the origin and character of that marvelous instrument, the screw; namely, the ratio in which this change of impetus along planes of different inclinations takes place. Given the inclined plane *AF*, for example, and taking as its elevation above the horizontal the line *FC*, along which the impetus of a heavy body and its momentum in descent is maximum, we seek the ratio that this momentum has to the momentum of the same moveable along the incline *FA*, which ratio, I say, is inverse to that of the said

27. This conception became a fundamental principle in Torricelli's continuation of Galileo's work; cf. E. Torricelli, *Opere* (Faenza, 1919), II, 105 ff. Comparison with Galileo's dictated text (note 26, above) suggests that this sentence was interpolated by Viviani when he put the argument in dialogue form.

28. Galileo's treatise *On Mechanics* was first published in a French translation by Marin Mersenne (1588–1648) in 1634. The original Italian, of which three manuscript forms exist (1593, 1594, and ca. 1600), was posthumously published in 1649.

lengths. This is the lemma to be put before the theorem that I hope then to be able to demonstrate.

It is manifest that the impetus of descent of a heavy body is as great as the minimum resistance or force that suffices to fix it and hold it [at rest]. I shall use the heaviness of another moveable for that force and resistance, and [as] a measure thereof. Let the moveable *G*, then, be placed on plane *FA*, tied with a thread which rides over *F* and is attached to the weight *H*; and let us consider that the space of the vertical descent or rise of this [*H*] is always equal to the whole rise or descent of the other moveable, *G*, along the incline *AF*—not just to the vertical rise or fall, through which the moveable *G* (like any other moveable) exclusively exercises its resistance. That much is evident. For consider the motion of the moveable *G* in the triangle *AFC* (for example, upward from *A* to *F*) as composed of the horizontal transversal *AC* and the vertical *CF*. As before, there is no resistance to its being moved along the horizontal, since by means of such a motion no loss or gain whatever is made with regard to its distance from the common center of heavy things, that being conserved always the same on the horizontal [as defined above]. It follows that the resistance is only with respect to compulsion to go up the vertical *CF*. Hence the heavy body *G*, moving from *A* to *F*, resists in rising only the vertical space *CF*; but that other heavy body *H* must descend vertically as much as the whole space *FA*. And this ratio of ascent and descent remains always the same, being as little or as great as the motion of the said moveables by reason of their connection together. Thus we may assert and affirm that when equilibrium (that is, rest) is to prevail between two moveables, their [overall] speeds or their propensions to motion—that is, the spaces they would pass in the same time—must be inverse to their weights [*gravità*], exactly as is demonstrated in all cases of mechanical movements.

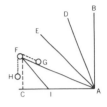

[217]

Thus, in order to hinder the descent of *G*, it will suffice that *H* be as much lighter than *G* as the space *CF* is proportionately less than the space *FA*. Hence if the heavy body *G* is, to the heavy body *H*, as *FA* is to *FC*, equilibrium will follow; that is, the heavy bodies *H* and *G* will be of equal moments, and the motion of these moveables will cease. Now, we have agreed that the impetus, energy, momentum, or propensity to motion of a moveable is as much as the minimum force or resistance that suffices to stop it; and it has been concluded that the heavy body *H* suffices to prohibit motion to the heavy body *G*; hence the lesser weight *H*, which exercises its total [static] moment in the vertical *FC*, will be the precise measure of the partial moment that the greater weight *G* exercises along the inclined plane *FA*. But the measure of the total moment of heavy body *G* is *G* itself, since to hinder the vertical descent of a heavy body, there is required the opposition of one equally heavy when both are free to move vertically. Therefore the partial impetus or momentum of *G* along the incline

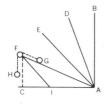

FA will be, to the maximum and total impetus of *G* along the vertical *FC*, as the weight *H* is to the weight *G*, which is (by construction) as the vertical *FC* (the height of the incline) is to the incline *FA* itself.

This is what was proposed to be demonstrated as the lemma; and as we shall see, it is assumed by our Author as known in the second part of Proposition VI of the present treatise.

Sagr. It seems to me that from what you have concluded thus far, it can be easily deduced, arguing by perturbed equidistance of ratios, that the momenta of the same moveable along differently inclined planes having the same height, such as *FA* and *FI*, are in the inverse ratio of those same planes.

218   *Salv.* A true conclusion. This established, I go on next to demonstrate the theorem itself; that is:

[ADDED THEOREM]

The degrees of speed acquired by a moveable in descent with natural motion from the same height, along planes inclined in any way whatever, are equal upon their arrival at the horizontal, all impediments being removed.

Here you must first note that it has already been established that along any inclinations, the moveable upon its departure from rest increases its speed, or amount of impetus, in proportion to the time, in accordance with the definition given by the Author for naturally accelerated motion. Whence, as he has demonstrated in the last preceding proposition, the spaces passed are in the squared ratio of the times, and consequently of the degrees of speed. Whatever the [ratio of] impetuses at the beginning [*nella prima mossa*], that proportionality will hold for the degrees of the speeds gained during the same time, since both [impetuses and speeds] increase in the same ratio during the same time.

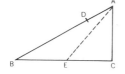

Now let the height of the inclined plane *AB* above the horizontal be the vertical *AC*, the horizontal being *CB*. Since, as we concluded earlier, the impetus of a moveable along the vertical *AC* is, to its impetus along the incline *AB*, as *AB* is to *AC*, [then] in the incline *AB* take *AD* as the third proportional of *AB* and *AC*; the impetus [to move] along *AC* is, to the impetus [to move] along *AB* (that is, [to move] along *AD*), as [*AB* is to *AC* or as] *AC* is to *AD*. Hence the moveable, in the same time that it passes the vertical space *AC*, would also pass the space *AD* along the incline *AB* (the momenta being as the spaces); and the degree of speed at *C* will have to the degree of speed at *D* the same ratio that *AC* has to *AD*. But the speed at *B* is to the speed at *D* as the time through *AB* is to the time through *AD*, by our definition of accelerated motion; and the time through *AB* is to the time through *AD* as *AC* (the mean proportional between *BA* and *AD*) is to AD, by the last corollary to Proposition II. Therefore the speeds at *B* and *C* [both] have to the speed at *D* the same ratio that *AC* has to *AD*, and hence [the speeds at *B* and *C*] are equal; which is the theorem intended to be demonstrated.

From this we may more conclusively prove the Author's ensuing Proposition III, in which he makes use of the [earlier] postulate; this [theorem] states that the time along the incline has to the time along the vertical the same ratio that the incline has to the vertical. So let us say: If *BA* is the time along *AB*,[29] the time along *AD* will be the mean proportional between these [*AB* and *AD*], that is, *AC*, by the second corollary to Proposition II. But if *AC* is the time along *AD*, it will also be the time along *AC*, since *AD* and *AC* are run through in equal times. And since if *BA* is the time along *AB*, *AC* will be the time along *AC*, then it follows that as *AB* is to *AC*, so is the time along *AB* to the time along *AC*. **[219]**

By the same reasoning it will be proved that the time along *AC* is to the time along some other incline, *AE*, as *AC* is to *AE*; therefore, by equidistance of ratios, the time along incline *AB* is to the time along incline *AE* homologously as *AB* is to *AE*, etc.

As Sagredo will readily see [later], the Author's Proposition VI could be immediately proved from the same application of this theorem. But enough for now of this digression, which has perhaps turned out to be too tedious, though it is certainly profitable in these matters of motion.

Sagr. And not only greatly to my taste, but most essential to a complete understanding of that principle.

Salv. Then I shall resume the reading of the text.

PROPOSITION III. THEOREM III   **215**

If the same moveable is carried from rest on an inclined plane, and also along a vertical of the same height, the times of the movements will be to one another as the lengths of the plane and of the vertical.

Let the inclined plane AC and the vertical AB each have the same altitude above the horizontal CB, that is, the line BA. I say that the time of descent along plane AC has, to the time of fall of the same moveable along the vertical AB, the same ratio that the length of plane AC has to the length of vertical AB. Assume any lines DG, EI, and FL parallel to the horizontal CB; it follows from our postulate that the degrees of speed acquired by the moveable from the first beginning of motion, A, to the points G and D, are equal, since their approaches to the horizontal are equal; likewise, the speeds at points I and E are the same, as are   **216**

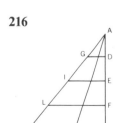

---

29. Galileo employs a single line to represent both distances and times frequently in the remaining propositions, using bisection for halving distances and mean proportionals for halving times, without further explanation; see, for example, Prop. XII, below, and see further at pp. **287–88**.

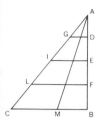

the speeds at L *and* F. *Now, if not only these, but parallels from all points of the line* AB *are supposed drawn as far as line* AC, *the momenta or degrees of speed at both ends of each parallel are always matched with each other. Thus the two spaces* AC *and* AB *are traversed at the same degrees of speed. But it has been*

217

shown that if two spaces are traversed by a moveable which is carried at the same degrees of speed, then whatever ratio those spaces have, the times of motion have the same [ratio].[30] *Therefore the time of motion through* AC *is to the time through* AB *as the length of plane* AC *is to the length of vertical* AB; *which was to be demonstrated.*

218

Sagr. It appears to me that the same can be very clearly and briefly concluded, since it has already been shown that the overall [*somma del*] accelerated motion[31] of passage through

219

AC (and *AB*) is that of the equable motion whose degree of speed is one-half the maximum degree, [at] *CB.* Therefore, the two spaces *AC* and *AB* being [considered as] passed with the same equable motion, it is manifest by Proposition I of Book I that the times of [these] passages will be as the spaces themselves.

[*Salv.* (resuming his reading):]

COROLLARY

From this it is deduced that the times of descent over differently inclined planes of the same height are to one another as their lengths. For if we suppose another plane AM *from* A, *terminated at the same horizontal* CB, *it will be proved likewise that the time of descent through* AM *is to the time through* AB *as line* AM *is to* AB; *also, as the time* AB *is to the time through* AC, *so is line* AB *to* AC; *therefore, by equidistance of ratios, as* AM *is to* AC, *so is*

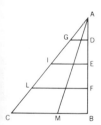

---

30. The plural, "degrees of speed," shows that reference is not directly to Prop. I on uniform motion; rather, this appears to be an extension of that proposition to instantaneous speeds by reasoning similar to the argument used in rejecting proportionality of speeds to distances in free fall (p. **203** and note 15, above). Strictly speaking, the extension had not been "shown," though it follows easily from such an argument and the general definition of equal speeds as those in which proportional distances are traversed in proportional times; cf. *Dialogue*, p. 24 (*Opere*, VII, 48).

31. The notion of a "total" or "overall" motion employed here in reference to Theorem I, above, had been preceded in Galileo's thought by a notion of total or overall speed; see *Dialogue*, p. 229 (*Opere*, VII, 256). Traces of that earlier concept survive in the scholium to Prop. XXIII, below; see also note 20 to the Added Day on percussion, below.

the time through AM *to the time through* AC.

PROPOSITION IV. THEOREM IV

The times of motion over equal planes, unequally inclined, are to each other inversely as the square root of the ratio of the heights of those planes.

Let BA *and* BC *be equal planes from the same terminus,* B, *but unequally inclined; and let horizontal lines* AE *and* CD *be drawn to the vertical* BD, *plane* BA *having height* BE *and plane* BC *height* BD. *And let* BI *be the mean proportional of these elevations* DB *and* BE; *it follows that the ratio of* DB *to* BI *is the square root of the ratio of* DB *to* BE. *I now say that the ratio of the times of descent or movement over planes* BA *and* BC *is the same as the inverse ratio of* DB *to* BI; *that is, the homologue of the time through* BA *is the height of the other plane,* BC, *which [height] is* BD, *and the homologue of the time through* BC *is* BI. *It is therefore to be demonstrated that the time through* BA *is to the time through* BC *as* DB *is to* BI.

220

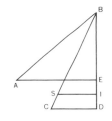

Draw IS *parallel to* DC; *as already demonstrated, the time of descent through* BA *is to the time of fall through the vertical* BE *as* BA *is to* BE, *while the time through* BE *is to the time through* BD *as* BE *is to* BI; *and the time through* BD *is to the time through* BC *as* BD *is to* BC, *or* BI *to* BS. *Therefore, by equidistance of ratios, the time through* BA *will be to the time through* BC *as* BA *is to* BS, *or* CB *to* BS, *and also* CB *is to* BS *as* DB *is to* BI; *therefore the proposition holds.*

PROPOSITION V. THEOREM V

The ratio of times of descent over planes differing in incline and length, and of unequal heights, is compounded from the ratio of lengths of those planes and from the inverse ratio of the square roots of their heights.[32]

Let planes AB *and* BC *be differently inclined, of unequal lengths, and of unequal heights; I say that the ratio of the time of descent through* AC *to the time through* AB *is compounded from the ratio of* AC *to* AB *and from [the ratio of] the square roots of their heights taken inversely.*

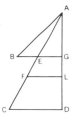

Draw the vertical AD, *meeting the horizontals* BG *and* CD, *and let* AL *be the mean proportional between heights* DA *and* AG. *From point* L *draw a parallel to the horizontal, meeting*

32. Mathematical functions of two variables had been similarly stated by Archimedes, using compound ratios; cf. Introduction, and note 9 to Second Day.

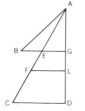

plane AC *at* F*; then* AF *will be the mean proportional between* CA *and* AE. *And since the time through* AC *is to the time through* AE *as line* FA *is to* AE, *and the time through* AE *is to to the time through* AB *as* AE *is to* AB, *it follows that the time through* AC *is to the time through* AB *as* AF *is to* AB. *Thus it remains to be proved that the ratio of* AF *to* AB *is compounded from the ratio of* CA *to* AB *and from the ratio of* GA *to* AL, *which* [*latter*] *is the ratio of the square roots of heights* DA *and* AG *taken inversely. But this is also evident: if* CA *is taken with respect to* FA *and* AB, *the ratio of* FA *to* AC *is the same as the ratio of* LA *to* AD, *or* GA *to* AL, *which is the square root of the ratio of the heights* GA *and* AD*; and the ratio of* CA *to* AB *is the ratio of the* [*corresponding*] *lengths; therefore the proposition holds.*

221

PROPOSITION VI. THEOREM VI

If, from the highest or lowest point of a vertical circle, any inclined planes whatever are drawn to its circumference, the times of descent through these will be equal.

Let a circle be erect to the horizontal GH, *and from its lowest point (that is, from its contact with the horizontal) let the diameter* FA *be erected. From the highest point,* A, *draw any inclined planes* AB *and* AC *out to the circumference; I say that the times of descent through these are equal.*

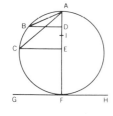

Draw BD *and* CE *perpendicular to the diameter, and let* AI *be the mean proportional between the heights of the planes* EA *and* AD. *Since the rectangles* FA–AE *and* FA–AD *are equal to the squares on* AC *and* AB*; and since also as rectangle* FA–AE *is to rectangle* FA–AD, *so* EA *is to* AD*; then as square* CA *is to square* AB, *so line* EA *is to* AD. *But as line* EA *is to* DA, *so square* IA *is to square* AD*; hence the squares on lines* CA *and* AB *are to each other as the squares on lines* IA *and* AD. *Therefore as line* CA *is to* AB, *so* IA *is to* AD. *Now, it was demonstrated in the preceding* [*proposition*] *that the ratio of the time of descent through* AC *to the time of descent through* AB *is compounded from the ratios of* CA *to* AB *and of* DA *to* AI, *which* [*latter*] *is the same as the ratio of* BA *to* AC*; therefore the ratio of the time of descent through* AC *to the time of descent through* AB *is compounded from the ratios of* CA *to* AB *and of* BA *to* AC. *Therefore the ratio of their times is the ratio of equality*[33]*; hence the proposition holds.*

33. The "ratio of equality" (x : x) remained a ratio and was not taken to

The same is demonstrated another way from mechanics; that in the next diagram the moveable passes through CA *and* DA *in equal times.*

Let BA *be equal to* DA, *and draw the verticals* BE *and* DF. **222** *From the elements of mechanics*[34] *it follows that the [static] moment of weight upon the plane elevated along line* ABC *is to the total moment [of that weight] as* BE *is to* BA; *and the similar moment of weight upon the incline* AD *is to its total moment as* DF *is to* DA, *or* BA; *therefore the moment of the same weight upon the plane inclined as* DA, *to its moment upon the incline* ABC, *is as line* DF *is to* BE. *Hence the spaces which the same weight passes through in equal times along the inclines* CA *and* DA *will be to each other as the lines* BE *and* DF, *by Proposition II of Book I.*[35] *It can indeed be demonstrated that as* BE *is to* DF, *so* AC *is to* DA; *therefore the same moveable passes through* CA *and* DA *in equal times.*

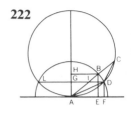

That CA *is to* DA *as* BE *is to* DF *is demonstrated thus: Join* C *and* D, *and through* D *and* B, *parallel to* AF, *draw* DGL *cutting* CA *at* I, *and [draw]* BH; *angle* ADI *will be equal to angle* DCA, *since they stand on the equal arcs* LA *and* AD. *Angle* DAC *is common to the similar triangles* CAD *and* DAI; *therefore the sides around equal angles in them will be proportional, and as* CA *is to* AD, *so* DA *is to* AI; *that is,* BA *is to* AI, *or* HA *to* AG, *which is [as]* BE *to* DF; *which was to be proved.*

This is also, and more quickly, demonstrated thus: Let there be a vertical circle whose diameter CD *is erect to the horizontal* AB, *and let there be any inclined plane* DF *from the highest point* D *to the circumference; I say that descent through plane* DF *and fall through the diameter* DC *will be finished* [absolvi] *in the same time by the same moveable.*

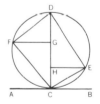

Draw FG *parallel to the horizontal* AB. *which will be per-*

be a unit or magnitude as we take it to be; no ratio was identified with any single number.

34. The proposition is found in Galileo's treatise *On Mechanics*, pp. 173–75 (*Opere*, II, 181–83), then unpublished in Italy (note 28, above). Here he speaks of the result as commonly known; it had been established during the Middle Ages. In the preceding paragraph, the word *Mechanicis* was capitalized, as if referring to a specific book, presumably Galileo's own, but here the phrase "elements of mechanics" is not capitalized, and no particular text seems to be meant.

35. This rather inconclusive argument "from mechanics" belongs to an early stage of Galileo's work, and is found in very similar form in a manuscript written at Padua probably about 1607. The reference to Prop. II of Bk. I was added for this book; cf. note 5, above.

pendicular to the diameter DC; *and join* F *and* C. *Since the*
223   *time of fall through* DC *is to the time of fall through* DG *as the*
mean proportional between CD *and* DG *is to* DG; *and since*
the mean proportional between CD *and* DG *is* DF; [*and*] *since*
angle DFC *is a right angle, being in a semicircle, and* FG *is*
perpendicular to DC, *the time of fall through* DC *is to the time*
of fall through DG *as line* FD *is to* DG. *But it was already*
demonstrated that the time of descent through DF *is to the time*
of fall through DG *as line* DF *is to* DG.[36] *Therefore the times*
of descent through DF *and of fall through* DC [*both*] *have*
the same ratio to the time of fall through DG, *and hence they*
are equal. Likewise it may be proved that if from the lowest
point C *the chord* CE *is raised,* EH *being drawn parallel to the*
horizontal and E *joined to* D, *the time of descent through* EC
will equal the time of fall through the diameter DC.

COROLLARY I

From this it is deduced that the times of descent through
all chords drawn from the terminals C *and* D *are equal*
to one another.

COROLLARY II

It is also deduced that if from the same point there descend
a vertical and an inclined plane, over which descents are
made in equal times, they are [*inscribable*] *in a semicircle*
of which the diameter is the vertical.

COROLLARY III

From this it is deduced that the times of movements over
inclined planes are equal when the heights of equal parts of
those planes are to one another as the lengths of the planes
themselves. For it has been shown that, in the penultimate
diagram, the times through AC *and* AD *are equal when*
the height of part AB *(which equals* AD*), or* BE, *is to*
height DF *as* CA *is to* DA.

Sagr. Please put off reading what follows for a time, until
I have resolved a certain idea that is now turning over in my
224   mind. If it is not a fallacy, then it borders on a sprightly prank,
as are all pranks of nature or necessity.

36. In fact this had not been proved up to this point, but is demonstrated
below (Prop. X). Probably the original order of propositions was changed,
this proof having first been placed with other theorems about vertical
circles which begin with Prop. XXX, below.

It is manifest that if, from a point marked in a horizontal plane, there were to be extended over that plane infinitely many straight lines, going in all directions, on each of which we imagine a point to move in equable motion, [and assuming that] all these points commence to move at the same instant of time from the designated point and that all their speeds are equal, these movable points would consequently mark out circumferences of ever-widening circles, all concentric around the original designated point. It is in just this way that we see little waves made in still water after a pebble has fallen into it from above, its impact serving to start motion in all directions and remaining as the center of all the circles that come to be made by these wavelets, ever larger and larger. But if we suppose a vertical plane in which some very high point is marked, from which are drawn infinitely many lines inclined in every direction; and upon these, we imagine heavy moveables descending, each with naturally accelerated motion at those speeds that suit the different slopes; then, supposing that these moveables are continually visible, in what sorts of lines would we see them continually arranged? This aroused my wonder when the preceding demonstrations assured me that they would all be seen ever in the same circumference of successively widening circles, in which the moveables would descend successively farther from the high point at which their fall began.

In order to explain myself better, I mark the high point *A*, from which lines *AF* and *AH* descend at whatever inclinations; and the vertical *AB*, in which points *C* and *D* are taken, around which are described circles passing through point *A* and cutting the inclined lines at points *F, H, B*, and *E, G, I*. It is evident from the preceding demonstrations that if moveables leave from terminus *A* at the same time and descend along these lines, then when one [moveable] is at *E*, another will be at *G*, and another at *I*; continuing to descend thus, they will [later] be found at the same instant of time at *F, H*, and *B*. **225** And these, together with infinitely many more, continuing to move along the infinitely many different inclinations, will ever be found successively on the same circumferences, which become greater and greater *in infinitum*.

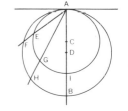

Thus from the two species of motion that nature employs,[37]

---

37. That is, the only entirely natural motions are uniform rectilinear motion and uniform acceleration, both of which are here made responsible for the production of circular effects. This passage throws light on Galileo's conception of inertia; cf. note 46, below.

there arises, with wonderfully corresponding diversity, the production of infinitely many circles. The first [generating motion] is situated at the center of infinitely many concentric circles, as its seat and originating principle; the second is located at the upper contact of infinitely many circumferences of circles, all of which are eccentric. The former are born from motions that are all equal and equable; the latter from motions always non-uniform within themselves, and each unequal to any of the rest that are carried out along infinitely many different inclinations.

But there is more. From the two points assigned for these emanations, let us imagine lines not only in two dimensions, horizontal and vertical, but in all directions, so that those which commenced at a single point and went to produce circles, from least to greatest, now commence from a single point and go to produce infinitely many spheres. Or let us say, one sphere might go amplifying itself into infinitely many magnitudes, and in two ways, by placing the origin of such spheres at the center, or else on the circumference.

Salv. The reflection is truly very beautiful, as befits the mind of Sagredo.

Simp. For my part, I can at least grasp this idea of the manner of production of circles and spheres by the two different natural motions, although I [still] do not completely understand some of the results that depend on accelerated motion, and some of its demonstrations. Yet since we can assign as the site of such emanations the lowest center, as well as the highest spherical surface, I believe that some great mystery may perhaps be contained in these true and admirable conclusions—I mean a mystery that relates to the creation of the universe, which is supposed to be spherical in shape, and perhaps [relates] to the residence of the first cause.[38]

Salv. I feel no repugnance to that same belief. But such profound contemplations belong to doctrines much higher than ours, and we must be content to remain the less worthy artificers who discover and extract from quarries that marble **226** in which industrious sculptors later cause marvelous figures

---

38. Simplicio speaks for those philosophers who admire mathematics but cannot quite follow its proofs. After Simplicio has thus deduced the First Cause from Sagredo's geometrical *tour de force*, Salviati, probably speaking for the physicist Galileo, replies sympathetically but claims no more than to have supplied raw material for those men of higher intelligence who create metaphysics and theology.

to appear that were lying hidden under those rough and formless exteriors.

Now, with your permission, we shall proceed.

PROPOSITION VII. THEOREM VII

If the heights of two planes have the squared ratio of their lengths, movements along these from rest will be made in equal times.

Let the planes AE *and* AB *be unequal and unequally inclined, having the heights* FA *and* DA *; and whatever ratio* AE *has to* AB, *let* FA *to* DA *have the square of that* [ratio] *; I say that the times of movements are equal from rest at* A *along planes* AE *and* AB. *Draw horizontal parallels* EF *and* DB *to the line of heights* [AF]; DB *cuts* AE *at* G. *Since the ratio of* FA *to* AD *is the square of the ratio of* EA *to* AB, *and* EA *is to* AG *as* FA *is to* AD, *the ratio of* EA *to* AG *is the square of the ratio of* EA *to* AB. *Hence* AB *is the mean proportional between* EA *and* AG. *And since the time of descent through* AB *is to the time through* AG *as* AB *is to* AG, *and the time of descent through* AG *is, to the time through* AE, *as* AG *is to the mean proportional between* AG *and* AE *(which is* AB*), then by equidistance of ratios the time through* AB *is to the time through* AE *as* AB *is to itself. Therefore the times are equal, which was to be proved.*

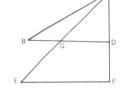

PROPOSITION VIII. THEOREM VIII

For planes cut by the same vertical circle, the times of movements in those that terminate at the upper or lower end of the [vertical] diameter are equal to the time of fall in that diameter; and in those [inclined planes] that do not reach the diameter, the times are shorter, while in those that cut the diameter they are longer [than the time through the diameter].

Let AB *be the vertical diameter of a circle erect to the horizontal. It has already been shown that the times of movements are equal along planes from terminals* A *and* B *to the circumference. That the time of descent is shorter in plane* DF, *which does not reach the diameter, is shown by drawing plane* DB, *which will be longer and less [steeply] inclined than* DF; *therefore the time through* DF *is shorter than [that] through* DB, *and hence through* AB. *But that the time of descent is longer in plane* CO, *cutting the diameter, follows in the same way; for this is longer and less [steeply] inclined than* CB. *Therefore the proposition holds.*

227

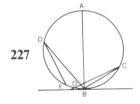

PROPOSITION IX. THEOREM IX

If any two planes are inclined from a point in a horizontal line, and are cut by a line that makes with them angles alternately equal to their angles with the horizontal, then movements in the parts cut off by the said line are made in equal times.

From point C *of horizontal line* X *let there be any two inclined planes* CD *and* CE. *At any point in line* CD *construct angle* CDF *equal to angle* XCE; *line* DF *cuts plane* CE *at* F *so that angles* CDF *and* CFD *equal angles* XCE *and* LCD, *taken alternately. I say that the times of descent through* CD *and* CF *are equal.*

It is manifest that angle CFD is also equal to angle DCL, angle CDF having been drawn equal to angle XCE. Take the common angle DCF from the three angles of triangle CDF (equal to two right angles, as are all [three] angles at point C on line LX), and there remain in the triangle two [angles], CDF and CFD, equal to the two [angles] XCE and LCD. But also CDF was put equal to XCE; therefore the remainder CFD [equals] the remainder DCL. Construct plane CE equal to plane CD, and drop perpendiculars DA and EB from points D and E to the horizontal XL; and from C to DF draw the perpendicular CG. Since angle CDG is equal to angle ECB, while DGC and CBE are right [angles], triangles CDG and CBE are similar; and as DC is to CG, so CE is to EB; also, DC is equal to CE; therefore CG will be equal to BE. And since in triangles DAC and CGF, angles [A]C[D] and [C]A[D] are [respectively] equal to angles [C]F[G] and [F]G[C], then CD will be to DA as FC is to CG; and by permutation, as DC is to CF, so DA is to CG or BE. *Thus the ratio of the heights of the equal planes* CD *and* CE *is the same as the ratio of lengths* DC *and* CF. *Therefore, by Corollary* I *to Proposition* VI *above, the times of descents in these will be equal; which was to be proved.*

228

Another [proof] of the same. Draw FS perpendicular to the horizontal AS. As triangle CSF is similar to triangle DGC, GC will be to CD as SF is to FC; and since triangle CFG is similar to triangle DCA, CD will be to DA as FC is to CG. Hence, by equidistance of ratios, as SF is to CG, so CG is to DA, whence CG is the mean proportional between SF and DA; and as DA is to SF, so the square of DA is to the square of CG. Further, since triangle ACD is similar to triangle CGF, GC will be to CF as DA is to DC; and by permutation, as DA is to CG, so DC is to CF; and as the square of DA is to the square *of* CG, *so is the square of* DC *to the square of* CF. *But it has been shown that the square of* DA *is to the square of* CG *as line* DA *is to line* FS, *whence as the square of* DC *is to the square of* CF, *so line* DA *is to* FS. *Hence, by Proposition VII above, since the heights* DA *and* FS *of planes* CD *and* CF *are in the squared ratio of those planes, the times of movements along these will be equal.*

PROPOSITION X. THEOREM X

Along planes of different inclines whose heights are equal, the times of movements are to each other as the lengths of those planes, whether [both] the movements start from rest or [both] are preceded by movement from the same height.

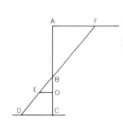

Let movements through ABC *and* ABD *to the horizontal* DC *be made in such a way that movement through* AB *precedes the movements through* BD *and through* BC; *I say that the time of movement through* BD *is [always] to the time through* BC *as length* BD *is to* BC.

Draw AF *parallel to the horizontal, meeting* DB *extended at* **229** F; *let* FE *be the mean proportional of* DF *and* FB. *Draw* EO *parallel to* DC, *whence* AO *will be the mean proportional between* CA *and* AB. *Now assuming that the time through* AB *is represented by [ut]* AB, *the time through* FB *will be as* FB, *and the time through all* AC *will be as the mean proportional* AO, *while [the time] through all* FD *will be [as]* FE. *Hence the time through the remainder* BC *will be* BO, *and [that] through the remainder* BD *will be* BE. *But as* BE *is to* BO, *so* BD *is to* BC. *Therefore the times through* BD *and* BC, *after fall through* AB *and* FB *(or, what is the same thing, through* AB *in common), will be to one another as lengths* BD *and* BC. *But it has been demonstrated above that the time through* BD *from rest at* B *will be to the time through* BC *as length* BD *is to* BC. *Therefore the times of movements through different planes of equal height are to one another as the lengths of the planes, whether motion is made in these from rest or whether another movement from a given height has preceded these movements; which was to be shown.*

PROPOSITION XI. THEOREM XI

If a plane in which motion is made from rest be divided in any way, the time of movement through the first part is, to the time of movement through that which follows, as the first part is to the excess by which that part is exceeded by the mean proportional between the whole plane and its first part.

Let there be movement from rest at A *through all* AB, *divided at any point* C, *and let* AF *be the mean proportional between all* AB *and its first part* AC. *The excess of the mean proportional* FA *over the part* AC *will be* CF; *I say that the time of movement through* AC *to the time of the subsequent movement through* CB *is as* AC *is to* CF. *This is clear because the time through* AC *is to the time through all* AB *as* AC *is to the mean proportional* AF; *therefore, by division [of ratio* AF:CF], *the time through* AC *is to the time through the remainder* CB *as* AC *is to* CF. *If, then, we assume the time through* AC *to be [represented by]* AC *itself, the time through* CB *will be* CF; *which is the proposition.*

230   *And if motion is made, not continually through* ACB, *but through the inflection* ACD *as far as [to] the horizontal* BD, *to which* FE *is drawn parallel from* F, *it is similarly demonstrated that the time through* AC *is, to the time through the diversion [reflexam]* CD, *as* AC *is to* CE. *For the time through* AC *is to the time through* CB *as* AC *is to* CF; *but it has been demonstrated that the time through* CB *after [fall through]* AC *is, to the time through* CD *after that same descent through* AC, *as* CB *is to* CD; *that is, as* CF *is to* CE. *Therefore, by equidistance of ratios, the time through* AC *will be to the time through* CD *as line* AC *is to* CE.

PROPOSITION XII. THEOREM XII

If a vertical and a plane however inclined intersect between given horizontal lines, and mean proportionals are taken between [each of] these and its part contained between the intersection and the upper horizontal, the time of movement in the vertical line will have, to the time of movement made in the upper part of the vertical and then in the lower part of the cutting plane, the same ratio as that which the entire vertical has to the line made up of the mean proportional taken in the vertical and the excess of the entire inclined plane over its mean proportional [of whole to upper part].

Let the horizontals be AF, *above, and* CD, *below; between these, the vertical* AC *and the inclined plane* DF *intersect at* B. *And let* AR *be the mean proportional between the entire vertical* CA *and its upper part* AB, *while* FS *is the mean proportional between all* DF *and its upper part* BF. *I say that the time of fall through the whole vertical* AC *has, to the time through its upper part* AB *plus the lower part of the plane,* BD, *the same ratio*

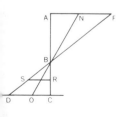

that AC *has to* AR *(the mean proportional in the vertical) plus* SD, *the excess of the whole plane* DF *over its mean proportional* FS.

Join R *and* S, *which [line] will be parallel to the horizontal. Since the time of fall through all* AC *is to the time through the part* AB *as* CA *is to the mean proportional* AR, *then if we assume* AC *to be the time of fall through* AC, *the time of fall through* AB *will be* AR, *and* RC *[will be] that through the remainder* BC. *But if the time through* AC *is assumed to be* AC *itself, as was done, then the time through* FD *will be* FD; *and it will likewise be concluded that* DS *is the time through* BD *after [fall through]* FB, *or after* AB.[39] *Therefore the time through all* AC *is* AR *plus* RC, *while that through the inflection* ABD *will be* AR *plus* SD; *which was to be proved.*

The same holds if, in place of the vertical, any other plane is assumed, as for example NO; *and the demonstration is the same.*

231

PROPOSITION XIII. PROBLEM I[40]

Given the vertical, to divert from it a plane having the same height as the given vertical, in which motion after fall in the vertical is made in the same time as [motion] in the given vertical from rest.

Let the given vertical be AB, *extended to* C *[by] an equal distance* BC, *and draw the horizontals* CE *and* AG; *it is required to divert from* B *a plane reaching to the horizontal* CE, *in which motion is made after fall from* A *in the same time as [motion] in* AB *from rest at* A.

Let CD *be equal to* CB, *and draw* BD; *let* BE *be constructed [applicetur] equal to the sum of [utrisque]* BD *and* DC; *I say that* BE *is the required plane.*

Extend EB *to meet the horizontal* AG *at* G, *and let* GF *be the mean proportional of* EG *and* GB; EF *will be to* FB *as* EG *is to* GF, *and the square of* EF *will be to the square of* FB *as the square of* EG *is to the square of* GF; *that is, as line* EG *is to* GB. *But* EG *is double* GB; *hence the square of* EF *is double the square of* FB. *But also the square of* DB *is double the square of* BC; *therefore as line* EF *is to* FB, *so* DB *is to* BC. *And by composition and permutation, as* EB *is to the sum of* DB *and* BC, *so* BF *is to* BC. *But* BE *is equal to the sum of* DB *and* BC,

---

39. Cf. note 29, above.

40. Problems, as distinguished from theorems, require the construction of an assigned unknown magnitude.

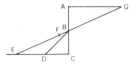

whence BF *is equal to* BC, *or* BA. *Hence if it is assumed that* AB *is the time of fall through* AB, *then* GB *will be the time through* GB, *and* GF *the time through all* GE. *Therefore* BF *will be the time through the remainder* BE, *after fall from* G, *or from* A; *which was proposed.*

232   PROPOSITION XIV PROBLEM II

Given a vertical and a plane inclined to it, to find the part in the upper vertical which is traversed from rest in a time equal to that in which, after fall in the part found in the vertical, the inclined plane is traversed.

Let the vertical be DB *and the plane inclined to it* AC; *it is required to find a part in the vertical* AD *which is traversed from rest in a time equal to that in which, after that fall* [post casum in ea], *the plane* AC *is traversed.*

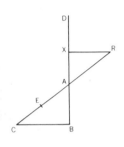

Draw the horizontal CB; *and as* BA *plus double* AC *is to* AC, *let* CA *be to* AE, *and let* EA *be to* AR *as* BA *is to* AC. *From* R, *draw* RX *perpendicular to* DB; *I say that* X *is the point sought. Since* BA *plus double* AS *is to* AC *as* CA *is to* AE, *then by division* [*of these ratios*] CE *will be to* EA *as* BA *plus* AC *is to* AC; *and since* EA *is to* AR *as* BA *is to* AC, *then by composition, as* BA *plus* AC *is to* AC, *so* ER *is to* RA. *But as* BA *plus* AC *is to* AC, *so* CE *is to* EA; *therefore, as* CE *is to* EA, *so* ER *is to* RA, *and the combined antecedents* [*are in this same ratio*] *to the combined consequents; that is, so* CR *is to* RE. *Hence* CR, RE, *and* RA *are* [*continued*] *proportionals.*

Next, since by construction EA *is to* AR *as* BA *is to* AC, *and by similarity of triangles,* XA *is to* AR *as* BA *is to* AC, *then as* EA *is to* AR, *so* XA *is to* AR; *and thus* EA *and* XA *are equal. But if we assume the time through* RA *to be as* RA, *the time through* RC *will be* [*as*] RE, *the mean proportional between* CR *and* RA; *and* AE *will be the time through* AC *after* [*fall through*] RA *or* XA. *But the time through* XA *is* XA *when* RA *is the time through* RA, *and it was shown that* XA *and* AE *are equal; therefore the proposition is evident.*

PROPOSITION XV. PROBLEM III

Given a vertical and a plane diverted from it, to find the part in the vertical extended downward which is traversed in the same time as the diverted plane is traversed after fall through [ex] *the given vertical.*

Let the vertical be AB *and the plane diverted therefrom be* BC; *it is required to find in the vertical, extended below* [B], *a*

part which is traversed after fall from [rest at] A in the same   **233**
time as is BC after the same fall from A.

*Draw the horizontal AD meeting CB extended at D, and
let DE be the mean proportional of CD and DB. Make BF
equal to BE; finally, let AG be the third proportional of BA
and AF; I say that BG is the space which, after the fall AB,
is traversed in the same time as [is] plane BC after the same
fall [AB]. For if we assume the time through AB to be as AB,
the time [through] DB will be as DB; and since DE is the
mean proportional between BD and DC, the time through all
DC will be DE, and BE [will be] the time through the remainder
BC from rest at D or after fall [through] AB. And it is similarly
concluded that BF is the time through BG after the same fall;
also, BF is equal to BE; therefore the proposition is evident.*

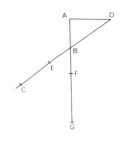

PROPOSITION XVI. THEOREM XIII

*Given portions of an inclined plane and of the vertical for
which the times of movements from rest are equal; if
these meet at the same point, a moveable coming from
any higher point will cover the portion of the inclined
plane more quickly than the portion of the vertical.*

*Let the vertical EB and the inclined plane CE meet at the
same point E, in which the times of movements from rest at
E are equal. Take any higher point in the vertical, A, from
which moveables are released; I say that after the fall AE, the
inclined plane EC is passed over in a shorter time than [is]
the vertical EB.*

*Join C and B, and draw the horizontal AD; let CE be ex-
tended to meet this at D, and let DF be the mean proportional
of CD and DE, while AG is the mean proportional of BA and*   **234**
*AE; draw FG and DG. Since the times of movements through
EC and EB from rest at E are equal, C will be a right angle,
by Corollary II of Proposition VI. Also A is a right angle,
and the vertex angles at E are equal, whence triangles AED
and CEB are similar, and the sides around [their] equal angles
are proportional; hence as BE is to EC, so DE is to EA. Therefore the rectangle BE–EA is equal to the rectangle CE–ED;
and since rectangle CD–DE exceeds rectangle CE–ED by
the square ED, while rectangle BA–AE exceeds rectangle
BE–EA by the square EA, the excess of rectangle CD–DE
over rectangle BA–AE (that is, the [excess of] square FD
over square AG) will be the same as the excess of square DE
over square AE, which excess is the square DA. Therefore*

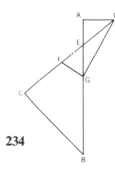

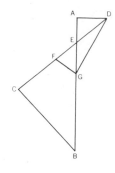

square FD *is equal to the two squares* GA *and* AD, *to which the square* GD *is also equal. Hence line* DF *is equal to* DG, *and angle* DGF *is equal to angle* DFG, *and angle* EGF *is less than angle* EFG, *and the opposite side* EF *is less than side* EG. *But if we assume the time of fall through* AE *to be as* AE, *the time through* DE *will be as* DE; *and since* AG *is the mean proportional between* BA *and* AE, *the time through all* AB *will be* AG, *and the remainder* EG *will be the time through the remainder* EB *from rest at* A. *And likewise it is concluded that* EF *is the time through* EC *after the descent* DE, *or after fall* AE. *But it has been shown that* EF *is less than* EG, *whence the proposition is evident.*

COROLLARY

It also follows from this and the preceding [proposition] that the space traversed in the vertical after fall from on high, during the same time as that in which the inclined plane is traversed, is less than that [space] traversed in that same time in the incline without a preceding fall from on high, though greater than the inclined plane itself.

It has been demonstrated that for moveables coming from the high point A, *the time accumulated [conversi] through* EC *is shorter than the time before [procedentis] through* EB; *hence it follows that the space traversed through* EB *in a time equal to the time through* EC *is less than the whole space* EB. *Now, that this same vertical space [*EB *in the previous diagram] is greater than* EC *is manifest from the diagram used for the preceding proposition, in which the part* BG *of the vertical is shown to be traversed in the same time as is* BC *after the fall*

235   AB. *But that* BG *is greater than* BC *is deduced as follows. Since* BE *and* FB *are equal, and* BA *is less than* BD, FB *has a greater ratio to* BA *than* EB *has to* BD; *and by composition,* FA *has a greater [ratio] to* AB *than* ED *has to* DB. *Hence as* FA *is to* AB, *so* GF *is to* FB *(for* AF *is the mean proportional between* BA *and* AG); *likewise as* ED *is to* BD, *so* CE *is to* EB; *therefore* GB *has a greater ratio to* BF *than* CB *has to* BE, *and hence* GB *is greater than* BC.

PROPOSITION XVII. PROBLEM IV

Given a vertical and a plane diverted from it, to mark a part in the given plane through which, after fall in the vertical, motion is made in a time equal to that in which the moveable traversed the vertical from rest.

Let AB *be the vertical and* BE *the plane diverted from it; it is required to mark in* BE *the space through which a moveable, after fall in* AB, *will be moved in a time equal to that in which it traversed the vertical* AB *from rest. Let* AB *be a horizontal line meeting the plane extended at* D, *and take* FB *equal to* BA; *and as* BD *is to* DF, *make* FD *to* DE; *I say that the time through* BE *after fall in* AB *will be equal to the time through* AB *from rest at* A.

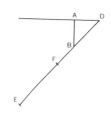

If AB *is assumed to be the time through* AB, *the time through* DB *will be* DB; *and since* FD *is to* DE *as* BD *is to* DF, *the time through the whole plane* DE *will be* DF, *and* BF [*will be the time*] *through the part* BE *from* D. *But the time through* BE *after* DB *is the same as* [*that*] *after* AB; *therefore the time through* BE *after* AB *will be* BF; *that is,* [*it will be*] *equal to the time* [*through*] AB *from rest at* A; *which was the problem.*

PROPOSTION XVIII. PROBLEM V

Given in the vertical any space marked from the beginning of movement, and given the time in which this is traversed, and given another smaller time, to find another space in the vertical which is traversed in the given smaller time.

Let the vertical be A[D], *in which let the space* AB *be given, for which the time from the beginning* A *is* AB; *and let the horizontal be* CBE; *and let a time less than* AB *be given, which is marked in the horizontal as equal to* BC. *It is required to find in the same vertical a space equal to* AB *which is traversed in the time* BC.

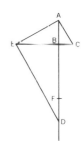

Draw line AC; *since* BC *is less than* BA, *angle* BAC *will be less than angle* BCA. *Draw* CAE *equal to the latter, and line* AE *meeting the horizontal at point* E; *perpendicular to this, draw* ED *cutting the vertical at* D, *and mark* DF *equal to* BA. *I say that* FD *is the part in the vertical which, in movement from the beginning of motion at* A, *is passed over in the given time* BC.

Since in the right triangle AED, EB *is drawn perpendicular to the side* AD *opposite the right angle at* E, AE *will be the mean proportional between* DA *and* AB, *and* BE *the mean proportional between* DB *and* BA, *or between* FA *and* AB *(for* FA *is equal to* DB*). And since* AB *is assumed to be the time through* A[B], *the time through all* AD *will be* AE *or* EC, *and* EB [*will be*] *the time through* AF. *Therefore the remainder* BC *will be the time through the remainder* FD; *which was the intent.*

237   PROPOSITION XIX. PROBLEM VI

*Given any space whatever in the vertical, run through
from the beginning of motion, and given the time of fall;
to find the time in which another equal space, taken some-
where in the same vertical, will be traversed by the same
moveable.*

Let any space AC be taken from the beginning of motion
at A in the vertical AB, to which is equal another space DB
taken anywhere; and given the time of motion through AC,
let this be AC; it is required to find the time of movement
through DB after fall from A. Describe the semicircle AEB
around all AB, and let CE be perpendicular to AB from C;
join A and E, which [line] will be longer than EC. Let EF be
cut equal to EC; I say that the remainder FA is the time of
movement through DB.

For since AE is the mean proportional between BA and AC,
and AC is the time of fall through AC, the time of fall through
all AB will be AE. And since CE is the mean proportional
between DA and AC (for DA is equal to BC), CE (that is,
EF) will be the time through AD. Therefore the remainder AF
is the time through the remainder DB; which is the proposition.

COROLLARY

*From this it is deduced that if any space is assumed, the
time through this after some adjoined space will be [as]
the excess of the mean proportional between the combined
spaces and the original space, over the mean proportional
between the original [space] and that added.*

Thus, supposing that the time through AB from rest at A is
AB, [then] when AS is added, the time through AB after SA
will be the excess of the mean proportional between SB and
SA over the mean proportional between BA and AS.

238   PROPOSITION XX. PROBLEM VII

*Given any space, and a part therein from [post] the
beginning of movement, to find another part at [versus] the
end which is traversed in the same time as the part first given.*

Let there be the space CB, and in this the part CD after the
beginning of movement at C; it is required to find another
part toward the end, B, which is traversed in the same time as
the given [part] CD.

Take the mean proportional between BC and CD, which is

to be put equal to BA, *and let* CE *be the third proportional to*
BC *and* CA; *I say that* EB *is the space which, after fall from*
C, *will be traversed in the same time as* CD. *For if we assume*
the time through all CB *to be as* CB, *then* BA *(that is, the*
mean proportional between BC *and* CD*) will be the time through*
CD, *and since* CA *is the mean proportional between* BC *and*
CE, *the time through all* CE *will be* CA; *also,* BC *is the time*
through all CB. *Therefore the remainder* BA *will be the time*
through the remainder EB *after fall from* C. *But* BA *was the*
time through CD; *therefore* CD *and* EB *will be traversed in*
equal times from rest at C;[41] *which was to be done.*

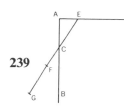

PROPOSITION XXI. THEOREM XIV

If fall in the vertical occurs from rest, of which a part
is taken at the beginning of movement that is run through
in a given time, after which there follows motion diverted
through a plane however inclined, the space which is
traversed in that plane during a time equal to the time of
fall already run through in the vertical will be more than
double, but less than triple, [that space already run].

    Below the horizontal [line] AE *let there be the vertical* AB,
in which let fall take place from the beginning, A, *in which there*
is taken a part, AC; *then from* C *let some plane* CG *be inclined*
on which motion continues after fall in AC. *I say that the*
space run through in that motion through CG, *in a time equal*   **239**
to the time of fall through AC, *is more than double but less*
than triple that same space AC.

    Take CF *equal to* AC, *and extending plane* CG *to the hori-*
zontal at E, *make* FE *to* EG *as* CE *is to* EF. *Then if the time*
of fall through AC *is taken as the line* AC, *the time through*
EC *will be* CE, *and* CF *(or* CA*) [will be] the time of motion*
through CG. *It is to be shown that space* CG *is more than*
double, but less than triple, CA. *Now since* FE *is to* EG *as*
CE *is to* EF, CF *will be [in] the same [ratio] to* FG; *but* EC *is*
less than EF, *whence* CF *will be less than* FG, *and therefore*
GC *will be more than double* FC *or* AC. *Further, since* FE *is*
less than double EC *(for* EC *is greater than* CA *or* CF*),* GF
will also be less than double FC, *and* GC *[will be] less than*
triple CF *or* CA; *which was to be demonstrated.*[42]

---

41. The 1638 text has *A* for *C*, an error corrected in Weston's translation.
42. This noteworthy theorem calls attention to the role of the integers
2 and 3 as limits governing speeds naturally conserved or acquired during
the second of two equal times. That Galileo did not insert any dialogue

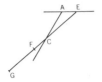

The same can be still more universally stated; for what happens in vertical and inclined planes also happens if, after motion in some inclined plane, there is deflection through a greater incline, as seen in this second diagram; and the demonstration is the same.

240

PROPOSITION XXII. PROBLEM VIII

Given two unequal times, and given the space traversed in the vertical (from rest) during the shorter of the given times, to divert a plane from the highest point of the vertical out to the horizontal, upon which the moveable descends in a time equal to the longer of the given [*times*].[43]

Let A *be the greater and* B *the lesser of the unequal times, and let* CD *be the space traversed from rest along the vertical in time* B; *it is required to divert from end* C *to the horizontal a plane which is traversed in time* A.

As B *is to* A, *let* CD *be to some other line equal to* CX, *descending from* C *to the horizontal. It is manifest that the plane* CX *is that along which the moveable descends in the given time* A, *for it was demonstrated* [*in Theorem III*] *that the time along an inclined plane has to the time along its height the ratio that the length of the plane has to its height. Therefore the time along* CX *is to the time along* CD *as* CX *is to* CD; *that is, as the time* A *is to the time* B. *But time* B *is that in which the vertical* CD *is traversed from rest; therefore time* A *is that in which the plane* CX *is traversed.*

PROPOSITION XXIII. PROBLEM IX

Given the space run through in any time along the vertical from rest, to divert a plane from the lower end of this space, upon which, after fall in the vertical and in equal time [*thereto*], *a given space is traversed that is more than double but less than triple the space run through in the vertical.*

241

---

discussing this is most surprising if, as some say, he had Platonist leanings. A manuscript copy exists in the hand of Mario Guiducci of a draft probably made in 1618, to which Galileo added this remark: "Note that if motion along the incline *CG* is accelerated *in infinitum*, it seems one might demonstrate that along the horizontal, [motion] must extend equably *in infinitum*; now it is also clear that if equable, it will also be infinite." Thus rectilinear inertia and an infinite universe must stand or fall together; cf. note 46, below.

43. This is one of the problems that Galileo had tried to solve nearly fifty years earlier, before he realized the importance of acceleration in fall; cf. *On Motion*, p. 69 (*Opere*, I, 301).

In the vertical AS *let space* AC *be run through in time* AC *from rest at* A, *to which* [*space* AC] IR *is more than double but less than triple; it is required to divert a plane from terminus* C, *upon which the moveable shall, in this same time* AC, *traverse a space equal to* IR *after fall through space* AC.

Let RN *and* NM *be equal to* AC; *and whatever ratio the remainder* IM *has to* MN, *let* AC *have the same to another line equal to* CF, *drawn from* C *to the horizontal* AE. *Extend this toward* O, *and let* CF, FG, *and* GO *be equal* [*respectively*] *to* RN, NM, *and* MI. *I say that the time along the diversion* CO, *after fall* AC, *is equal to the time* [*through*] AC *from rest at* A. *For since* FC *is to* CE *as* OG *is to* GF, *then by composition, as* OF *is to* FG *or* FC, *so* FE *will be to* EC, *and as one antecedent is to one consequent, so is the sum of* [omnia] *antecedents to the sum of consequents; hence the whole of* OE *is to* EF *as* FE *is to* EC. *And thus* OE, EF, *and* FC *are continued proportionals; and since it was assumed that the time through* AC *is as* AC, *the time through* EC *will be* CE, *and* EF [*will be*] *the time through all* EO, *and the remainder* CF [*will be that*] *through the remainder* CO. *But* CF *is equal to* CA; *therefore what was required has been done. For time* CA *is the time of fall through* AC *from rest at* A, *while* CF *(which equals* CA*) is the time through* CO *after descent through* EC, *or after fall through* AC; *which was the thing proposed.*

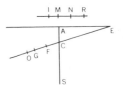

It is to be noted here that the same happens if the preceding motion is made not vertically, but along an inclined plane, as in the next diagram; there the initial [praecedens] *motion is made along the inclined plane* AS, *below the horizontal line* AE, *and the demonstration is exactly the same.*   **242**

SCHOLIUM

If due attention is paid, it will be manifest that the less the given line IR *falls short of triple* AC, *the closer the diverted plane* CO, *on which the second motion is made, comes to the vertical, in which ultimately, in a time equal to* AC, *a space triple* AC *is run through. For if* [cum] IR *is almost triple* AC, IM *will be nearly equal to* MN; *and since, by construction,* AC *is made to* CE *as* IM *is to* MN, *it is clear that* CE *will be found to be little more than* CA, *and consequently that the point* E *will be found close to point* A, *while* CO *and* CS *will contain a very acute angle, and will nearly coincide. On the other hand, if* IR *is the minimum that is greater than double* AC, *then* IM *will be a very short line, whence it comes about*

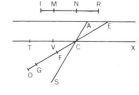

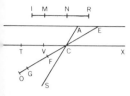

that AC *will be very short with respect to* CE, *which will become very long, and nearly parallel to the horizontal drawn through* C. *And we may then deduce that if, in the above diagram, after descent through the inclined plane* AC, *there is diversion along a horizontal line such as* CT, *the space through which the moveable will next* [consequentur] *be moved, in a time equal to that of descent through* AC, *would be exactly double the space* AC.

Further, it is seen that this fits with [*other*] *like reasoning. For from the fact that* OE *is to* EF *as* FE *is to* EC, *it appears that* FC *determines the time through* CO. *For if the horizontal part* TC, *double* CA, *is bisected at* V, *its extension toward* X *will be prolonged indefinitely in seeking to meet with* AE *produced; and the ratio of an infinite* TX *to an infinite* VX *will not be different from the ratio of an infinite* VX *to an infinite* XC.

We may reach the same conclusion by another approach, taking up again an argument like that which we used in the demonstration of Proposition I.[44] *For take again the triangle*

243   ABC, *and by its parallels to the base* BC *let us represent to ourselves the degrees of speed continually increased according to the increments of time. From those, which are infinitely many (as the points in line* AC *are infinitely many, and* [*so are*] *the instants in any* [*interval of*] *time), there arises* [exurget] *the surface of the triangle* [ABC]; *and if we assume the motion to be continued for another equal time, no longer in accelerated but in equable motion at the maximum degree of speed acquired (which degree is represented by line* BC), *then from these degrees* [*of speed*] *a like parallelogram* ABCD *will be produced* [conflabitur], *double the triangle* ABC. *Hence the space which is traversed in the same time with similar degrees* [*of speed*] *will be double the space run through with the degrees of speed represented by triangle* ABC.

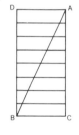

But motion in the horizontal plane is equable, as there is no cause of acceleration or retardation; therefore it is to be concluded that the space CT,[45] *run through in time equal to the*

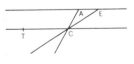

44. See note 31, above.
45. With the exception of the 1638 original and Weston's translation, all editions erroneously read *CD* here in place of *CT*. Galileo's reference is to *CT* in the previous diagram, here (though not in the original) repeated in part for convenience of reference. The erroneous *CD*, a plausible editorial emendation made in 1655, has ever since made it seem that Galileo in this one place had had used an area to represent a distance traversed. We do this, but it would be incongruous for a strict Euclidean mathematician. Compare

time AC, *is double the space* AC. *Indeed, the former motion, accelerated, is made from rest according to the parallels of the triangle, while the latter [equable, is made] according to the parallels of the parallelogram, which, being infinitely many, are doubles to the infinitely many parallels of the triangle.*

It may also be noted that whatever degree of speed is found in the moveable, this is by its nature [suapte natura] *indelibly impressed on it when external causes of acceleration or retardation are removed, which occurs only on the horizontal plane; for on declining planes there is cause of more* [maioris] *acceleration, and on rising planes, of retardation. From this it likewise follows that motion in the horizontal is also eternal, since if it is indeed equable it is not [even] weakened or remitted, much less removed.*[46]

Furthermore, one must consider the existing degree of speed acquired by the moveable in natural descent to be naturally indelible and eternal; but if after descent along a declining plane it is diverted through another upward plane, a cause of retardation presents itself there, for on such a plane the same moveable would naturally descend. Wherefore a certain mixture of contrary influences [affectionum] *arises—that of the degree of speed acquired in the preceding descent, which by itself would carry the moveable away uniformly* in infinitum, *and [that of] a natural propensity to downward motion according to that same ratio of acceleration in which it is always moved. Whence it is seen to be quite reasonable if, in inquiring what events take place when a moveable is diverted through some rise after descent through some inclined plane, we assume that that maximum degree acquired in descent is in itself perpetually kept* [servari] *in the ascending plane, but [that] in the ascent there supervenes the natural tendency downward; that is, to a motion from rest accelerated in the ratio always assumed.*   **244**

Theorem I, above, where a seemingly superfluous line was introduced into Galileo's diagram to represent distances traversed. The area which for us would represent distance was seen by Galileo only as representing an overall speed.

46. Cf. note 42, above. Here, and in the ensuing argument, is to be found everything of significance in Galileo's restricted inertial concept, which he limited to phenomena of heavy bodies near the earth's surface. His remark that interference with uniformity of motion is always present except in supported horizontal motion is a simple statement of fact, as is his subsequent declaration that no truly horizontal plane exists, though there are surfaces on earth on which uniform motion would ideally be conserved. Cf. *Dialogue*, p. 148 (*Opere*, VII, 174), and see further, pp. **273–75**, below.

*But lest perhaps this be understood but darkly, it is explained
more clearly by another depiction.*

*Let it be understood, then, that descent is [first] made through
the downward plane AB, from which reflected motion is continued upward through BC; and first let these be equal planes,
elevated at equal angles to the horizontal GH. It follows that
the moveable descending along AB from rest at A acquires
speed according to the increase of time itself; the degree at
B is the maximum acquired, and this is naturally impressed
immutably—any cause of new acceleration or retardation
being removed. [There would be a cause] of acceleration, I
say, if it were to continue its progress further on the [same]
plane extended, and of retardation if diverted to the rising
plane BC. But on the horizontal plane GH, it would go in*
infinitum *in equable motion at the degree of speed acquired
[in descent] from A to B, and this speed would be such that,
in a time equal to the time of descent along AB, it would traverse
in the horizontal a space double the space AB.*

*However, let us suppose the same moveable to be moved
equably at the same degree of speed along plane BC, so that
also in this [case], in a time equal to the time of descent along
AB, it would traverse on BC extended a space double that of
AB. Truly, we understand that as soon as the ascent begins,
there naturally supervenes that which happens to it from A
on the plane AB; namely, a certain descent from rest according
to those same degrees of acceleration, by force of which [vi
quorum],[47] as happened on AB, it descends the same amount
in the same time on the diverted plane [BC] that it descended
along AB. It is manifest that from this kind of mixture of
equable ascending and accelerated descending motion, the
moveable is carried to terminus C, along plane BC, at the same
degrees of speed, which will be equal [in ascent and descent].
And indeed, we can deduce that assuming any two points D
and E, equally distant from angle B, the transit through DB
is made in a time equal to the time of reflection through BE.
Draw DF parallel to BC; it is evident that the descent through
AD will be reflected through DF; now if, after [reaching] D,
the moveable is carried along the horizontal DE, its impetus
at E will be the same as its impetus at D; therefore from E*

47. It was unusual for Galileo to introduce acceleration in terms of force,
his customary procedures being kinematic rather than dynamic. The essential idea here is that inertial motion is found in bodies supported on planes
other than the horizontal, but is cloaked by continual deceleration.

it ascends to C, *whence the degree of speed at* D *is equal to the degree at* E.

From this we may therefore reasonably assert that if descent is made through some inclined plane, after which there follows reflection through some rising plane, the moveable ascends, by the impetus received, all the way to the same altitude or height from the horizontal. Thus if the descent is along AB, the moveable is carried along the diverted plane BC to the horizontal ACD; and not only if the inclinations of the planes are equal, but also if they are unequal, as is plane BD. For it was assumed earlier that the degrees of speed acquired over unequally inclined planes are equal whenever the planes are of the same height above the horizontal. But if the same inclination exists for planes EB and BD, descent through EB suffices to impel the moveable along plane BD all the way to D, as such an impulse is made on account of the received impetus of speed at point B; and there is the same impetus at B whether the moveable descends through AB or through EB. It follows that the moveable is pushed out likewise along BD after descent along AB or along EB. It happens indeed that the time of ascent through BD will be longer than that through BC, inasmuch as descent through EB also takes a longer time than through AB; and the ratio of these times has already been shown to be the same as that of the lengths of the planes.

245

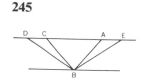

Next we shall inquire into the ratio of the spaces passed in equal times on planes of whatever different inclinations, but of the same heights; that is, those which are included between the same horizontal parallels. And this takes place according to the following ratio.

PROPOSITION XXIV. THEOREM XV

Given a vertical, and a plane elevated from its lower end, lying between given horizontal parallels; the space traversed by a moveable on the inclined plane, after fall through the vertical, in a time equal to its time of [vertical] fall, is greater than in the vertical but less than double that in the vertical.

Between the same horizontal parallels BC and HG let there be the vertical AE and the inclined plane EB; over EB, after fall along the vertical AE, let deflection be made toward B from the point E; I say that the space through which the moveable ascends, in a time equal to the time of descent AE, is greater than AE but less than double AE. Take ED equal to AE, and

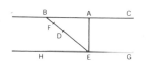

make DB *to* BF *as* EB *is to* BD*; it is to be shown, first, that* F
marks the point to which the moveable comes, moved by

246 *reflection along* EB, *in a time equal to the time of* AE*; and
next, that* EF *is greater than* EA *but less than its double. If
we understand that the time of descent along* AE *is as* AE,
the time of descent along BE, *or of ascent along* EB, *will be*

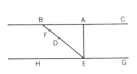

as the line BE*; and since* DB *is the mean proportional between*
EB *and* BF, *and* BE *is the time of descent through all* BE,
BD *will be the time of descent through* BF, *and the remainder*
DE *[will be] the time of descent through the remainder* FE.
But the time through FE *from rest at* B *is the same as the time
of ascent through* EF, *when the degree of speed at* E *is that
acquired through the descent* BE *(or* AE*). Therefore the
same time* DE *will be that in which the moveable arrives at
point* F *after fall from* A *through* AE *and reflected motion
along* EB. *But it was assumed that* ED *is equal to* AE*; whence
the first [part] has been shown.*

And since as all EB *is to all* BD, *so the removed part* DB
is to the removed part BF, *[likewise] as all* EB *is to all* BD, *so
will the remainder* ED *be to* DF*; whence* EB *is greater than*
BD. *Therefore* ED *is greater than* DF, *and also* EF *is less than
double* DE, *or* AE*; which was to be shown. The same holds if
the precedent motion is made not in the vertical, but on an
inclined plane; and the proof is the same when the plane of
reflection is less steep and hence longer than the declining plane.*

PROPOSITION XXV. THEOREM XVI

*If, after fall through some inclined plane, there follows
motion through the horizontal plane, the time of fall through
the inclined plane will be, to the time of motion through
any horizontal line, as double the length of the inclined
plane is to that horizontal line.*

Let CB *be the horizontal line and* AB *the inclined plane; and,
after fall through* AB, *let motion follow through the horizontal,
in which any space* BD *is taken. I say that the time of fall
through* AB *is to the time of motion through* BD *as double*
AB *is to* BD. *For take* BC *double* AB, *and it follows from
what was demonstrated above that the time of fall through*
AB *equals the time of motion through* BC*; but the time of
motion through* BC *is to the time of motion through* DB *as*

247 *line* CB *is to line* BD. *Therefore the time of motion through*
AB *is to the time through* BD *as double* AB *is to* BD*; which
was to be proved.*

PROPOSITION XXVI. PROBLEM X

Given a vertical between parallel horizontal lines, and a space greater than that vertical, but less than its double; to raise a plane from the lower terminus of the vertical between the parallels, upon which in reflected motion after descent in the vertical, a moveable will traverse a space equal to that given, in a time equal to the time of descent through the vertical.

Let the vertical be AB, and between the parallels AO and BC, let FE be greater than BA but less than its double; it is required to erect a plane from B, between the horizontals, on which the moveable, after fall from A to B, in reflected motion during a time equal to the time of descent through AB, traverses in ascent a space equal to EF. Make ED equal to AB; the remainder DF will be less [than DE], since all EF is less than double AB. Let DI equal DF, and as EI is to ID, make DF be to FX; and from B reflect BO equal to EX. I say that the plane through BO is that on which, after fall AB, in a time equal to the time of fall through AB, the moveable in rising will pass through a space equal to the given space EF. Put BR and RS equal to ED and DF; then since as EI is to ID, so DF is to FX; and by composition, as ED is to DI, DX will be to XF; that is, as ED is to DF, DX is to XF, and EX is to XD, whence as BO is to OR, RO is to OS. Now if we assume the time through AB to be AB, the time through OB will be OB, and RO will be the time through OS; and the remainder BR will be the time through the remainder SB in descent from O to B. But the time of descent through SB from rest at O is equal to the time of ascent from B to S after the descent AB; therefore BO is the plane, raised from B, upon which, after descent through AB, the space BS, equal to the given space EF, is traversed in the time BR, or BA; which was required to be done.

248

PROPOSITION XXVII. THEOREM XVII

If a moveable descends on unequal planes of the same height, the space traversed in the lower part of the longer [plane], in a time equal to that in which all the shorter plane is traversed, is equal to the space made up of the shorter plane and that length to which the shorter plane has the same ratio as the longer plane has to the excess by which the longer surpasses the shorter.

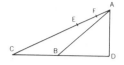

Let AC *be the longer plane and* AB *the shorter, of which each is of height* AD*; and in the lower part of* AC*, take* CE *equal to* AB*. Let the ratio of all* CA *to* AE *(that is, to the excess of plane* CA *over* AB*) be* [*the ratio*] *of* CE *to* EF*; I say that the space* FC *is that which is traversed after departure from* A *in a time equal to the time of descent through* AB*. For since all* CA *is to all* AE *as the part* CE *is to the part* EF*, the removed part* EA *will be to the removed part* AF *as all* CA *is to all* AE*; and thus the three* [*magnitudes*] CA, AE, *and* AF *are in continued proportion. Now, if the time through* AB *is assumed to be as* AB*, the time through* AC *will be as* AC*; but the time through* AF *will be as* AE*, and* [*that*] *through the remainder* FC*, as* EC*; whence* EC *is equal to* AB*; therefore the proposition is evident.*

249

PROPOSITION XXVIII. PROBLEM XI

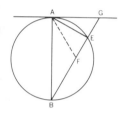

Let the horizontal line AG *be tangent to a circle at its diameter* AB*, and draw any two chords,* AE *and* EB*; it is required to find the ratio of the time of fall through* AB *to the time of descent through both* AE *and* EB [*combined*]*.*

Extend BE *to the tangent at* G*, and draw* AF*, bisecting angle* BAE*; I say that the time through* AB *is, to the time through* AE *and* EB*, as* AE *is to* AE *plus* EF*. Since angle* FAB *is equal to angle* FAE*, and angle* EAG *to angle* ABF*, all* GAF *will be equal to the sum of* FAB *and* ABF*, to which also angle* GFA *is equal; therefore line* GF *is equal to* GA*. And since the rectangle* BG–GE *is equal to the square on* GA*, it is also equal to the square on* GF*; and the three lines* BG, GF, *and* GE *are* [*continued*] *proportionals. Now, if* AE *is assumed to be the time through* AE*, GE *will be the time through* GE*; GF *the time through all* GB*; and* EF *the time through* EB *after descent from* G *(or from* A *through* AE*); therefore the time through* AE *(or through* AB*) is, to the time through* AE *and* EB*, as* AE *is to* AE *and* EF*; which was to be found.*

Otherwise, more briefly. Make GF *equal to* GA*; it is evident that* GF *is the mean proportional between* BG *and* GE*. The rest as above.*

PROPOSITION XXIX. THEOREM XVIII

Given any horizontal distance, from the end of which is erected a perpendicular, in which is taken a part equal to one-half the distance given in the horizontal; a moveable descending from this altitude and being turned to the horizontal will traverse the [given] horizontal space **250** *together with the vertical in a shorter time than any other vertical distance together with that same horizontal space.*

Take any distance BC in the horizontal plane, and from B let there be the vertical in which BA is one-half of BC; I say that the time in which a moveable sent from A will traverse both distances, AB and BC, is the shortest time of all during which the same distance BC is traversed together with any part of the vertical, whether greater or less than the part AB. Let EB be taken as greater [than AB], as in the first diagram, or as less, as in the second. It is to be shown that the time in which distances EB and BC are traversed is longer than the time in which AB and BC are traversed. It is assumed that the time through AB is as AB, and [that this is] also the time of motion in the horizontal BC, since BC is double AB; and through both distances, AB and BC, the time will be double BA. Let BO be the mean proportional between EB and BA; BO will be the time of fall through EB. Further, let the horizontal distance BD be double BE; it follows that the time after fall EB is BO. Make OB to BN as DB is to BC, or as EB is to BA; and since motion in the horizontal is equable, and OB is the time through BD after fall from E, NB will be the time through BC after fall from the same height E. From this it follows that OB plus BN is the time through EB and BC; and since double BA is the time through AB and BC, it remains to be shown that OB plus BN is more than double BA. But since OB is the mean proportional between EB and BA, the ratio of EB to BA is the square of the ratio of OB to BA; and since EB is to BA as OB is to BN, the ratio of OB to BN will also be the square of the ratio of OB to BA. Now, the ratio of OB to BN is compounded from the ratios of OB to BA and of AB to BN; therefore the ratio of AB to BN is the same as the ratio of OB to BA. Hence BO, BA, and BN are three [magnitudes] in continued proportion, and OB plus BN is greater than double BA; from which the proposition is evident.

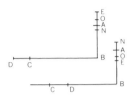

PROPOSITION XXX. THEOREM XIX   **251**

If a vertical is let fall from some point of a horizontal line, and from another point in the same horizontal a plane is drawn that meets the vertical, along which [plane] a moveable will descend in the shortest time [from that point] to the vertical, that plane will be such that it cuts off on

*the vertical a distance equal to that from the other point
in the horizontal to the origin of the vertical.*

Drop the vertical BD *from a point* B *in the horizontal line*
AC, *in which take any point* C; *and in the vertical, take a
distance* BE *equal to the distance* BC, *drawing* CE. *I say that
of all inclined planes from point* C *to the vertical* [BD], CE
is that along which descent will be made to the vertical [BD]
in the shortest time of all. For take planes CF *and* CG, *above
and below it* [CE], *and draw* IK *tangent at* C *to the radius of
the circle* BC, *which* [*tangent*] *will be parallel to the vertical.
Draw* CF *parallel to* EK, *which extends to the tangent and
which cuts the circumference of the circle at* L. *It is evident
that the time of fall through* LE *is equal to the time of fall
through* CE; *but the time through* KE *is longer than that
through* LE; *therefore the time through* KE *is longer than that
through* CE. *But the time through* KE *equals the time through*
CF, *as these are equal and are drawn at the same inclination;
likewise, since* CG *and* EI *are equal and at the same inclination,
the times of movements through them will be equal. But the
time through* HE, *shorter than* IE, *is briefer than the time
through* IE; *whence also the time through* CE *(which is equal
to the time through* HE*) is briefer than the time through* IE.
Hence the proposition holds.

PROPOSITION XXXI. THEOREM XX

*If to a straight line inclined in any way above the horizontal
there is drawn to the incline, from any given point in the
horizontal, that plane on which descent [from any point
on the incline] is made in the shortest time of all, this
[plane] will bisect the angle between two perpendiculars
from the given point, one [perpendicular] to the horizontal,
and the other to the inclined [line].*

Let CD *be a line inclined in any way above the horizontal*
AB; *and given in the horizontal some point,* A, *draw from this*
AC, *perpendicular to* AB, *and* AE, *perpendicular to* CD;
bisect the angle CAE *by line* FA. *I say that of all inclined
planes from any point on line* CD *to point* A, *that incline which
is extended through* FA *is the one in which the time of descent
will be briefest of all. Draw* FG *parallel to* AE; *the alternate
angles* GFA *and* FAE *will be equal, and also* EAF *is equal
to* FAG; *therefore the sides* FG *and* GA *of the triangle* [FGA]
will be equal. Now, if with center G *and radius* GA *a circle is
described, it will pass through* F *and will be tangent to the*

252

horizontal and to the inclined [line] at points A *and* F *[respectively], whence angle* GFC *is a right angle, since* GF *is parallel to* AE. *From this it follows that all lines from point* A *to the inclined [line] extend beyond the circumference, and consequently, movements through these are passed over in longer times than through* FA; *which was to be demonstrated.*

LEMMA

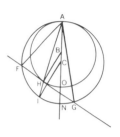

If two circles are internally tangent, and the inner [circle] is tangent to some straight line that cuts the outer [circle]; and if three lines are drawn from the point of tangency of the circles to three points of the tangent line, [one] to its point of tangency with the inner circle, and [the others] to its intersections with the outer [circle]; these [lines] will contain equal angles at their contact with the circles.

Let two circles be tangent internally at point A, *the center of the smaller circle being* B, *and that of the larger,* C; *and let the inner circle be tangent at point* H *to some straight line* **253** *FG which cuts the larger [circle] at points* F *and* G. *Draw lines* AF, AH, *and* AG; *I say that angles* FAH *and* GAH *contained by these [lines] are equal. Extend* AH *to the circumference at* I, *and from the centers draw* BH *and* CI. *Draw* BC *through the centers, which extended will fall on the point of tangency* A *and on the circumferences of the circles at* O *and* N. *Since angles* ICN *and* HBO *are equal, each being double the angle* IAN, *lines* BH *and* CI *will be parallel. And since* BH, *from center to point of tangency, is perpendicular to* FG, CI *will also be perpendicular to the same [FG]; and arc* FI *is equal to arc* IG, *and consequently angle* FAI *[is equal] to angle* IAG; *which was to be shown.*

PROPOSITION XXXII. THEOREM XXI

If two points are taken in the horizontal, and from one of them some line is inclined toward the other, from which [in turn] a line is drawn that cuts off on it a part equal to the distance between the two points on the horizontal, then fall is finished more swiftly through this drawn [line] than through any other line extending from the same point to the same incline. Furthermore, in other lines made at equal angles on either side of this line, falls take place in equal times.

Let A *and* B *be two points on the horizontal, and from* B *let the straight line* BC *be inclined, in which* BD *is taken from*

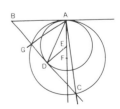

B, *equal to* BA; *and* A *and* D *are joined. I say that fall is made more swiftly through* AD *than through any* [*other*] *line extending from* A *to the incline* BC.

254   *For from points* A *and* D *draw* AE *and* DE, *intersecting at* E *and perpendicular* [*respectively*] *to* BA *and* BD. *Since in the isosceles triangle* ABD, *angles* BAD *and* BDA *are equal, their complements,* DAE *and* EDA, *will be equal; hence if a circle is described with center* E *and radius* EA, *it will pass through* D *and will be tangent to lines* BA *and* BD *at points* A *and* D. *And since* A *is an end of the vertical* AE, *fall will be finished more swiftly through* AD *than through any other* [*line*] *from the end* A *to the line* BC *extended beyond the circumference of the circle; which was to be shown first.*

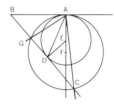

Now if the vertical AE *is extended, and some center* F *is taken for a circle* AGC *described with radius* FA *and cutting the tangent line at points* G *and* C, *then when* AG *and* AC *are drawn, they will be divided into equal angles by the median* AD, *as previously demonstrated. Hence the times of motions along these* [*lines*] *will be equal, since they are bounded by the high point* A *and the circumference of the circle* AGC.

PROPOSITION XXXIII. PROBLEM XII

Given a vertical and a plane inclined to it, of the same height and having the same upper terminus; to find a point, vertically above the common point, from which a moveable, falling and then deflected along the inclined plane, consumes the same time in this plane as [*in fall*] *from rest through the* [*given*] *vertical.*

Let AB *be the vertical and* AC *the inclined plane having the same height; it is required to find in the vertical,* BA, *extended in the direction of* A, *a point from which the descending moveable traverses the space* AC *in the same time as it traverses the given vertical* AB *from rest at* A.

Draw DCE *at right angles to* AC, *and* CD *equal to* AB, *and join* A *and* D; *angle* ADC *will be greater than angle* CAD, *since* CA *is greater than* AB *or* CD. *Make angle* DAE *equal to angle* ADE, *and extend* EF *perpendicular to* AE *until it*

255   *meets the inclined plane at* F. *Make both* AI *and* AG *equal to* CF, *and draw* GH *through* G *parallel to the horizontal. I say that* H *is the point sought. Let* AB *be assumed to be the time of fall through the vertical* AB, *and the time through* AC *from rest at* A *will be* AC; *and since in the right triangle* AEF, EC *is drawn from the right angle* E *perpendicular to the base* AF, AE *will be the mean proportional between* FA *and* AC, *while* CE *is the mean proportional between* AC *and* CF *(that is, between* CA *and* AI*). And since* AC *is the time from* A *through* AC, *the time through all* AF *will be* AE, *and* EC [*will be*] *the time through* AI. *But since in the isosceles triangle* AED, *the side* AE *is equal to the side* ED, *the time through* AF *will be* ED, *and* CE *is the time through* AI. *Hence* CD *(that is,* AB*) will be the time through* IF *from rest at* A, *which is the same as to say that* AB *is the time through* AC *from* G *or from* H; *which was to be done.*

PROPOSITION XXXIV. PROBLEM XIII

Given an inclined plane and a vertical, both with the same high point; to find in the vertical extended upward a higher point from which a moveable that descends and is then deflected into the inclined plane traverses both in the same time as the inclined plane alone [*is traversed*] *from rest at its high point.*

Let AB *and* AC *be the inclined plane and the vertical whose* [*common*] *terminus is* A; *it is required to find a higher point in the vertical extended beyond* A, *from which the moveable falling, and being deflected through the plane* AB, *traverses that part of the vertical and* [*all*] *the plane* AB *in the same time as* AB *alone from rest at* A.

Let BC *be a horizontal line, and draw* AN *equal to* AC; *and as* AB *is to* BN, *make* AL *to* LC. *Draw* AI *equal to* AL, *and let* CE *be the third proportional to* AC *and* BI, *marked in the vertical* AC *extended. I say that* CE *is the distance sought, such that the vertical being extended above* A, *and a part* AX [*being*] *taken* [*in it*] *equal to* CE, *the moveable from* X *would traverse both distances* XA *and* AB [*combined*] *in equal time with* AB *alone from* A. *Draw the horizontal* XR *parallel to* BC, *meeting* BA *extended at* R; *then,* AB *being extended to* D, *draw* ED *parallel to* CB, *and describe a semicircle on* AD. *From* B, *perpendicular to* DA, *draw* BF *to the circumference. It is evident that* FB *is the mean proportional between* AB *and* BD, *and* FA [*is that*] *between* DA *and* AB. *Draw* BS *equal to* BI, *and* FH *equal to* FB. *Since* AC *is to* CE *as* AB *is to* BD, *and since* BF *is the mean proportional between* AB *and* BD *(as is* BI *between* AC *and* CE*),* FB *will be to* BS *as* BA *is to* AC. *And since* BA *is to* AC *(or to* AN*)*

256

257

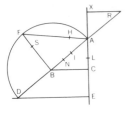

as FB *is to* BS, *then by conversion of ratios* BF *will be to* FS *as* AB *is to* BN; *that is, as* AL *is to* LC. *Hence rectangle* FB–CL *is equal to rectangle* AL–SF. *But rectangle* AL–SF *is the excess of rectangle* AL–FB *(or* AI–BF) *over rectangle* AI–BS *(or* AI–IB), *while rectangle* FB–LC *is the excess of rectangle* AC–BF *over rectangle* AL–BF. *And rectangle* AC–BF *will equal rectangle* AB–BI, *for as* BA *is to* AC, *so* FB *is to* BI. *Therefore the excess of rectangle* AB–BI *over rectangle* AI–BF *(or* AI–FH) *will equal the excess of rectangle* AI–FH *over rectangle* AI–IB. *Hence twice rectangle* AI–FH *equals the two* [*rectangles*] AB–BI *and* AI–IB; *that is, twice* AI–IB *plus the square on* BI. *Make the square on* AI *common to both, and twice rectangle* AI–IB *plus the squares on* AI *and* IB *(that is, the square of* AB *itself) will be equal to twice the rectangle* AI–FH *plus the square on* AI. *Again, make the square on* BF *common to both, and the squares of* AB *and* BF *(or the square of* AF) *will equal twice the rectangle* AI–FH *plus the squares on* AI *and* FB; *that is,* [*the squares on*] AI *and* FH. *But the square of* AF *is equal to twice the rectangle* AH–HF *plus the squares on* AH *and* HF, *whence twice the rectangle* AI–FH *plus the squares on* AI *and* FH *will be equal to twice the rectangle* AH–HF *plus the squares on* AH *and* HF. *Taking away the common square on* HF, *twice the rectangle* AI–FH *plus the square on* AI *will equal twice the rectangle* AH–HF *plus the square on* AH. *And since* FH *is a common side to all the rectangles, line* AH *will equal line* AI; *for if it were greater or less, the* [*double*] *rectangle* FH–HA *plus the square on* HA *would be greater or less than the* [*double rectangle*] FH–IA *plus the square on* IA, *against that which has been demonstrated.*

Now assuming the time of fall through AB *to be as* AB, *the time through* AC *will be as* AC; *and* IB, *the mean proportional between* AC *and* CE, *will be the time through* CE, *or through* XA *from rest at* X. *And since* AF *is the mean proportional between* DA *and* AB *(or* RB *and* BA), *and* BF *is the mean proportional between* AB *and* BD *(that is,* RA *and* AB), *and* [BF] *is equal to* FH, *then, from the above, the excess* AH *will be the time through* AB *from rest at* R, *or after fall from* X; *while the time through* AB *from rest at* A *is* AB. *Therefore the time through* XA *is* IB, *but through* AB *after* RA *or* XA *it is* AI; *whence the time through* XA *and* AB *will be as* AB; *that is, the same as through* AB *alone from rest at* A; *which was proposed.*

PROPOSITION XXXV. PROBLEM XIV   **258**

Given a line at an angle [inflexa] *to a given vertical, to find a part therein through which alone, from rest, motion is made in the same time as through it and the vertical together.*

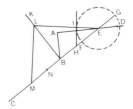

Let AB *be the vertical, and* BC *be at an angle with it; it is required to find a part in* BC *through which motion is made from rest in the same time as through that part together with vertical* AB. *Draw the horizontal* AD, *which the incline* CB *extended meets at* E, *and draw* BF *equal to* BA; *with center* E *and radius* EF, *describe the circle* FIG, *extending* FE *to meet the circumference at* G; *make* BH *to* HF *as* GB *is to* BF. *Draw* HI *tangent to the circle at* I; *then from* B, *erect* BK *perpendicular to* FC, *meeting line* EIL *at* L; *finally, draw* LM *perpendicular to* EL, *meeting* BC *at* M. *I say that motion in line* BM *from rest at* B *will be made in the same time as* [motion] *from rest at* A *through both* AB *and* BM. *Take* EN *equal to* EL; *and since* BH *is to* HF *as* GB *is to* BF, *then by permutation, as* GB *is to* BH, BF *is to* FH; *and by division,* GH *is to* HB *as* BH *is to* HF. *Hence rectangle* GH–HF *equals the square on* HB; *but the same rectangle also equals the square on* HI; *therefore* BH *is equal to* HI. *And since in the quadrilateral* ILBH *the sides* HB *and* HI *are equal, and angles* B *and* I *are right angles, side* BL *is also equal to* [side] LI. *Then* EI *equals* EF, *whence all* LE *(or* NE) *is equal to* LB   **259** *plus* EF. *Take away the common* [part] EF, *and the remainder* FN *will be equal to* LB. *But* FB *was assumed equal to* BA, *whence* LB *is equal to* AB *plus* BN. *Further, if it is assumed that the time through* AB *is* AB *itself, the time through* EB *will be equal to* EB; *but the time through all* EM *will be* EN, *that is, the mean proportional between* ME *and* EB, *whence the time of fall through the remainder* BM *after* EB *(or* AB) *will be* BN. *Now it was assumed that the time through* AB *is* AB; *therefore the time of fall through both* AB *and* BM *is* AB *plus* BN. *But since the time through* EB *from rest at* E *is* EB, *the time through* BM *from rest at* B *will be the mean proportional between* BE *and* EM, *which is* BL. *Therefore the time through both* AB *and* BM *from rest at* A *is* AB *plus* BN, *while the time through* BM *alone from rest at* B *is* BL. *But it was shown that* BL *is equal to* AB *plus* BN; *whence the proposition holds.*

Another, more direct, proof: Let BC *be the inclined plane and* BA *the vertical. Draw a perpendicular through* B *to* EC,

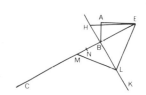

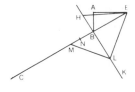

and on its extension mark BH *equal to the excess of* BE *over* BA. *Draw angle* BHE *equal to angle* HEL, *and let* EL *extended meet* BK *at* L; *from* L, *erect a perpendicular* LM *to* EL, *meeting* BC *at* M. *I say that* BM *is the distance sought in the plane* BC. *For since* MLE *is a right angle,* BL *is the mean proportional between* MB *and* BE, *as* LE *is between* ME *and* EB. *Make* EN *equal to* EL; *the three lines* NE, EL, *and* LH *will be equal, and* HB *will be the excess of* NE *over* BL. *But the same* HB *is also the excess of* NE *over* NB *plus* BA; *therefore* NB *plus* BA *is equal to* BL. *Now if* EB *is assumed to be the time through* EB, *the time through* BM *from rest at* B *will be* BL, *and* BN *will be the time of the same after* EB *(or* AB *); and* AB *will be the time through* AB. *Therefore the time through* AB *plus* BM *is equal to* AB *plus* BN, *which is equal to the time through* BM *alone from rest at* B; *which was the intent.*

260

[FIRST] LEMMA

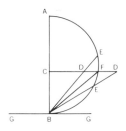

Let DC *be perpendicular to the diameter* BA, *and from end* B *draw* BED *in any direction, and connect* F *and* B. *I say that* FB *is the mean proportional between* DB *and* BE. *Connect* E *and* F, *and through* B *draw the tangent* BG, *which will be parallel to* CD, *whence angle* DBG *will be equal to angle* FDB. *But* GBD *is also equal to angle* EFB *in the alternate part; therefore the triangles* FBD *and* FEB *are similar; and as* BD *is to* BF, FB *is to* BE.

[SECOND] LEMMA

Let line AC *be greater than* DF, *and let* AB *have to* BC *a greater ratio than* DE *has to* EF; *I say that* AB *is greater than* DE. *For since* AB *has to* BC *a greater ratio than* DE *has to* EF, *whatever ratio* AB *has to* BC, DE *will have [that ratio] to some smaller [magnitude] than* EF. *Let it have this to* EG. *Since* AB *is to* BC *as* DE *is to* EG, *then by composition and conversion of ratios, as* CA *is to* AB, GD *will be to* DE; *but* CA *is greater than* GD; *hence* BA *is greater than* DE.

[THIRD] LEMMA

Let ACIB *be one quadrant of a circle, and from* B *draw* BE *parallel to* AC. *From some center taken in this* [BE], *describe the circle* BOES, *tangent to* AB *at* B *and*

cutting the arc of the quadrant [ACIB] at I. Join C and B,   **261**
and extend CI to S; I say that CI is always less than line
CO. Draw AI, tangent to circle BOE. Now if DI is drawn,
it will be equal to DB, since indeed DB is tangent to the
quadrant, and DI is also tangent to it and perpendicular
to the diameter AI, for AI is also tangent to circle BOE
at I. And since angle AIC is greater than angle ABC,
because it stands on a longer arc, angle SIN will also be
greater than ABC; whence arc IES is greater than arc
BO. And line CS, closer to the center, is greater than CB,
so that CO is greater than CI, for SC is to CB as OC is
to CI.

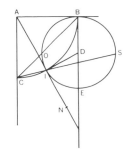

   The same would apply even more if, as in the second
diagram, BIC were less than a quadrant. For the vertical
DB will cut the circle CIB, whence DI also [would cut it],
being equal to DB, and angle DIA would be obtuse, and
AIN would cut the circumference BIE. And since angle
ABC is less than angle AIC, which equals SIN, this
[angle] is still less than [that which] SI would form with
the tangent at I; hence the arc SEI is much longer than
the arc BO; therefore, etc.; which was to be demonstrated.

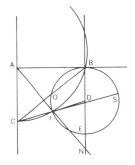

PROPOSITION XXXVI. THEOREM XXII

From the lowest point of a vertical circle, let an inclined
plane be raised, subtending an arc no greater than one
quadrant, from the ends of which two other planes are   **262**
inclined, meeting at any point on the arc; descent in both
these planes will be finished in a shorter time than [descent]
in the first inclined plane alone, or in only the lower of
these planes.

Let circumference CBD be no more than one quadrant of
the vertical circle with its lowest point at C, to which is raised
the plane CD; and let two planes be deflected from the ends
D and C to some point B taken on the circumference; I say
that the time of descent through both the planes DB and BC
is briefer than the time of descent through DC alone, or through
BC alone from rest at B. Draw the horizontal MDA through
D, meeting CB extended at A; make DN and MC perpendicular
to MD, and BN [perpendicular] to BD. Around the right
triangle DBN describe the semicircle DFBN, cutting DC at
F; DO is the mean proportional of CD and DF, while AV
is the same of CA and AB. Let PS be the time of running
through all DC, or BC (it is evident that these times of traversal

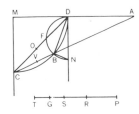

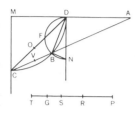

are equal) and let the time SP *have to the time* PR *the ratio that* CD *has to* DO *;* PR *will be the time in which a moveable from* D *runs through* DF, *and* RS [*will be*] *that in which the remainder* FC [*is passed*]. *Since* PS *is also the time in which a moveable from* B *runs through* BC, *if* SP *is made to* PT *as* BC *is to* CD, *then* PT *will be the time of fall from* A *to* C, *since* DC *is the mean proportional between* AC *and* CB, *as shown above. Finally, make* TP *to* PG *as* CA *is to* AV *;* PG *will be the time in which a moveable from* A *comes to* B, *and* GT *the time of the remaining motion* BC *after motion from* A *to* B. *Now, since* DN *(the diameter of circle* DFN*) is vertical, lines* DF *and* DB *are run through in equal times, so if it is*

263   *proved that the moveable goes more swiftly through* BC *after fall* DB *than* [*through*] FC *after running through* DF, *we have our goal.*

But the moveable coming from D *through* DB *will traverse* BC *with the same swiftness of time as if it came from* A *through* AB, *since in either fall (*DB *or* AB*) it would receive equal momenta of speed; therefore it will be demonstrated that it runs through* BC *after* AB *in shorter time than* [*through*] FC *after* DF. *But it has further been made clear that* GT *is the time in which* BC *is run through after* AB, *while* RS *is the time* [*through*] FC *after* DF *; and thus it is to be shown that* RS *is greater than* GT. *This is shown thus: since* CD *is to* DO *as* SP *is to* PR, *then by conversion of ratios and inverting, as* RS *is to* SP, OC *is to* CD *; but as* SP *is to* PT, *so* DC *is to* CA *; and since* CA *is to* AV *as* TP *is to* PG, *then by conversion of ratios,* AC *will be to* CV *as* PT *is to* TG. *Therefore, by equidistance of ratios, as* RS *is to* GT, *so* OC *is to* CV *; but* OC *is greater than* CV, *as is presently to be shown. Therefore time* RS *is greater than time* GT, *which was to be demonstrated.*

But since CF *is greater than* CB *and* FD *is less than* BA, *the ratio of* CD *to* DF *is greater than the ratio of* CA *to* AB *; but as* CD *is to* DF, *so the square on* CO *is to the square on* OF, *because* CD, DO, *and* DF *are* [*continued*] *proportionals. And as* CA *is to* AB, *so the square on* CV *is to the square on* VB *; therefore* CO *has to* OF *a greater ratio than* CV *has to* VB. *Hence, by the foregoing lemmas,* CO *is greater than* CV. *Moreover it is evident that the time through* DC *is to the time through* DB–BC *as* D–O–C *is to* DO *plus* CV.

SCHOLIUM

From the things demonstrated, it appears that one can deduce that the swiftest movement of all from one terminus to the other is not through the shortest line of all, which is the straight line [AC], *but through the circular arc.*[48] *For in the quadrant BAEC, of which side BC is the vertical, arc AC [may be] divided into any equal parts AD, DE, EF, FG, and GC, and straight lines [may be] drawn from C to points A, D, E, F, and G, as well as straight lines AD, DE, EF, FG, and GC; and it is manifest that movement through the two [lines] AD–DC is finished more quickly than through AC alone, or [through] DC from rest at D. But from rest at A, DC is finished more quickly than the two AD–DC. Yet it seems true that from rest at A, descent is finished more quickly through the two DE–EC than through CD only; therefore descent through the three AD–DE–EC is finished more quickly than through the two AD–DC. It is likewise true that with prior descent through AD–DE, movement is made more quickly through the two EF–FC than through EC alone; hence motion is swifter through the four AD–DE–EF–FC than through the three AD–DE–EC. And ultimately, after prior descent through A–D–E–F, movement is finished more quickly through the two FG–GC than through FC alone. Therefore descent is made in still shorter time through the five AD–DE–EF–FG–GC than through the four AD–DE–EF–FC. Hence motion between two selected points, A and C, is finished the more quickly, the more closely we approach the circumference through inscribed polygons.*

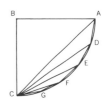

264

What has been explained for the quadrant happens also in arcs less than the quadrant; and the reasoning is the same.

PROPOSITION XXXVII. PROBLEM XV

Given a vertical and an inclined plane of the same height, to find a part in the incline that is equal [in length] to the vertical and is traversed in the same time as this vertical [always starting from rest at the intersection].

Let AB be the vertical and AC the inclined plane; it is required to find a part in AC, equal to the vertical AB, which is traversed from rest at A in the same time as that in which the vertical is traversed.

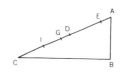

Take AD equal to AB, and bisect the remainder DC at I; as AC is to CI, make CI to some other [line] AE, put [in turn] equal to DG. It is evident that EG equals AD plus AB. I say, moreover, that EG is that [part] which is traversed by a

---

48. All that could properly be deduced was that the shortest descent is along some kind of curve. The curve is in fact only approximately circular, and was later shown to be cycloidal; cf. note 21, above.

moveable coming from rest at A *in a time equal to the time in which the moveable falls through* AB. *For since* CI *is to* AE, **265** *or* ID *to* DG, *as* AC *is to* CI, *then by inversion of ratios, as* CA *is to* AI, *so* DI *is to* IG; *and since as all* CA *is to all* AI, *so the removed part* CI *is to the removed part* IG, *the remainder* IA *will be to the remainder* AG *as all* CA *is to all* AI. *Thus* AI *is the mean proportional between* CA *and* AG, *as is* CI *between* CA *and* AE. *Thus if we put the time through* AB *to be as* AB, *the time through* AC *will be* AC, *and* CI *(or* ID*) will be the time through* AE. *And since* AI *is the mean proportional between* CA *and* AG, *and* CA *is the time through all* AC, *the time through* AG *will be* AI, *and the remainder* IC [*will be the time*] *through the remainder* GC. *But* DI *was the time through* AE, *so* DI *and* IC *are the times through both* AE *and* CG. *Therefore the remainder* DA *will be the time through* EG, *equal to the time through* AB; *which was to be done.*

COROLLARY

From this it is evident that the required distance lies between those upper and lower parts [*of the inclined plane*] *which are traversed in equal times.*

PROPOSITION XXXVIII. PROBLEM XVI

Given two horizontal planes cut by a vertical; to find a point on high in the vertical from which falling moveables, deflected into horizontal planes, will, in times equal to their times of [*vertical*] *fall, traverse distances in these horizontals (that is, in both upper and lower) that have to one another any given ratio of lesser to greater.*[49]

Let the horizontal planes CD *and* BE *be cut by the vertical* ACB, *and let the given ratio of lesser to greater be* N *to* FG. *It is required to find a high point in the vertical* AB *from which a moveable, falling and deflected into plane* AC, *will, in a time equal to its time of fall, traverse a distance which, compared with the distance traversed by a second moveable coming from the same high point* [*and moving*] *through plane* BE [*for*] *a time equal to the time of its fall, shall have the same ratio as the given* N *to* FG.

266   *Make* GH *equal to* N, *and make* BC *to* CL *as* FH *is to* HG; *I say that* L *is the required high point. For assume that* CM *is double* CL *and draw* LM *meeting plane* BE *at* O; BO *will*

---

49. Cf. Fourth Day, pp. **283**–**85**, below.

be double BL. *And since* BC *is to* CL *as* FH *is to* HG, *then by composition and conversion, as* HG (*that is,* N) *is to* GF, CL *will be to* LB; *that is,* CM [*will be*] *to* BO. *But since* CM *is double* LC, *it follows that the distance* CM *is that which will be traversed in plane* CD *by a moveable coming from* L *after fall* LC; *and for the same reason,* BO *is that which is traversed after fall* LB, *in a time equal to the time of fall through* LB, *since* BO *is double* BL; *whence the proposition is evident.*

Sagr. It appears to me that we may grant that our Academician was not boasting when, at the beginning of this treatise, he credited himself with bringing to us a new science concerning a most ancient subject. When I see with what ease and clarity, from a single simple postulate, he deduces the demonstrations of so many propositions, I marvel not a little that this kind of material was left untouched by Archimedes, Apollonius, Euclid, and so many other illustrious **267** mathematicians and philosophers; especially seeing that many and thick volumes have been written on motion.

Salv. There is a little fragment of Euclid concerning motion,[50] but in it one finds no indication that he went on to investigate the ratio of acceleration, and of its diversities along different slopes. Thus one may truly say that only now has the door been opened to a new contemplation, full of admirable conclusions, infinite in number, which in time to come will be able to put other minds to work.

Sagr. Truly, I believe that just as those few properties of the circle (I mean this by way of example) demonstrated by Euclid in the third book of his *Elements* are the gateway to innumerable others, more recondite, so these [of motion] which have been produced and demonstrated in this brief treatise, when they have passed into the hands of others of a speculative turn of mind, will become the path to many others, still more marvelous. This is likely to be the case because of the preëminence of this subject above all the rest of physics.

This has been a long and laborious day, in which I have enjoyed the bare propositions more than their demonstrations, many of which I believe are such that it would take me more than an hour to understand a single one of them. That study

50. The fragment, of very doubtful authenticity, was known in the Middle Ages and was printed with Euclid's works in many editions beginning in 1537.

I reserve to carry out in quiet, if you will leave the book in my hands after we have seen this part that remains, which concerns the motion of projectiles. This will be done tomorrow, if that suits you.

   Salv. I shall not fail to be with you.

The Third Day Ends

Salv. Simplicio is just arriving now, so let us begin on motion without delay. Here is our Author's text:

On the Motion of Projectiles

We have considered properties existing in equable motion, and those in naturally accelerated motion over inclined planes of whatever slope. In the studies on which I now enter, I shall try to present certain leading essentials, and to establish them by firm demonstrations, bearing on a moveable when its motion is compounded from two movements; that is, when it is moved equably and is also naturally accelerated. Of this kind appear to be those which we speak of as projections, the origin of which I lay down as follows.

I mentally conceive of some moveable projected on a horizontal plane, all impediments being put aside. Now it is evident from what has been said elsewhere at greater length that equable motion on this plane would be perpetual if the plane were of infinite extent;[1] but if we assume it to be ended, and [situated] on high, the moveable (which I conceive of as being endowed with heaviness), driven to the end of this plane and going on further, adds on to its previous equable and indelible motion that downward tendency which it has from its own heaviness. Thus there emerges a certain motion, compounded from equable horizontal and from naturally accelerated downward [motion], which I call "projection." We shall demonstrate some of its properties [accidentia], of which the first is this:

PROPOSITION I.   THEOREM I   **269**

When a projectile is carried in motion compounded from equable horizontal and from naturally accelerated downward [motions], it describes a semiparabolic line in its movement.

Sagr. As a favor to me, Salviati, and I believe also to Simplicio, it is necessary to pause here for a moment. I did not go so deeply into geometry as to make a study of

1. See pp. **243**–**45**.

Apollonius,[2] beyond my knowing that he deals with these parabolas and other conic sections. Without my knowing about these and their properties, I do not believe that the demonstrations of further propositions pertaining to them can be understood. And since already in this very first proposition proposed to us by the Author, it must be demonstrated that the line described by a projectile is parabolic, then even if we need not deal with lines other than those, I suppose it is absolutely necessary to have a complete understanding, if not of all the properties of such figures demonstrated by Apollonius, at least of those that are required for the present science.

Salv. You are too humble, wishing thus to make something new of things [*cognizioni*] that you assumed not long ago as well known. I refer to the matter of resistances, for when we there needed knowledge of a certain proposition of Apollonius, you made no difficulty concerning it.[3]

Sagr. I may have happened to know that one, or may have assumed it for the moment because it was required of me throughout that treatment. But here, where I suppose that all the demonstrations I am about to hear concern such lines, there is no point in my gulping them down, as people say, throwing away time and effort.

Simp. And for my part, I believe that although Sagredo is well enough supplied for his needs, these very first terms already begin to strike me as novel. For although our philosophers have treated this matter of projectile motion, I don't recall that they felt themselves obliged to define the lines described thereby, other than in very general terms— that these are curved lines, except for things thrown vertically

270   upward. And unless that little that I have learned of geometry from Euclid, after our other discussions a long time ago, will be sufficient to render me capable of what is required for understanding the demonstrations to come, I shall have to content myself with merely believing the propositions without comprehending them.

Salv. But I want you to know them through the Author of the treatise himself. Now, when he allowed me to see this work of his, neither had I at that time mastered the books of Apollonius; but he took the trouble to demonstrate to me

2. Apollonius of Perga (262–190 B.C.) was the author of the most complete ancient treatise on conic sections.
3. See Second Day, Prop. [XII] (p. **177**)

two principal properties of the parabola without [assuming] any previous knowledge [on my part]. Those are all we will need in the present treatise. They are indeed also proved by Apollonius, after many preliminaries that it would take a long time to see. My wish is that we much shorten the journey, deducing the first [proposition] immediately from the pure and simple generation of the parabola, from which in turn immediately [follows] the demonstration of the second.

Taking up the first, then, imagine a right cone whose base is the circle *IBKC* and whose vertex is the point *L*. When cut by a plane parallel to the side *LK*, this yields the section *BAC*, called a parabola, whose base cuts at right angles the diameter *IK* of circle *IBKC*. The axis *AD* of the parabola is parallel to the side *LK*. Taking any point *F* in line *BFA*, draw the straight line *FE* parallel to *BD*. Now I say that:

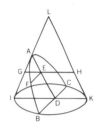

[FIRST LEMMA]

The square of *BD* has to the square of *FE* the same ratio that the axis *DA* has to the part *AE*.

Suppose a plane parallel to the circle *IBKC* and passing through point *E*; this makes a circular section in the cone, of which the diameter will be the line *GEH*. Since *BD* is perpendicular to the diameter *IK* of circle *IBK*, the square of *BD* will be equal to the rectangle of sides *ID* and *DK*. Likewise in the upper circle, assumed to pass through points *G, F*, and *H*, the square of line *FE* is equal to the rectangle of sides *GE–EH*. Therefore the square of *BD* has to the square of *FE* the same ratio that rectangle *ID–DK* has to rectangle *GE–EH*. And since line *ED* is parallel to *HK, EH* will be equal to *DK* [these being] also parallel; whence rectangle *ID–DK* will have to rectangle *GE–EH* the same ratio that *ID* has to *GE*, that is, that which *DA* has to *AE*.   **271** Hence rectangle *ID–DK* has to rectangle *GE–EH* (that is, the square *BD* has to the square *FE*) the same ratio that the axis *DA* has to the part *AE*; which was to be proved.

The other proposition necessary to the present treatise we shall make manifest thus.

[SECOND LEMMA]

Draw a parabola with its axis *CA* extended to *D*, and take any point *B* [on the parabola], drawing through this the line *BC* parallel to the base of this parabola. Take *DA* equal to the part *CA* of the axis, and draw a

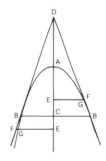

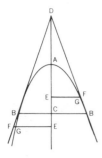

straight line from *D* to *B*. I say that line *DB* does not
fall within the parabola, but touches it on the outside
only, at point *B*.

For if possible, let it [*DB*] fall within and cut [the parabola]
above [*B*], or below when extended. Take in it [extended]
some point *G*, through which draw the line *FGE*. Since the
square of *FE* is greater than the square of *GE*, the square of
FE will have a greater ratio to the square of *BC* than the
square of *GE* has to [that of] *BC*. And since by what preceded,
the square of *FE* is to the square of *BC* as line *EA* is to *AC*,
then *EA* has a greater ratio to *AC* than the square of *GE*
has to the square of *BC* (that is, than the square of *ED* has
to the square of *DC*), seeing that in triangle *DGE, ED* is to
DC as *GE* is to the parallel *BC*. But line *EA* has to *AC* (that
is, to *AD*) the same ratio that four rectangles *EA–AD* have
to four squares on *AD*—or to the square of *CD*, which is
equal to four squares on *AD*. Then four rectangles *EA–AD*
would have a greater ratio to the square of *CD* than the
square of *ED* has to the square of *DC*; and four rectangles
EA–AD would be greater than the square of *ED*. But this
is false; these four [rectangles] are less, inasmuch as parts
EA and *AD* of line *ED* are not equal. Hence line *DB* touches
the parabola at *B* and does not cut it; which was to be
demonstrated.

Simp. You proceed too grandly in your demonstrations;
272 it seems to me that you always assume that all Euclid's
propositions are as familiar and ready at hand for me as his
very first axioms. That is not the case. You have just now
tossed it at me, over your shoulder, that four rectangles
EA–AD are less than the square of *DE* because the parts
EA and *AD* of line *ED* are unequal. This does not satisfy
me, but leaves me up in the air.

Salv. Well, all mathematicians worthy of the name take
it for granted that the reader has ready at hand at least the
Elements of Euclid. Here, to supply your need, it will suffice
to remind you of a proposition in the second [book],[4] where
it is proved that when a line is cut into equal and unequal
parts, the rectangle of unequal parts is less than the rectangle
of equal parts—that is, [less] than the square of the half[-line]
—by as much as the square of the line comprised between
the points of section. Hence it is manifest that the square of

4. Euclid, *Elements* II.5

the whole [line], which contains four squares of the half[-line], is greater than four rectangles of the unequal parts.

Now, we must keep in mind these two propositions just demonstrated, taken from the elements of conics, in order to understand the things to follow in the present treatise; for the Author makes use of these, and no more. So let us take up his text again, and see how he demonstrates his first proposition, in which his purpose is to prove to us that:

[THEOREM I, restated]

The line described by a heavy moveable, when it descends with a motion compounded from equable horizontal and natural falling [motion], is a semiparabola.[5]

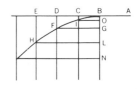

Imagine a horizontal line or plane AB *situated on high, upon which the moveable is carried from* A *to* B *in equable motion, but at* B *lacks support from the plane, whereupon there supervenes in the same moveable, from its own heaviness, a natural motion downward along the vertical* BN. *Beyond the plane* AB *imagine the line* BE, *lying straight on, as if it were the flow or measure of time, on which there are noted any equal parts of time* BC, CD, DE; *and from points* B, C, D, *and* E *imagine lines drawn parallel to the vertical* BN. *In the first of these, take some part* CI; *in the next, its quadruple* DF; *then its nonuple* EH, *and so on for the rest according to the rule of squares of* CB, DB, *and* EB; *or let us say, in the duplicate ratio of those lines.*

273

If now to the moveable in equable movement beyond B *toward* C, *we imagine to be added a motion of vertical descent according to the quantity* CI, *the moveable will be found after time* BC *to be situated at the point* I. *Proceeding onwards, after time* DB *(that is, double* BC*), the distance of descent will be quadruple the first distance,* CI; *for it was demonstrated in the earlier[6] treatise that the spaces run through by heavy things in naturally accelerated motion are in the squared ratio of the times. And likewise the next space,* EH, *run through in time* BE, *will be as nine times* [CI]; *so that it manifestly*

5. This was discovered by Galileo late in 1608 in connection with a very precise experimental test of his belief that horizontal motion would remain uniform in the absence of resistance. The test required observations like those described in the ensuing paragraph. Cf. note 30 to First Day.

6. The text reads *primo*, but the proposition meant is Theorem II in the second Latin treatise of the Third Day.

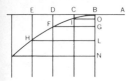

appears that spaces EH, DF, *and* CI *are to one another as the squares of lines* EB, DB, *and* CB. *Now, from points* I, F, *and* H, *draw straight lines* IO, FG, *and* HL *parallel to* EB; *line by line,* HL, FG, *and* IO *will be equal to* EB, DB, *and* CB *respectively, and* BO, BG, *and* BL *will be equal to* CI, DF, *and* EH. *And the square of* HL *will be to the square of* FG *as line* LB *is to* BG, *while the square of* FG [*will be*] *to the square of* IO *as* GB *is to* BO; *therefore points* I, F, *and* H *lie in one and the same parabolic line.*

And it is similarly demonstrated, assuming any equal parts of time, of any size whatever, that the places of moveables carried in like compound motion will be found at those times in the same parabolic line. Therefore the proposition is evident.

Salv. This conclusion is deduced from the converse of the first of the two lemmas given above. For if the parabola is described through points *B* and *H*, for example, and if either of the two [points], *F* or *I*, were not in the parabolic line described, then it would lie either inside or outside, and consequently line *FG* would be either less or greater than that which would go to terminate in the parabolic line. Whence the ratio that line *LB* has to *BG*, the square of *HL* would have, not to the square of *FG*, but to [the square of] some other [line] greater or less [than *FG*]. But it [the square of *HL*] does have [that ratio] to the square of *FG*. Therefore point *F* is on the parabola; and so on for all the others, etc.

Sagr. It cannot be denied that the reasoning is novel, ingenious, and conclusive, being argued *ex suppositione*; that is, by assuming that the transverse motion is kept always equable, and that the natural downward [motion] likewise maintains its tenor of always accelerating according to the squared ratio of the times; also that such motions, or their speeds, in mixing together do not alter, disturb, or impede one another. In this way, the line of the projectile, continuing its motion, will not finally degenerate into some other kind **274** [of curve]. But this seems to me impossible; for the axis of our parabola is vertical, just as we assume the natural motion of heavy bodies to be, and it goes to end at the center of the earth. Yet the parabolic line goes ever widening from its axis, so that no projectile would ever end at the center [of the earth],[7] or if it did, as it seems it must, then the path of

7. In general it would not reach the center, but would take an elliptical path around it. In the ensuing discussion, Galileo wishes to distinguish

the projectile would become transformed into some other line, quite different from the parabolic.

Simp. To these difficulties I add some more. One is that we assume the [initial] plane to be horizontal, which would be neither rising nor falling, and to be a straight line—as if every part of such a line could be at the same distance from the center, which is not true. For as we move away from its midpoint towards its extremities, this [line] departs ever farther from the center [of the earth], and hence it is always rising. One consequence of this is that it is impossible that the motion is perpetuated, or even remains equable through any distance; rather, it would be always growing weaker. Besides, in my opinion it is impossible to remove the impediment of the medium so that this will not destroy the equability of the transverse motion and the rule of acceleration for falling heavy things.[8] All these difficulties make it highly improbable that anything demonstrated from such fickle assumptions can ever be verified in actual experiments.

Salv. All the difficulties and objections you advance are so well founded that I deem it impossible to remove them. For my part, I grant them all, as I believe our Author would also concede them. I admit that the conclusions demonstrated in the abstract are altered in the concrete, and are so falsified that horizontal [motion] is not equable; nor does natural acceleration occur [exactly] in the ratio assumed; nor is the line of the projectile parabolic, and so on. But on the other hand, I ask you not to reject in our Author what other very great men have assumed, despite its falsity. The authority of Archimedes alone should satisfy everyone; in his book *On Plane Equilibrium* [*Mecaniche*], as in the first book of his *Quadrature of the Parabola*, he takes it as a true principle that the arm of a balance or steelyard lies in a straight line equidistant at all points from the common center of heavy things, and that the cords to which [balance-]weights are attached hang parallel to one another. These liberties are pardoned to him by some for the reason that in using our instruments, the distances we employ are so small in com-   **275** parison with the great distance to the center of our terrestrial

---

sharply between purely speculative results and actual phenomena near the surface of the earth.

8. Note that while Sagredo had objected to a theoretical implication of Salviati's assumption, Simplicio rejects that assumption as departing from actual conditions realizable in practice.

globe that we could treat one minute of a degree at the
equator as if it were a straight line, and two verticals hanging
from its extremities as if they were parallel. Indeed, if such
minutiae had to be taken into account in practical operations,
we should have to commence by reprehending architects,
who imagine that with plumb-lines they erect the highest
towers in parallel lines.

Here I add that we may say that Archimedes and others
imagined themselves, in their theorizing, to be situated at
infinite distance from the center. In that case their said
assumptions would not be false, and hence their conclusions
were drawn with absolute proof.[9] Then if we wish later to put
to use, for a finite distance [from the center], these conclusions
proved by supposing immense remoteness [therefrom], we
must remove from the demonstrated truth whatever is
significant in [the fact that] our distance from the center is
not really infinite, though it is such that it can be called
immense in comparison with the smallness of the devices
employed by us. The greatest among these will be the
shooting of projectiles, and in particular, artillery shots; and
[even] these, though great, do not exceed four miles, in
comparison with about that many thousand miles for our
distance from the center. And these shots coming to end on
the surface of the terrestrial globe may alter in parabolic
shape only insensibly, whereas that shape is conceded to be
enormously transformed in going on to end at the center.

Next, a more considerable disturbance arises from the
impediment of the medium; by reason of its multiple varieties,
this [disturbance] is incapable of being subjected to firm rules,
understood, and made into science. Considering merely the
impediment that the air makes to the motions in question
here, it will be found to disturb them all in an infinitude of
ways, according to the infinitely many ways that the shapes
of the moveables vary, and their heaviness, and their speeds.
As to speed, the greater this is, the greater will be the
opposition made to it by the air, which will also impede
bodies the more, the less heavy they are. Thus the falling
heavy thing ought to go on accelerating in the squared ratio
of the duration of its motion; yet, however heavy the moveable may be, when it falls through very great heights the
impediment of the air will take away the power of increasing

9. Cf. note 8, above; Salviati stresses the validity of an argument inde-
pendently of the truth of the assumptions behind it.

its speed further, and will reduce it to uniform and equable   **276**
motion. And this equilibration will occur more quickly and
at lesser heights as the moveable shall be less heavy.

Also that motion in the horizontal plane, all obstacles
being removed, ought to be equable and perpetual; but it
will be altered by the air, and finally stopped; and this again
happens the more quickly to the extent that the moveable
is lighter.

No firm science can be given of such events of heaviness,
speed, and shape, which are variable in infinitely many ways.
Hence to deal with such matters scientifically, it is necessary
to abstract from them. We must find and demonstrate conclusions abstracted from the impediments, in order to make
use of them in practice under those limitations that experience
will teach us. And it will be of no little utility that materials
and their shapes shall be selected which are least subject to
impediments from the medium, as are things that are very
heavy, and rounded. Distances and speeds will for the most
part not be so exorbitant that they cannot be reduced to
management by good accounting [*tara*]. Indeed, in projectiles
that we find practicable, which are those of heavy material
and spherical shape, and even in [others] of less heavy
material, and cylindrical shape, as are arrows, launched
[respectively] by slings or bows, the deviations from exact
parabolic paths will be quite insensible.[10]

Indeed I shall boldly say that the smallness of devices
usable by us renders external and accidental impediments
scarcely noticeable. Among them that of the medium is the
most considerable, as I can make evident by two experiences.
I shall consider movements made through air, since it is
principally of these that we shall be speaking. The air
exercises its force against them in two ways: one is by impeding less heavy moveables more than [it does] the heaviest
ones; the other is by opposing a greater speed more than a
lesser speed in the same body.

As to the first, experience shows us that two balls of equal
size, one of which weighs ten or twelve times as much as
the other (for example, one of lead and the other of oak),
both descending from a height of 150 or 200 braccia, arrive
at the earth with very little difference in speed. This assures
us that the [role of] the air in impeding and retarding both

10. Circumstances in which even this is not true are discussed later on;
see p. **279**.

is small; for if the lead ball, leaving from a height at the same moment as the wooden ball, were but little retarded, and the other a great deal, then over any great distance the lead ball should arrive at the ground leaving the wooden ball far behind, being ten times as heavy. But this does not happen at all; indeed, its victory will not be by even one percent of the entire height; and between a lead ball and a stone ball that weighs one-third or one-half as much, the difference in time of arrival at the ground will hardly be observable. Now the impetus that a lead ball acquires in falling from a height of 200 braccia is so great that, continuing in equable motion, it would run 400 braccia in as much time as it spent in falling, a very considerable speed with respect to that which we confer on our projectiles with bows or other devices (except for impetus depending on firing). Hence we can conclude without much error, by treating as absolutely true those propositions that are to be demonstrated without taking into account the effect of the medium.

As to the other point, we must show that the impediment received from the air by the same moveable when moved with great speed is not very much more than that which the air opposes to it in slow motion.[11] The following experiment gives firm assurance of this. Suspend two equal lead balls from two equal threads four or five braccia long. The threads being attached above, remove both balls from the vertical, one of them by 80 degrees or more, and the other by no more than four or five degrees, and set them free. The former descends, and passing the vertical describes very large [total] arcs of 160°, 150°, 140°, etc., which gradually diminish. The other, swinging freely, passes through small arcs of 10°, 8°, 6°, etc., these also diminishing bit by bit. I say, first, that in the time that the one passes its 180°, 160°, etc., the other will pass its 10°, 8°, etc. From this it is evident that the [overall] speed of the first ball will be 16 or 18 times as great as the [overall] speed of the second; and if the greater speed were to be impeded by the air more than the lesser, the oscillations in arcs of 180°, 160°, etc. should be less frequent

11. The statement is false, and the experiment adduced in its support is fictitious; see note 12, below. Galileo probably deduced the result from his mistaken assumption that resistance of the medium is proportional to velocity, rather than to its square; see notes 52 and 61 to First Day, and Sagredo's conclusion, below.

than those in the small arcs of 10°, 8°, 4°, and even 2° or one degree. But experiment contradicts this, for if two friends shall set themselves to count the oscillations, one counting the wide ones and the other the narrow ones, they will see that they may count not just tens, but even hundreds, without disagreeing by even one, or part of one.[12]

This observation assures us of both [the above] proposi- **278** tions at once; that is, that the greatest and least oscillations all are made, swing by swing, in equal times, and that the impediment and retardation of the air does no more in the swiftest [of these] motions than in the slowest, contrary to what all of us previously believed.

Sagr. Yet since it cannot be denied that the air does impede both, because both [motions] go weakening and finally stop, we must say that the retardations are made in the same ratio in both cases. But how? How can the air make greater resistance at one time than another? Can this happen except by its being assailed at one time with greater impetus and speed, and at another with less? Now if that is the case, the amount of speed of the moveable is itself both the cause and the measure of the amount of resistance. Thus all motions, slow or fast, are retarded and impeded in the same ratio [to their speeds], which seems to me an idea not to be scorned.

Salv. Hence we can conclude, in this second case, that the [practical] fallacies in conclusions that are to be demonstrated by abstracting from external accidents, matter little respecting motions of great speed in devices which we usually deal with, or over distances that are small in relation to the radius of the earth.[13]

Simp. I should like to hear your reason for sequestering things projected by the impetus of firing [*fuoco*], which I take it is the force of gunpowder, from other projections as by slings, bows, or catapults, [treating these] as not subject in the same way to alteration and impediment by the air.

12. A disagreement of about one beat in thirty should occur with pendulums of the length and amplitudes described here as isochronous. Writing this passage in his old age, Galileo may have recalled his valid experiments with pendulums weighted with cork and lead as described in the First Day (p. **129**), and confused them with the quite different result incorrectly asserted here.

13. Literally, "in relation to the magnitude of the semidiameter of great circles of the terrestrial globe."

Salv. Here I am influenced by the excessive, or I might say supernatural,[14] violence [*furia*] with which these projectiles are shot. It seems to me no exaggeration to say that the speed with which the ball is shot from musket or cannon may be called "supernatural," for in natural fall through air from some immense height, the speed of the ball—thanks to opposition from the air—will not go on increasing forever. Rather, what will happen is seen in bodies of very little weight falling through no great distance; I mean, a reduction to equable motion, which will occur also in a lead or iron ball after the descent of some thousands of braccia. This bounded terminal speed may be called the maximum that such a heavy body can naturally attain through air, and I deem this speed to be much smaller than that which can be impressed on the same ball by exploding powder.

279

A very suitable [*acconcia*] experiment can assure us of this. From a height of one hundred braccia or more, fire a lead bullet from an arquebus vertically downward on a stone pavement.[15] Then shoot with the same [gun] against a like stone from a distance of one or two braccia, and see which of the two bullets is more badly smashed. If the one which came from on high is found to be less flattened than the other, it will be a sign that the air impeded the first [bullet] and diminished the speed conferred on it at the beginning of its motion by the firing, and that consequently the air does not permit this [second] speed ever to be gained by a bullet coming from as great a height as you please. For if the speed impressed on it by firing did not exceed that which it could acquire by itself in falling naturally, its thrust [*botta*] downward ought to be more effective, rather than less so. I have not made such an experiment, but I am inclined to believe that the ball from an arquebus or a cannon, coming from any height whatever, will not strike the blow that it makes against a wall a few braccia distant; that is, when so close that the short penetration, or we might say "cut," to be made in the air is insufficient to take away the excess of the supernatural violence impressed on it by firing.

---

14. The word "supernatural" does not mean miraculous, but simply "not natural"; that is, incapable of being acquired in natural acceleration (free fall) through a given medium, from any height whatever. Thus the muzzle velocity of a cannonball may be greater than the speed it gains in any fall through air, where a terminal velocity is ultimately reached.
15. See note 65 to First Day.

This excessive impetus of violent shots can cause some deformation in the path of a projectile, making the beginning of the parabola less tilted and curved than its end. But this will prejudice our Author little or nothing in practicable operations, his main result being the compilation of a table of what is called the "range" of shots, containing the distances at which balls fired at [extremely] different elevations will fall. Since such shots are made with mortars charged with but little powder, the impetus is not supernatural in these, and the [mortar] shots trace out their paths quite precisely.[16]

Meanwhile we have got ahead of the treatise, where the Author [next] wants to introduce to us the contemplation and investigation of the impetus of a moveable when it moves with one motion compounded from two; and first, of the composition of two equable [motions], one horizontal and the other vertical.[17]

PROPOSITION II. THEOREM II.   **280**

If some moveable is equably moved in double motion, that is, horizontal and vertical, then the impetus or momentum of the movement compounded from both will be equal in the square[18] to both momenta of the original motions.

Let some moveable be equably moved in double motion, the vertical displacements [mutationi] *corresponding to space* AB, *and let the horizontal movement traversed in the same time correspond to* BC. *Since spaces* AB *and* BC *are traversed in the same time, in equable motions, the momenta of those motions will be to one another as* AB *is to* BC, *and the moveable that is moved according to these displacements will describe the diagonal* AC, *while its momentum of speed will be as* AC.

16. It had long been known that artillery shots descend more sharply near the end to their flight; cf. Tartaglia's diagrams (1537), *Mechanics in Italy*, pp. 78–94 *passim*. Galileo here restricts his later tables (pp. **304**, **307**) to low-speed mortar shots on the grounds that long-range artillery is never fired at great elevations.

17. The composition of velocities of this type had been noted in antiquity, at least so far as the direction of the resultant motion was concerned; see *Questions of Mechanics*, 1 (Loeb ed., pp. 337–39). But strict Aristotelians clung to the idea that only one motion could act on a body at one time, whence disparate motions must impede one another (*Physica* 202a.34). A moving body was thought to obey the motion that was the more powerful at a given instant; cf. *Mechanics in Italy*, p. 80, and Galileo's counter arguments, *Dialogue*, pp. 176–79 (*Opere*, VII, 203–5).

18. See Glossary. The terminology is antiquated, but the idea is that of vector addition. See further, pp. **288–89**.

Truly AC *is equal in the square to* AB *and* BC*; therefore the momentum compounded from both momenta of* AB *and* BC *is equal in the square to both of them taken together; which was to be shown.*

Simp. You must remove a little doubt that is aroused in me. It seems to me that what has just been concluded contradicts another proposition, in the foregoing treatise, where it was affirmed that the impetus of the moveable coming from *A* to *B* is equal to that [of a moveable] coming from *A* to *C*.[19] But now it is concluded that the impetus at *C* is greater than that at *B*.

Salv. Both propositions are true, Simplicio, but quite different from one another. Here, a single moveable is spoken of, moved by a single motion compounded from two, both equable; whereas there, two moveables were concerned, moved by naturally accelerated motions, one through the vertical *AB* and the other through the inclined plane *AC*. Besides, there the times were not assumed to be equal, but the time through the incline *AC* was greater than the time through the vertical *AB*. In the motion of which we are now speaking, all motions through *AB, BC*, and *AC* are meant to be equable and to be made in the same time.

Simp. Excuse me, and go on, for I am satisfied.

281   *Salv.* The Author next undertakes to lead us to understand what happens with the impetus of a moveable moved by one motion compounded from two, one horizontal and equable, and the other vertical and naturally accelerated; from which finally are compounded the motion of the projectile, and the parabolic line is described. It is sought to demonstrate the magnitude of the impetus of the projectile at every point in this [curve].[20] For an understanding of this, the Author shows us the way, or let us say gives us the method, of finding a rule for and measuring that impetus along the line in which a falling heavy body descends with naturally accelerated motion starting from rest. He says:

PROPOSITION III.   THEOREM III

Let motion take place from rest at A *through line* AB,

19. Simplicio refers to the Postulate, Third Day (pp. **205, 218**).

20. An earlier attempt to analyze the impetus of a projectile (as distinguished from its path) was made by Tartaglia; cf. *Mechanics in Italy*, pp. 81, 91–93.

and take therein some point C. *Assume* AC *to be the time, or measure of the time, of fall through space* AC, *as well as the measure of the impetus or momentum at point* C *acquired from descent* AC. *And take in the same line* AB *some other point such as* B, *at which is to be determined the impetus acquired by the moveable from descent* AB, *as a ratio to the impetus obtained at* C, *of which the measure was assumed to be* AC. *Let* AS *be the mean proportional between* BA *and* AC; *we shall demonstrate that the impetus at* B *is to the impetus at* C *as line* SA *is to* AC.

Draw the horizontal lines CD *(double* AC*) and* BE *(double* BA*); it follows from what has been demonstrated that* [*a moveable*] *falling through* AC, *turned into the horizontal* CD *and carried in equable motion according to the impetus acquired* **282** *at* C, *traverses space* CD *in a time equal to that in which* AC *was traversed in accelerated motion. Similarly,* BE *is traversed in the same time as* AB. *But the time of fall* AB *is* AS; *therefore the horizontal* BE *is traversed in time* AS. *Make* EB *to* BL *as time* SA *is to time* AC. *Since motion through* BE *is equable, space* BL *will be run through in time* AC *with the momentum of swiftness* [*acquired*] *at* B. *But in the same time,* AC, *the space* CD *is traversed with the momentum of swiftness* [*acquired*] *at* C; *and the momenta of swiftness are to each other as the spaces traversed in the same time with these momenta. Therefore the momentum of swiftness at* C *is to the momentum of swiftness at* B *as* DC *is to* BL. *Now, as* DC *is to* BE, *so are their halves, that is, so* CA *is to* AB, *while as* EB *is to* BL, *so* BA *is to* AS. *Hence, by equidistance of ratios, as* DC *is to* BL, *so* CA *is to* AS; *that is, as the momentum of swiftness at* C *is to the momentum of swiftness at* B, *so* CA *is to* AS; *that is, as the time through* CA *is to the time through* AB.

This makes evident the rule for measuring the impetus or momentum of swiftness over the line in which motion of descent takes place, which impetus is assumed to increase in the ratio of the times. But here, before we proceed, it is first to be noted that we are going to speak of motion compounded from equable horizontal and naturally accelerated downward [*motions*]. From such a mixture is produced [conflatur] *the path of projectiles; that is, the parabola is traced; and we must define some common standard according to which we may measure the speed, impetus, or momentum of both motions. Now, there*

*are innumerable degrees of speed for equable motions, and
one of these is to be taken, not at random from the indefinitely
many, but rather one [is to be selected] that is compatible and
connected with a degree of speed acquired through naturally
accelerated motion.*

*I can think of no easier way to select and determine this
than by taking some motion of that same kind. To explain
myself more clearly; I imagine the vertical* AC *and the horizontal* CB, AC *being the altitude and* BC *the amplitude of the
semiparabola described by the compounding of two motions,
one of which is that of a moveable descending through* AC *in
naturally accelerated motion from rest at* A, *and the other is
that of equable transverse motion through the horizontal* AD.
The impetus acquired at C *by descent through* AC *is determined
by the quantity of height* AC; *for the impetus of a moveable
falling from the same height is ever one and the same. But
innumerable degrees of speed may be assigned to equable
motion in the horizontal, not just one. From that multitude I
may select one and segregate it from the rest, as if pointing
a finger at it, by extending upward the altitude* CA, *in which,
whenever necessary, I shall fix the "sublimity"* AE.[21]

283

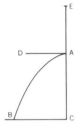

Now if I mentally conceive something falling from rest at
E, *it is evident that the impetus it acquires at terminus* A *is
identical with that with which I conceive the same moveable
to be carried when [it is] turned through the horizontal* AD.
*This is that degree of swiftness with which, in the time of fall
through* EA, *it would traverse double that distance* EA *in the
horizontal. This prefatory remark I consider necessary.*

It is further to be noted that I call the horizontal [line] CB
the "amplitude" of semiparabola AB; [I call] *the axis of this
parabola,* AC, *its "altitude"; and the line* EA, *from descent
through which the horizontal impetus is determined, I call the
"sublimity" [of the parabola].*

*These things explained and defined, I now go on to the
things to be demonstrated.*

Sagr. Pause, I pray you, because it seems to me proper to
adorn the Author's thought here with its conformity to a
conception of Plato's regarding the determination of the
various speeds of equable motion in the celestial motions of

---

21. Galileo's "sublimity" is equivalent to the distance from apex to the
point of intersection of axis and directrix of a vertical parabola.

revolution. Perhaps entertaining the idea that a moveable cannot pass from rest to any determinate degree of speed, in which it must then equably perpetuate itself, except by passing through all the other lesser degrees of speed (or let us say of greater slowness) that come between the assigned degree and the highest [degree] of slowness, which is rest, he said that God, after having created the movable celestial bodies, in order to assign to them those speeds with which they must be moved perpetually in equable circular motion, made them depart from rest and move through determinate spaces in that natural straight motion in which we sensibly see our moveables to be moved from the state of rest, successively accelerating. And he added that these having been made to gain that degree [of speed] which it pleased God that they should maintain forever, He turned their straight motion into circulation, the only kind [of motion] **284** that is suitable to be conserved equably, turning always without retreat from or approach toward any pre-established goal desired by them. The conception is truly worthy of Plato, and is to be the more esteemed to the extent that its foundations, of which Plato remained silent, but which were discovered by our Author in removing their poetical mask or semblance, show it in the guise of a true story.

And since, through very competent astronomical doctrines, we have data about the sizes of the planetary orbs and the distances from the center about which they turn, as well as about their speeds, it seems very credible to me that our Author, from whom the Platonic concept did not remain hidden, may at some time have had the curiosity to try whether he could assign a determinate sublimity from which the bodies of the planets left from a state of rest, and were moved through certain distances in straight and naturally accelerated motion, and were then turned at the acquired speeds into equable motions. This might be found to correspond with the sizes of their orbits and the times of their revolutions.

Salv. Indeed, I seem to remember that he told me he had once made the computation, and also that he found it to answer very closely to the observations.[22] But he did not

---

22. See *Dialogue*, pp. 20–21, 29–30 (*Opere*, VII, 44–45, 53–54). Galileo's cosmogonical speculation could not be reconciled with astronomical data, as shown by Marin Mersenne (*Harmonie Universelle*, Paris, 1636–37, Bk. 2, pp. 103 ff.). The passage in Plato behind which Galileo

want to talk about it, judging that he had discovered too many novelties that have provoked the anger of many, and others might kindle still more sparks. But if anyone should have a similar wish, he may satisfy his taste for himself through the teachings of the present treatise. So let us get on with our material, which is to demonstrate:

PROPOSITION IV.   PROBLEM I

How to determine the impetus at any given point [punctis singulis] *in a given parabola described by a projectile.*

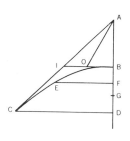

Let BEC *be the semiparabola whose amplitude is* CD *and whose altitude is* DB, *which* [*latter*] *extended upward meets the tangent* CA *to the parabola at* A. *Draw the horizontal* BI *through the vertex* B *and parallel to* CD. *Now if the amplitude* CD *is equal to the whole altitude* DA, *then* BI *will equal* BA *and* BD; *and if the time of fall through* AB *and the momentum of speed acquired at* B *through descent* AB *from rest at* A *are* [*both*] *assumed to be measured by this same* AB, *then* DC, *double* BI, *will be the space traversed in the same time when* [*the moveable is*] *turned through the horizontal with impetus* AB. *But in the same time, falling through* BD *from rest at* B, *it traverses the altitude* BD; *therefore the moveable falling from rest at* A *through* AB, *and turned through the horizontal with impetus* AB, *traverses a space equal to* DC. *Fall through* BD *supervening, the altitude* BD *is traversed and the parabola* BC *is traced, in which the impetus at terminus* C *is made up of the equable transverse* [*motion*] *whose momentum is as* AB, *and the other momentum acquired in descent* BD *at terminus* D *(or* C*), which momenta are equal. Hence if we assume* AB *to be the measure of either of these, say of the equable transverse* [*momentum*], *while* BI *(which is equal to* BD*) is the measure of the impetus acquired at* D *(or* C*), then the subtended* [*line*] IA *will be the quantity of the momentum compounded from both. This will therefore be*

285

---

claimed to have found his mathematical rule was probably *Timaeus*, 38–39, beginning: "Now, when all the stars . . ." There seems to be a contradiction in the discussion here. The higher speed of the inner planets, if acquired in this way, ought to give them larger orbits than the outer planets, even allowing that the Divine will converted Galileo's parabolas into circles. Probably what Galileo had noted was that the planetary speeds are inversely as the square roots of their distances from the sun, a relation strikingly like the law of free fall. But the inverse proportionality is not made a direct one by reversing the direction of fall.

the quantity or measure of the combined [integri] *momenta with which impetus is made at* C *by the projectile coming through parabola* BC.

Keeping this in mind, take in the parabola any point E *at which the impetus of the projectile is to be determined. Draw the horizontal* EF, *and take* BG *as the mean proportional between* BD *and* BF; *since it was assumed that* AB *(or* BD*) is the measure of the time and of the momentum of speed in fall* BD *from rest at* B, *then* BG *will be the time, or measure of time and impetus, at* F, *coming from* B. *If, therefore,* BO *is taken equal to* BG, *the added diagonal* AO *will be the quantity of impetus at point* E; *for* AB *is assumed to determine the time and impetus at* B, *which* [impetus] *turned through the horizontal continues always the same, while* BO *determines the impetus at* F *(or* E*) through fall from rest at* B *through altitude* BF. *But* AO *is equal in the square to both* AB *and* BO *together; what was sought is therefore evident.*

Sagr. The theory of compounding these different impetuses and of the quantity of impetus that results from such mixing is so new to me as to leave no little confusion in my mind. I speak not of the mixing of two equable movements, one along the horizontal line and the other along the vertical, even though unequal to one another; for as to this, I quite understand that a motion results which is equal in the square to both components of it. But I am confused by the mixture of equable horizontal and naturally accelerated vertical [motion]. So I should appreciate it if we were to digest this matter a bit more thoroughly.

Simp. I am even more in need of that, since I am still not **286** entirely satisfied in my own mind, as is necessary, about these propositions which are to be the essential foundations of others to follow. What I mean is that even as to the mixing of two equable motions, horizontal and vertical, I need to understand better that "power" [that "in the square"] of their compounding. You, Salviati, surely understand our need and our wishes.

Salv. The wish is very reasonable, and since I have had a longer time than you to think it over, I shall try to ease your understanding if I can. But you must bear with me, and excuse me if in the discussion of it I repeat a good deal of what was already set forth by the Author.

We can reason definitively about movements and their

speeds or impetuses (whether these are equable or naturally accelerated) only if we first determine some standard [*misura*] that we can use to measure such speeds, as also some measure of time. As to the measure of time, we already have universal agreement on hours, minutes, seconds, etc.; and just as the measure of time is for us that one in common use, accepted by everybody, so it is necessary to assign some measure for speeds to be commonly understood and accepted by all; that is, one that will be the same for everyone.

As explained previously, the Author deemed suitable for such a purpose the [accelerated] speed of naturally falling heavy bodies, of which the growing speeds keep the same tenor everywhere in the world.[23] Thus, for example, the speed acquired by a one-pound lead ball starting from rest and falling from a height of one pikestaff [*picca*] is always and everywhere the same, and for that reason it is very well suited to stand for [*explicar*] the impetus deriving from natural descent. It now remains to find the method of determining also the quantity of impetus in an equable motion, in such a way that everyone who reasons about this may form the same conception of its magnitude and speed. In this way, one man will not imagine it faster and another slower, with the result that in conjoining and mixing this [motion], originally conceived as equable, with the established [*statuito*] accelerated motion, different men shall not form different ideas involving divergent magnitudes of [compound] impetuses.

To determine and represent this unique [*particolare*] impetus and speed, our Author has found no better means than to make use of the impetus acquired by the moveable in [a specified] naturally accelerated motion. Any acquired **287** momentum, turned to equable motion, retains its limited speed precisely, and it is such that in another time equal to that of the descent, it will pass through exactly twice the distance of the height from which fall took place. Since this is a principal point in the matter we are discussing, it is good to make it completely understood by means of some particular example.

Consider again, therefore, the speed and the impetus acquired by a heavy body falling from a height of one

23. Galileo had no reason to suspect that altitude and latitude affect the acceleration of free fall. He was certainly the first to propose a world-wide standard of measure based on a universally familiar phenomenon.

pikestaff; we wish to make use of this speed as a measure of other speeds and impetuses on other occasions. Assume, for example, that the time of such a fall is four seconds.[24] Now, in order to find, by this measure, the impetus of the body falling from any other height, greater or smaller, we must not argue from the ratio of this new height to the height of one pikestaff, and deduce the amount of impetus acquired through the second height by thinking, for example, that in falling from a quadruple height the body would acquire four times the speed, because that is false.[25] For the speed of naturally accelerated motion increases or diminishes not in the ratio of the distances, but in that of the times; and the ratio of distances is greater than this in a squared ratio, as was already demonstrated. Thus when we have taken one part in a straight line for the measure of speed, and also for the time and for the space passed in that time—since, for the sake of brevity, all three of these magnitudes are often represented on the same line—then, in order to find the quantity of time and the degree of speed that the same moveable will have acquired at some other distance, we do not obtain this immediately from that second distance, but from the line that is the mean proportional between the two distances.

I can explain myself better by an example. In the vertical line AC, the part AB is assumed to be the space passed by a heavy body falling naturally in accelerated motion. I can represent the time of that passage by any line I please, and to be brief I wish to represent this to be as much as the line AB. Likewise, for a measure of the impetus and of the speed acquired through such motion, I still take the line AB, so that the measure of all the spaces to be considered in the course of the reasoning is this segment AB. Having established these three measures at pleasure, under a single magnitude AB, [these being measures] of such diverse quantities as spaces, times, and impetuses, let it be proposed to determine **288** the assigned distance and height AC, the amount of time of fall from A to C, and the amount of impetus found to be acquired at the terminus C, [all] in relation to the time and

24. The *picca* being about 12 feet, and free fall covering 16 feet in the first second, Galileo's assumption is clearly arbitary and intended only for the purpose of illustration, as was his earlier assumption of fall through 100 braccia in 5 seconds (*Dialogue*, p. 233 (*Opere*, VII, 250); cf. *Opere*, XVIII, 77.
25. Cf. note 10 to Third Day.

impetus measured by *AB*. Both questions will be determined by taking the mean proportional *AD* of the two lines *AC* and *AB*, affirming that the time of fall through the whole distance *AC* is as much as the time *AD* in relation to the [unit] time *AB*, which was assumed at the outset to be the quantity of time in the fall *AB*. Likewise, we shall say that the impetus, or degree of speed, that the falling body will attain at the terminus *C*, in relation to the impetus that it had at *B*, is this same line *AD* in relation to *AB*, seeing that the speed increases in the same ratio as does the time. This conclusion was taken as a postulate; yet the Author wanted to explain its application above, in Proposition III.

This point being well understood and established, we come to the consideration of the impetus deriving from two motions compounded, one [instance] of which shall be [motion] compounded of horizontal and always equable [motion] together with a vertical [motion] when this is also always equable. The other [instance] will be [motion] compounded from the horizontal, still always equable, and the vertical naturally accelerated [motion].

When both are equable, it has already been seen that the resultant impetus from a compounding of both is equal in the square to both [components]. This, for clear understanding, we shall exemplify thus: It is assumed that the moveable falling through the vertical *AB* has, for example, three degrees of equable impetus, while carried along *BC* [text: *AB*] toward *C* its speed and impetus are four degrees, so that in the time that in falling it would pass three braccia, for example, in the vertical, it would pass four in the horizontal. But in the compounding of both speeds, it comes in the same time from point *A* to terminus *C*, traveling always in the diagonal *AC*. This is not of length 7, as would be the compound of the two, *AB*, 3, and *BC*, 4; but it is [of length] 5, which 5 is equal in the square to the two, 3 and 4; for if the squares of 3 and 4 are taken, which are 9 and 16, and they are added together, they make 25 as the square of *AC*, and this is equal to the two squares of *AB* and *BC*; whence *AC* will be as much as the side, or let us call it the root, of the square 25, which is 5.

289   For a firm and secure rule, then, when one must designate the quantity of impetus resulting from two given impetuses, one horizontal and the other vertical, both being equable, one must take the squares of both, add these together, and

extract the square root of their combination; this will give us the quantity of the impetus compounded from both. And thus in the example given, the moveable that in virtue of its vertical motion would have struck on the horizontal with three degrees of force, and with its horizontal motion alone would have struck at *C* with four degrees, in striking then with both impetuses combined, the blow will be that of a striker moved with five degrees of speed and force. And such a stroke would be of the same value at any point of the diagonal *AC*, the impetuses compounded being always the same and never increased or diminished.

Now let us see what happens in compounding equable horizontal motion with a vertical motion, starting from rest, goes naturally accelerating. It is already manifest that the diagonal which is the line of the motion compounded from these two is not a straight line, but a semiparabola, as has been shown, in which the impetus goes always growing, thanks to the continual growth of speed in the vertical motion. Hence in order to determine the impetus at an assigned point of this parabolic diagonal, it is first necessary to assign the quantity of uniform horizontal impetus, and then to investigate the impetus due to falling at the assigned point, which cannot be determined without consideration of the time elapsed from the beginning of the compounding of the two motions. This consideration of time is not required in the compounding of equable motions, the speeds and impetuses of which are always the same. But here, where there enters a mixing of motion that starts from the greatest slowness and increases its speed in accordance with the continuation of time, it is necessary that the quantity of time shall manifest to us the quantity of the degree of speed at the given point. As for the rest, the impetus compounded from these two is (as in uniform motions) equal in the square to both components.

Here again it is best that I explain by an example. In the vertical *AC* take any part *AB*, which I imagine to serve as a measure of the space of natural motion in this vertical, and likewise as a measure of the time, and also of the degree of speed, or let us say of the impetus. Now first, it is evident that if the impetus at *B* of the [body] falling from rest at *A* shall be **290** turned upon *BD* parallel to the horizontal, in equable motion, then the quantity of its speed will be such that in the time *AB* it will pass a distance twice the distance *AB*; and so much is the line *BD*. Next, taking *BC* equal to *BA*, draw *CE* parallel

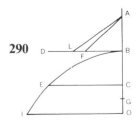

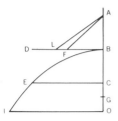

and equal to *BD*, marking the parabolic line *BEI* through points *B* and *C*. Since the horizontal *BD* (or *CE*), double *AB*, is passed in time *AB* with impetus *AB*, and in equal time the vertical *BC* is passed with the impetus acquired at *C*, equal to that same horizontal [*BD*], the moveable will be found to have come in a time equal to *AB* through the parabola *BE* from *B* to *E*, with a single impetus compounded from two, each equal to impetus *AB*. And since one of these is horizontal and the other is vertical, the impetus compounded from them will be equal in the square to both of them; that is, [its square will be] twice the square of either one. Whence *BF* being taken equal to *BA*, and the diagonal *AF* being drawn, the impetus and blow at *E* will be greater than the blow at *B* [dealt] by a body falling from height *A*, or than the blow of the horizontal impetus through *BD*, in the ratio of *AF* to *AB*.

But, always keeping *BA* as the measure of the distance of fall to *B* from rest at *A*, and as a measure of the time and of the impetus of the falling [body] acquired at *B*, if the height *BE* is not equal to, but is greater than, *BA*, then take the mean proportional *BG* between *AB* and *BO* as the measure of time and impetus acquired at *O* by fall through height *BO*. The distance through the horizontal, passed with impetus *AB* in time *AB*, will be double *AB*; but during the time *BG* it will be greater [than *AB*] in proportion as *BG* is greater than *BA*. Next, taking *LB* equal to *BG*, draw the diagonal *AL*, from which we shall have the compounded quantity of the two impetuses, horizontal and vertical, by which the parabola is described; of these, the horizontal [impetus] is equable and is acquired at *B* by the fall *AB*, while the other is that acquired at **291** *O*, or let us say at *I*, by the fall *BO* in time *BG*, which [latter] is also the quantity of its momentum. By similar reasoning we may investigate the impetus at the extreme end of the parabola when its altitude is less than the sublimity *AB*, taking the mean proportional between the two. Setting this along the horizontal in place of *BF*, and drawing the diagonal *AF*, we shall have from this the quantity of the impetus at the extreme end of the parabola.

To that which has been said up to this point about these impetuses, blows, or let us say impacts of projectiles, we should add one other very necessary consideration. This is that it is not sufficient to have in mind just the speed of the projectile, in order to determine fully the force and energy of its impact, but it is further necessary to specify separately the state and condition of that which receives the impact, in the effectiveness of which this [condition] has a great share and contribution in several respects. First, everyone understands that the thing struck suffers violence thereby from the speed of the thing striking [only] to the extent that it opposes this and entirely or partly restrains [*frena*] its motion. For if a blow arrives on that which yields to the speed of the striker without any resistance at all, there will be no blow. And he who runs to strike an enemy with his lance, if it happens that as he overtakes him the enemy moves in flight with like speed, will effect no blow, and the action will be a simple touching without wounding. If an impact is received in an object that does not yield to the striker entirely, but only partly, the impact will [do] damage, not with all its impetus, but only with the excess of speed of the striker over the speed of retirement and yielding of the thing struck. For example, if the striker arrives with ten degrees of speed upon the thing struck, which, by yielding partly, retires with four degrees, then the impetus and impact will be as of six degrees. And finally, the impact on the part of the striker will be entire and maximal when the struck does not yield at all, but entirely opposes itself and stops all the motion of the striker, if indeed this can happen.[26]

I said [impact] "on the part of the striker," because if the struck moves with contrary motion against the striker, the blow and the encounter will be made so much the more strongly, as the two contrary speeds united are greater than that of the striker alone. Moreover, it must also be noticed that to yield more, or less, may derive not only from the quality of the material, harder or less hard, as of iron, or lead, or wool, etc., but also from the placement of the body that   **292**
receives the impact. If this placement is such that the motion of the striker comes against it at right angles, the impetus of the impact will be maximal; but if the motion comes obliquely, and gives a slanting blow as we say, the blow will be weaker, and the more so according to the greater obliquity. For any object obliquely situated, though of very solid material, does not remove and stop all the impetus and motion of the striker, which escapes and passes on beyond, continuing (at least in part) to be moved over the surface of the opposed resistent.

---

26. A perspicuous question which later was to become central in the development of conservation laws in physics. That Galileo rejected the possibility of impact without effect is evident in the Added Day; see p. **337**.

If therefore the magnitude of the impetus of the projectile at the extremity of its parabolic line is determined as above, it must be understood as being the impact received on a line at right angles to the parabolic [line], or rather to its tangent at the said point; for although the motion is compounded of a horizontal and a vertical [motion], the impetus is not maximal either upon the horizontal or upon the vertical, being received obliquely on both.

Sagr. Your bringing up of these blows and impacts has awakened in my mind a problem, or call it a question, of mechanics; one to which I have not found the answer in any writer, nor anything that lessens my marvel at it, or even partially satisfies my mind. This doubt and puzzlement resides in my inability to understand the origin and principle of the immense energy and force that is seen to exist in impact, when, with a simple blow of a hammer that weighs no more than eight or ten pounds, we see resistances overcome that would not yield to the weight of a body exerting its impetus on it without impact, by merely weighing down on and pressing it, though this heaviness may amount to many hundreds of pounds. I should still like to find a way of measuring this force of impact, which I do not believe to be infinite, but rather think that it has its limit of equalization with, and finally of control by, other forces—pressure, heaviness, levers, screws, and other mechanical instruments, the multiplication of force by which I quite understand.

Salv. You are by no means alone in marveling at the effect of such puzzling events, and at the obscurity of their cause. I thought about these things for some time in vain, my confusion merely growing, until finally, meeting with our Academician, **293** I received double consolation from him—first, by hearing that he, too, had long remained in the same shadows, and second, by his telling me that after he had spent thousands of hours during his life in theorizing and philosophizing about this, he had arrived at some ideas very distant from our first conceptions, and hence novel, and admirable for their novelty. And, since I now know that your curiosity will make you glad to hear those thoughts—which are far from easy to believe—I shall not await your request, but shall tell you of them. As soon as we have read this treatise on projectiles, I shall explain to you all those fantasies, or let us say extravagancies, that stick in my memory from the reasonings of the Academician. Meanwhile let us go on with the Author's propositions.

PROPOSITION V. PROBLEM [II]

In the axis of a given parabola extended [upward], to find a high point from which a falling body describes this same parabola [when deflected horizontally at its vertex].
Let there be a parabola AB *whose amplitude is* HB *and whose axis extended is* HE. *We seek the sublimity from which a falling body, being turned horizontally with the impetus aquired at* A, *describes the said parabola. Draw the horizontal* AG *parallel to* BH, *and putting* AF *equal to* AH, *draw the straight line* FB *tangent to the parabola at* B, *which intersects the horizontal line* AG *at* G. *Take* AE, *the third proportional to* FA *and* AG *; I say that* E *is the high point sought, from which a body falling from rest at* E, *and turned into the horizontal with the impetus acquired at* A, *where there supervenes the impetus of fall to* H *[as if] from rest at* A, *will describe the parabola* AB. *If we assume* EA *to be the measure of time of fall from* E *to* A *and of the impetus acquired at* A, *then* AG, *(that is, the mean proportional between* EA *and* AF*) will be the time and impetus of fall from* F *to* A *or from* A *to* H. *And since [the moveable] coming from* E *in time* EA *with the impetus acquired at* A *will traverse twice* EA *in equable horizontal movement [in time* EA*], movement at that same impetus would also traverse twice* GA *in time* AG, *one-half of* BH *(for the spaces traversed in equable motion are to one another as their times of motion); and in vertical motion from rest, during the same time* GA, *it would traverse* AH. *Therefore the amplitude* HB *and the altitude* AH *would be traversed by the moveable in the same time. Thus the parabola* AB *is described by fall from sublimity* E*; which was to be found.*

294

COROLLARY

From this it follows that one-half the base, or amplitude, of a semiparabola (which is one-quarter the amplitude of the whole parabola) is a mean proportional between its altitude and the sublimity from which a falling [body] would describe it.

PROPOSITION VI. PROBLEM [III]

Given the sublimity and the altitude of a semiparabola, to find its amplitude.
Let DC *be a horizontal line and* AC *a vertical to it, in which are given the altitude* CB *and the sublimity* BA*; it is required to find in the horizontal* CD *the amplitude of the semiparabola*

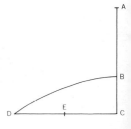

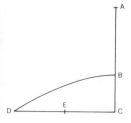

which is determined by the sublimity BA *and the altitude* BC. *The mean proportional between* CB *and* BA *is to be taken, of which the double is put as* CD; *I say that* CD *is the required amplitude. This is manifest from the preceding* [*corollary*].

PROPOSITION VII. THEOREM [IV]

In projectiles by which semiparabolas of the same amplitude are described, less impetus is required for the describing of one whose amplitude is double its altitude than for any other.

Let semiparabola BD *be one whose amplitude* CD *is double its altitude* CB; *and in the axis extended upward, take* BA *equal to the altitude* BC. *Draw* AD, *which will be tangent to the semi-parabola at* D *and will intersect the horizontal* BE *at* E, *while* BE *will be equal to* BC *(or* BA*). It follows that this* [*curve*] *will be described by a projectile whose equable horizontal impetus is that of fall from rest at* A *to* B, *and whose natural downward impetus is that of fall to* C *from rest at* B. *From this it is evident that the impetus compounded from these and impinging on point* D, *is as the diagonal* AE, *equal in the square to both* [CD *and* DB].

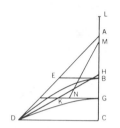

295

Now let some other semiparabola GD *be taken, whose amplitude is the same* CD, *but whose altitude* CG *is less (or greater) than the altitude* BC. *Let* HD *be tangent to this, intersecting the horizontal through* G *at point* K. *Make* KG *to* GL *as* HG *is to* GK; *the sublimity* [altitudo] GL, *as previously demonstrated, will be that from which a falling* [*body*] *describes parabola* GD. *Let* GM *be the mean proportional between* AB *and* GL; *then* GM *will be the time and the momentum or impetus at* G [*after*] *fall from* L; *for it is assumed that* AB *is the measure of time and impetus. Let* GN *be the mean proportional between* BC *and* CG; *this will be the measure of the time and impetus of fall from* G *to* C. *Therefore, joining* MN, *this will be the measure of impetus of a projectile through the parabola* DG, *striking at* D. *This impetus, I say, is greater than the impetus of a projectile through parabola* BD, *of which the quantity was as* AE. *For since* GN *was taken as the mean proportional between* BC *and* CG, *and* BC *is equal to* BE *or* KG *(for each is one-half* DC*), NG is to GK as CG is to GN; and as CG (or HG) is to GK, so the square of NG is to the square of GK. But as HG is to GK, so, by construction, is KG to GL; therefore as NG is to the square of GK, so KG is to GL. But as KG is to GL, so the square of KG is to the square of GM, since GM is the mean proportional between KG and GL. Thus the three squares of NG, KG, and*

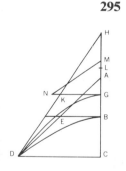

GM *are in continued proportion, and the* [*product of the*] *two extremes* NG *and* GM *(that is, the square of* MN*) is greater than twice the square of* KG*, of which the square is twice* AE. *Hence the square of* MN *is greater than the square of* AE*, and line* MN *is greater than line* EA *; which was to be demonstrated.*

COROLLARY   **296**

From this it is clear that in reverse [*direction*] *through the semiparabola* DB*, the projectile from point* D *requires less impetus that through any other* [*semiparabola*] *having greater or smaller elevation than semiparabola* BD*, which* [*elevation*] *is according to the tangent* AD *and contains one-half a right angle with the horizontal. Hence it follows that if projections are made with the same impetus from point* D*, but according to different elevations, the maximum projection, or amplitude of semiparabola (or whole parabola) will be that corresponding to the elevation of half a right angle. The others, made according to larger or smaller angles, will be shorter* [*in range*].

Sagr. The force of necessary demonstrations is full of marvel and delight; and such are mathematical [demonstrations] alone. I already knew, by trusting to the accounts of many bombardiers, that the maximum of all ranges of shots, for artillery pieces or mortars—that is, that shot which takes the ball farthest—is the one made at elevation of half a right angle, which they call "at the sixth point of the [gunner's] square."[27] But to understand the reason for this phenomenon infinitely surpasses the simple idea obtained from the statements of others, or even from experience many times repeated.

Salv. You say well. The knowledge of one single effect acquired through its causes opens the mind to the understanding and certainty of other effects without need of recourse to experiments. That is exactly what happens in the present instance; for having gained by demonstrative reasoning the certainty that the maximum of all ranges of shots is that of elevation at half a right angle, the Author demonstrates to us something that has perhaps not been observed through

27. An instrument devised by Tartaglia for measuring the elevation of a cannon; see *Mechanics in Italy*, p. 64. It consisted of a rigid right angle having a plumb line suspended from the inside corner, and was read along a quadrant graduated into twelve equal arcs of $7\frac{1}{2}$ degrees each, called "points." A horizontal shot was accordingly called "point blank."

experiment; and this is that of the other shots, those are equal [in range] to one another whose elevations exceed or fall short of half a right angle by equal angles.[28] Thus two balls shot, one at an elevation of $52\frac{1}{2}°$ [7 *punti*] from the horizon, and the other at $37\frac{1}{2}°$ [5 *punti*], strike the ground at equal distances; as do those shot at 60° and 30°, or at $67\frac{1}{2}°$ and $22\frac{1}{2}°$, and so on. Now let us hear the proof.

297

PROPOSITION VIII. THEOREM [V]

The amplitudes of parabolas described by projectiles sent forth with the same impetus, according to elevations having angles equidistant above and below half a right angle, are equal to one another.

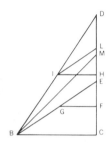

In triangle MCB, *around its right angle* C, *let the horizontal* BC *and the vertical* CM *be equal; angle* MBC *will then be half a right angle. Extend* CM *to* D *and form at* B *two equal angles,* MBC *and* MBD, *above and below the diagonal* MB. *It is to be demonstrated that the amplitudes of the parabolas of projectiles sent forth from point* B *with the same impetus, according to elevations at angles* EBC *and* DBC, *are equal. Indeed, the external angle* BMC *is equal to the [sum of the] internal angles* MDB *and* DBM, *and angle* MBC *will also be equal to these; for if we put angle* MBE *in place of [angle]* DBM, *angle* MBC *will equal the two [angles]* MBE *and* BDC. *And subtracting the common [angle]* MBE, *the remainder* BDC *will equal the remainder* EBC. *Hence triangles* DCB *and* BCE *are similar. Bisect line* DC *at* H, *and* EC *at* F; *and draw* HI *and* FG *parallel to the horizontal* CB. *As* DH *is to* HI, *so* IH *is to* HL; *and triangle* IHL *will be similar to triangle* IHD, *to which* EGF *is also similar. And as* IH *and* GF *are equal (that is, are half of* BC), FE *(that is,* FC*) will be equal to* HL. *Adding* FH *in common,* CH *will be equal to* FL. *Through* H *and* B *describe a semiparabola of altitude* HC *and sublimity* HL; *its amplitude will be* CB, *which is twice* HI, *the mean proportional between* DH *(or* CH*) and* HL; *and* DB *will be tangent to it, since* CH *is equal to* HD. *If we further describe the parabola through* FB *with sublimity* FL *and altitude* FC, *between which* FG *is the*

298   *mean proportional, then the horizontal* CB *being twice this,* CB *will likewise be its amplitude, and* EB *[will be] tangent to it,*

28. Tartaglia had implied a knowledge of this, however, when he declared the maximum range to be attained at elevation of 45°; see *Mechanics in Italy*, pp. 85–86, 91–94.

since EF *and* FC *are equal. But angles* DBC *and* EBC, *which are the elevations, are equidistant from half a right angle; therefore the proposition is evident.*

PROPOSITION IX. THEOREM [VI]

The amplitudes of parabolas are equal when their altitudes and sublimities are inversely proportional.

Let the altitude GF *of parabola* FH *have to altitude* CB *of parabola* BD *the same ratio that the sublimity* BA *[of the latter] has to the sublimity* FE *[of the former]; I say that amplitudes* HG *and* DC *are equal. For since the first,* GF, *has to the second,* CB, *the same ratio that the third,* BA, *has to the fourth,* FE, *the rectangle* GF–FE *of the first and fourth will be equal to the rectangle* CB–BA *of the second and third. Therefore the squares to which these rectangles are [respectively] equal are equal to each other. But rectangle* GF–FE *is equal to the square of one-half* GH, *and rectangle* CB–BA *is equal to the square of one-half* CD; *hence these squares, and their sides, and the doubles of their sides, are [respectively] equal. But those [last named] are the amplitudes* GH *and* CD; *therefore the proposition is evident.*

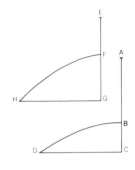

LEMMA FOR THE NEXT [THEOREM]

If a straight line is cut anywhere, the squares of the mean proportionals between the whole and the parts are equal [in sum] to the square of the whole.

Let AB *be cut anywhere at* C; *I say that the squares of the mean proportional lines between the whole* AB *and its parts* AC *and* CB, *taken together, are equal to the square of the whole* AB. *This is evident; for describe a semicircle on the whole of* AB, *and and from* C *erect the perpendicular* CD. *Join* DA *and* DB. *Then* **299** DA *is the mean proportional between* BA *and* AC, *and* DB *is the mean proportional between* AB *and* BC; *and the squares of lines* DA *and* DB *taken together are equal to the square of all* AB, *angle* ADB *being a right angle inscribed in a semicircle; whence the proposition is evident.*

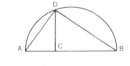

PROPOSITION X. THEOREM [VII]

The impetus or momentum of [fall through] any semi-parabola is equal to the momentum of natural vertical fall to the horizontal through the combined sublimity and altitude of the semiparabola.

Let there be the semiparabola AB *with sublimity* DA *and*

altitude AC, *which* [*together*] *make up the vertical* DC; *I say that the impetus of the semiparabola at* B *is equal to the momentum of natural fall from* D *to* C.

Take DC *as the measure of the time and impetus* [*of the said fall*], *and assume* CF *to be the mean proportional between* CD *and* DA; *let* CE *be the mean proportional between* DC *and* CA; *then* CF *will be the measure of the time and of the momentum of that which descends through* DA *from rest at* D, *while* CE *will be the time and momentum of that which descends through* AC *from rest at* A; *and the diagonal* EF *will be the momentum at* B (*compounded from these*) *of* [*fall through*] *the semiparabola. Since* DC *was cut anywhere at* A, *while* CF *and* CE *are mean proportionals between all* CD *and its parts* DA *and* AC, *the squares of these, taken together, will equal the square of the whole, by the foregoing lemma. But the same squares are also equal to the square of* EF; *hence line* EF *is equal to* DC. *From this it is evident that the momenta through* DC *at* C, *and through the semiparabola* AB *at* B, *are equal; which was required.*

COROLLARY

From this it is evident that for all semiparabolas whose combined altitudes and sublimities are matched, the [*related*] *impetuses* [*of fall*] *are also equal.*

PROPOSITION XI. PROBLEM [IV]

Given the impetus and the amplitude of a semiparabola, to find its altitude.

Let the given impetus be defined by the vertical AB, *and let the amplitude be* BC *in the horizontal; it is required to find the*

300   *sublimity of the semiparabola whose impetus is* AB *and whose sublimity is* BC. *It is evident from what has been demonstrated that one-half the amplitude* BC *will be the mean proportional between the altitude and the sublimity of the semiparabola, the impetus of which (from the preceding) is the same as that of fall from rest at* A *through all* AB. *For that reason, cut* BA *so that the rectangle contained by the parts shall be equal to the square of one-half* BC, *which is* BD. *From this it is clearly necessary that* DB *cannot exceed one-half of* BA; *for the maximum rectangle contained by parts* [*of a line*] *is* [*obtained*] *when the line is bisected. Bisect* BA *at* E. *Now, if* BD *is equal to* BE, *the work is finished,* [*and*] *the altitude of the semiparabola will be* BE *and its sublimity* EA. *The amplitude of such a parabola, of elevation one-half a right angle, is the maximum of*

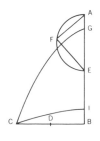

*all those that are described with the same impetus, as demon-
strated above.*

But let BD *be less than one-half* BA, *which is so cut that the
rectangle of its parts is equal to the square of* BD. *Describe a
semicircle on* EA, *take* AF *equal to* BD, *and connect* FE *equal to
one part cut,* EG. *The rectangle* BG–GA *plus the square of* EG
now equals the square EA, *which also equals square* AF *plus
[square]* FE. *Subtracting the equal squares* GE *and* EF, *there
remains the rectangle* BG–GA *equal to the square of* AF *(that
is, of* BD *) ; and line* BD *is the mean proportional between* BG
and GA. *It follows that the semiparabola whose amplitude is* BC,
and whose impetus [of fall] is AB, *has the altitude* BG *and the
sublimity* GA. *Now if* BI, *equal to* GA, *is taken below, [then]*
BI *will be the altitude and* IA *the sublimity of semiparabola* IC.

From what has been demonstrated, we can now:

PROPOSITION XII. PROBLEM [V]

*Calculate and compile tables of all amplitudes of semi-
parabolas described by projectiles sent forth with the same
impetus.*

*It follows from what has been demonstrated that when
parabolas are described by projectiles having the same [initial]
impetus, the sublimities and altitudes [thereof], added together,
comprise equal verticals; whence those verticals must be included
between the same horizontal parallels. Thus take the horizontal*
CB, *equal to the vertical* BA, *and draw the diagonal* AC; *angle*
ACB *will be half a right angle, or* 45°. *Bisect the vertical* BA *at*
D; *the semiparabola* DC *will be that which is determined by the
sublimity* AD *plus the altitude* DB, *and its impetus at* C *will be as
much as that of a moveable at* B *coming from rest at* A *through
line* AB. *If* AG *is drawn parallel to* BC, *the combined altitudes
and sublimities of all the rest of the semiparabolas whose impetus
is the same as that just described must lie between the parallels*
AG *and* BC. *Furthermore, as already demonstrated, the semiparabolas of which the tangents [at the base] are equidistant
above and below the elevation of half a right angle are equal in
amplitude, so that the calculation listed for the greater [of such
matched] elevations will serve also for the smaller.*

*Let us select ten thousand (10000) for the number of parts in
the maximum amplitude of projection for a parabola of elevation*
45°, *and assume that line* BA *and amplitude* BC *of the semiparabola are of that many parts. (We chose the number 10000
because in our calculations we use a table of tangents in which*

301

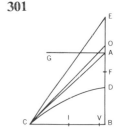

*that number corresponds to the tangent of 45°). Now, commenc-
ing the task, draw* CE *at angle* ECB, *greater than* ACB *though
still acute; and, tangent to* EC, *let a semiparabola be drawn
whose sublimity and altitude together equal* BA. *From our table
of tangents, for the given angle* BCE, *we find the tangent* BE;
bisect this at F, *and then find the third proportional to* BF *and*
BI *(one-half* BC*), which will necessarily be greater than* FA.

302 *Let it be* FO. *Then, for the semiparabola inscribed in triangle*
ECB *and tangent to* CE, *of which the amplitude is* CB, *we find
the altitude* BF *and the sublimity* FO.

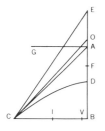

  But the whole line BO *goes above the parallels* AG *and* CB,
*whereas we require [a line] confined between them; for only in
that way will both this [line] and the semiparabola* DC *be traced
out by projectiles sent forth from* C *with the same impetus.
Hence another, similar to this, is to be found among the innumerable [semiparabolas], larger and smaller, that can be
designed within angle* BCE, *whose combined sublimity and
altitude (homologous, that is, with* BC*) are equal to* BA. *Therefore let amplitude* BC *be to* CR *as* OB *is to* BA, *and* CR *will be
found to be the amplitude of the semiparabola having elevation
at angle* BCE, *while its combined sublimity and altitude are
comprised within and are equal to the space between the parallels*
GA *and* CB; *which is what was sought. The operation is as
follows.*

  Take the tangent of the given angle BCE; *to one-half of this,
add the third proportional to this [half] and one-half* BC, *which
[third proportional] is* FO. *Then, as* OB *is to* BA, *make* BC *to
some other [line]* CR, *and this is the amplitude sought.*

  Let us do an example. Let angle ECB *be 50°; its tangent is
11918, of which one-half (or* BF*) is 5959; one-half* BC *is 5000.
To these [two] halves, the third proportional is 4195, which added
to* BF *gives 10154 for* BO. *Next, as* OB *is to* BA *(that is, as
10154 is to 10000), make* BC, *or 10000 (for both [*BA *and* BC*]
are [equal to] the tangent of 45°), to some other [magnitude],
and we shall find the required amplitude,* RC, *to be 9848, the
maximum amplitude (*BC*) being 10000. The amplitudes of the
whole parabolas are the doubles of these; that is, 19696 and
20000. This [calculation] also gives the amplitude of the parabola
of elevation 40°, which is at the same distance from 45°.*

303 Sagr. For complete understanding of this demonstration, I
still lack knowledge of why the Author says it is true that the
third proportional to *BF* and *BI* must be greater than *FA*.

[TABLE 1]   [TABLE 2]   **304**

Amplitudes of semiparabolas described with the same initial speed.

Altitudes of semiparabolas described with the same initial speed.

Angle of Elevation		Angle of Elevation
45°	10000	
46	9994	44°
47	9976	43
48	9945	42
49	9902	41
50	9848	40
51	9782	39
52	9704	38
53	9612	37
54	9511	36
55	9396	35
56	9272	34
57	9136	33
58	8989	32
59	8829	31
60	8659	30
61	8481	29
62	8290	28
63	8090	27
64	7880	26
65	7660	25
66	7431	24
67	7191	23
68	6944	22
69	6692	21
70	6428	20
71	6157	19
72	5878	18
73	5592	17
74	5300	16
75	5000	15
76	4694	14
77	4383	13
78	4067	12
79	3746	11
80	3420	10
81	3090	9
82	2756	8
83	2419	7
84	2079	6
85	1736	5
86	1391	4
87	1044	3
88	698	2
89	349	1

Angle of Elevation		Angle of Elevation	
1°	3	46°	5173
2	13	47	5346
3	28	48	5523
4	50	49	5698
5	76	50	5868
6	108	51	6038
7	150	52	6207
8	194	53	6379
9	245	54	6546
10	302	55	6710
11	365	56	6873
12	432	57	7033
13	506	58	7190
14	585	59	7348
15	670	60	7502
16	760	61	7649
17	855	62	7796
18	955	63	7939
19	1060	64	8078
20	1170	65	8214
21	1285	66	8346
22	1402	67	8474
23	1527	68	8597
24	1685	69	8715
25	1786	70	8830
26	1922	71	8940
27	2061	72	9045
28	2204	73	9144
29	2351	74	9240
30	2499	75	9330
31	2653	76	9415
32	2810	77	9493
33	2967	78	9567
34	3128	79	9636
35	3289	80	9698
36	3456	81	9755
37	3621	82	9806
38	3793	83	9851
39	3962	84	9890
40	4132	85	9924
41	4302	86	9951
42	4477	87	9972
43	4654	88	9987
44	4827	89	9998
45	5000	90	10000

Salv. I think that consequence may be deduced in this manner. The square of the mean of three proportional lines is equal to the rectangle of the extremes, whence the square of *BI* (or of its equal, *BD*) must be equal to the rectangle of the first, *FB*, and the third, to be found; and this third must necessarily be greater than *FA*, because the rectangle *BF–FA* is less than the square of *BD*, the deficiency being the square of *DF*, as Euclid proves in a proposition of his second [book].[29]

You must also note that point *F*, which bisects the tangent *EB*, will as often fall above point *A* [as beneath it], and also, once, at *A* itself; in the former cases, it is self-evident that the third proportional to the half-tangent and *BI*, which gives the sublimity, is entirely above *A*. But the Author has taken the case in which it was not manifest that the said third proportional is always greater than *FA*, and hence passes beyond the parallel *AG* when added above point *F*. But now let us go on.

It will be useful, with the help of this tabulation, to complete another in which are compiled the altitudes of the same semiparabolas of projectiles sent forth with the same impetus. The construction of this [other table] follows.

305   PROPOSITION XIII. PROBLEM [VI]

From the amplitudes of the semiparabolas gathered in the previous table, and preserving the common impetus with which each is described, to obtain the respective altitudes of individual semiparabolas.

Let the given amplitude be BC, *and the measure of impetus (assumed to be always the same),* OB *; that is, the sum of [each] altitude and [the associated] sublimity; it is required to find and distinguish the altitude itself. This will be done when* BO *is so divided that the rectangle of its parts shall be equal to the square of one-half the amplitude* BC. *Let this division fall at* F, *and bisect both* OB *and* BC, *at* D *and* I. *Then the square of* IB *is equal to the rectangle* BF–FO, *and the square of* DO *is equal to the same rectangle plus the square of* FD; *if therefore from the square of* DO *there is taken the square of* BI *(equal to rectangle* BF–FO*), the square of* FD *will remain. The side of this,* DF, *added to line* BD, *will give the required altitude,* BF. *This is arranged from the things given, as follows:*

From the square of one-half BO, *take the square of* BI, *also*

29. *Elements* II.5.

known; extract the square root of the remainder, and add BD,
[which is] known, and you will have the required altitude, BF.

*To be found is the altitude of the semiparabola described
at elevation 55°. The amplitude, from the preceding tabulation,
is 9396; half of this is 4698, of which the square is 22071204.
Subtract this from the square of half* BO, *which is 25000000*   **306**
*(and is always the same); the remainder is 2928796, of which
the square root is approximately 1710. This, added to half* BO
(that is, 5000), gives 6710, which is the altitude BF.

*It will be useful to give a third table, containing the altitudes
and sublimities of semiparabolas of which the amplitude will be
be the same.*

Sagr. I shall be happy to see this, since by it I may come to
know the difference of the impetuses and forces required in
shooting the projectile to the same distance, using what are
called "ranging shots." I believe that this difference is very
great for the various elevations, so that, for instance, if we
wished to use an elevation of three or four degrees, or of 87°
or 88°, and [still] make the ball fall where it went when shot at
an elevation of 45°, which has been shown to require the
minimum impetus, then I believe that an immense excess of
force would be required.

Salv. You are quite right, and you will see that in order to
carry out the entire operation, for all elevations, one is rapidly
driven toward infinite impetus. Now let us look at the construction of the [ensuing] table.

PROPOSITION XIV. PROBLEM [VII]   **308**

*To find the altitudes and sublimities of semiparabolas of
which the amplitudes shall be equal, for individual degrees
of elevation.*

We shall obtain all this by means of an easy procedure.
Let the amplitude of the semiparabolas be always 10000 parts;
then one-half the tangent, for any degree of elevation, gives the
altitude. For example, let the elevation of the semiparabola be
30°, and its amplitude, as assumed, 10000 parts; its altitude
will be 2887, for that is approximately one-half the tangent
[of 30°]. And the altitude having been found, we get the sublimity as follows. As has been demonstrated, half the amplitude
of a semiparabola is the mean proportional between the altitude

307   [TABLE 3]

Giving the altitudes and sublimities of parabolas of constant amplitude, namely 10000, computed for each degree of elevation.

Angle of Elevation	Altitude	Sublimity	Angle of Elevation	Altitude	Sublimity
1°	87	286533	46°	5177	4828
2	175	142450	47	5363	4662
3	262	95802	48	5553	4502
4	349	71531	49	5752	4345
5	437	57142	50	5959	4196
6	525	47573	51	6174	4048
7	614	40716	52	6399	3906
8	702	35587	53	6635	3765
9	792	31565	54	6882	3632
10	881	28367	55	7141	3500
11	972	25720	56	7413	3372
12	1063	23518	57	7699	3247
13	1154	21701	58	8002	3123
14	1246	20056	59	8332	3004
15	1339	18663	60	8600	2887
16	1434	17405	61	9020	2771
17	1529	16355	62	9403	2658
18	1624	15389	63	9813	2547
19	1722	14522	64	10251	2438
20	1820	13736	65	10722	2331
21	1919	13024	66	11230	2226
22	2020	12376	67	11779	2122
23	2123	11778	68	12375	2020
24	2226	11230	69	13025	1919
25	2332	10722	70	13237	1819
26	2439	10253	71	14521	1721
27	2547	9814	72	15388	1624
28	2658	9404	73	16354	1528
29	2772	9020	74	17437	1433
30	2887	8659	75	18660	1339
31	3008	8336	76	20054	1246
32	3124	8001	77	21657	1154
33	3247	7699	78	23523	1062
34	3373	7413	79	25723	972
35	3501	7141	80	28356	881
36	3633	6882	81	31569	792
37	3768	6635	82	35577	702
38	3906	6395	83	40222	613
39	4049	6174	84	47572	525
40	4196	5959	85	57150	437
41	4346	5752	86	71503	349
42	4502	5553	87	95405	262
43	4662	5362	88	143181	174
44	4828	5177	89	286499	87
45	5000	5000	90	infinity	[zero]

and the sublimity. Hence, the altitude having been already found, and half the amplitude being always the same (that is, 5000 parts), then if the square of this is divided by the given altitude, the required sublimity results.

As in our example above, the altitude was 2887; the square of 5000 parts is 25000000; divided by 2887, this gives approximately 8659 for the sublimity sought.

Salv. Here we see, first of all, how true is that conception mentioned earlier: that in different elevations, the farther we depart from the middle one, whether [by going] higher or lower, the greater is the impetus and violence required for shooting the projectile to the same distance. For the impetus consists of the mixture of two motions, an equable horizontal motion and a vertical, naturally accelerated; and this impetus, coming to be measured by the sum of the altitude and the sublimity, you see from the table that this sum is a minimum at the elevation of 45°, where the altitude and sublimity are equal, each being 5000 and their sum being 10000. For if we look at some other, greater, altitude, say for example [at elevation] of 50°, we shall find this altitude to be 5959, and the sublimity 4196, which added together make 10155. And that much, likewise, we shall find to be the impetus for [an elevation of] 40°, the two elevations being equally distant from the middle [elevation of 45°].

In the second place we should note that it is true that equal impetuses are required for elevations equally distant, two by two, from the middle; and with this pleasing additional alternation, that the altitudes and sublimities of the upper elevations are inverse to the sublimities and altitudes of the lower. Thus in the example given, for an elevation of 50° **309** the altitude is 5959 and the sublimity 4196, while for an elevation of 40°, it turns out the other way; the altitude is 4196, and the sublimity 5959. The same happens with all the rest, without any difference except where, in order to escape the tedium of calculation, we leave fractions out of account; in sums so great, these are of no moment or prejudice whatever.

Sagr. I observe that with regard to the two impetuses, horizontal and vertical, as the projectile is made higher, less is required of the horizontal, but much of the vertical. On the other hand, in shots of low elevation there is need of great force in the horizontal impetus, since the projectile is shot to so small a height. But if I understand correctly, at full elevation of 90°, all the force in the world would not suffice to shoot the projectile one single inch out of the vertical, and it must necessarily fall back at the same place from which it was shot. Yet I dare not affirm with equal certainty that a projectile, even at zero elevation, which is to say in the horizontal line, could not be shot to some [little] distance by some [great] force, or that infinite force would be required—as if, for example, not even a culverin had the power to shoot an iron ball horizontally, or "point blank" as they say (that is, at no point [on the gunner's square]), where there is zero elevation. I say that in this case there remains some ambiguity, and that I am unable to deny resolutely either fact, for the reason that another event seems no less strange, though I have a logically conclusive demonstration of it. This is the impossibility of stretching a rope so [tightly] that it shall be pulled straight, and [held] parallel to the horizontal; for it always sags and bends, nor is there any force that will suffice to hold it straight.

Salv. Well, Sagredo, in this matter of the rope, you may cease to marvel at the strangeness of the effect, since you have a proof of it; and if we consider well, perhaps we shall find some relation between this event of the rope and that of the projectile [fired horizontally].

The curvature of the line of the horizontal projectile seems to derive from two forces, of which one (that of the projector) drives it horizontally, while the other (that of its own heaviness) draws it straight down. In drawing the rope, there is [likewise] the force of that which pulls it horizontally, and also that of the weight of the rope itself, which naturally inclines it downward. So these two kinds of events are very similar. Now, if you give to the weight of the rope such power and energy as to be able to oppose and overcome any immense force that wants to stretch it straight, why do you want to deny this [power] to the weight of the ball?

But I wish to cause you wonder and delight together by telling you that the cord thus hung, whether much or little stretched, bends in a line that is very close to parabolic. The similarity is so great that if you draw a parabolic line in a vertical plane surface but upside down—that is, with the vertex down and the base parallel to the horizontal—and then hang a little chain from the extremities of the base of the parabola thus drawn, you will see by slackening the little chain now more and now less, that it curves and adapts itself to the parabola; and the agreement will be the closer, the less curved and the more extended the parabola drawn shall be. In parabolas described with an elevation of less than 45°, the chain will go almost exactly along the parabola.[30]

Sagr. Then with a chain wrought very fine, one might speedily mark out many parabolic lines on a plane surface.

Salv. That can be done, and with no little utility, as I am about to tell you.

Simp. But before you go on, I also wish at least some assurance about that proposition of which you said that there is a necessarily conclusive demonstration—I mean of the impossibility, by any immense force, of making a rope stay stretched straight and parallel to the horizon.

Sagr. Let me see if I remember the demonstration. To understand it, Simplicio, it is necessary that you assume as true something which is verified in all mechanical instruments, and not only by experience, but by demonstration as well. This is that the speed of the moving thing, though it be [one] of very weak force, can overcome the resistance, though great, of something that can be moved only slowly, provided only that the speed of the moving thing have a greater ratio to the slowness of the resistent than the resistance of that which is to be moved has to the force of the moving thing.

Simp. This is well known to me, and is demonstrated by Aristotle in his *Questions of Mechanics*,[31] and it is plainly seen in the lever. Again, in the steelyard, the counterweight that weighs no more than four pounds will lift a weight of 400, when the distance of the counterweight from the center on   **311** which the steelyard turns is more than one hundred times the distance from that center to the point from which the great weight hangs. This happens because the counterweight, in its descent, goes more than one hundred times the distance through which the great weight rises in the same time.[32] This is the same as saying that the little counterweight moves with more than one hundred times the speed of the great weight.

30. The ensuing discussion more or less duplicates that toward the end of the Second Day, suggesting that this section of the final dialogue may have been originally intended to conclude the discussion of strength of materials; cf. note 42 to Second Day.

31. *Questions of Mechanics*, 20 (Loeb ed., pp. 375–77).

32. It was this emphasis on operation "in the same time" that separated Galileo's approach on the one hand from medieval statics and on the other hand from the mechanics of Descartes, who considered displacements alone to be truly relevant in mechanical theory and who ridiculed those who, like Galileo, considered the role of speeds essential in the theory of simple machines; cf. p. **329**.

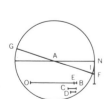

Sagr. You reason very well, and doubtless you will grant that however small the force of the mover, it will overcome any resistance, however great, whenever the mover exceeds the resistance in speed by more than it falls short of it in vigor and in heaviness.

Now we come to the case of the rope, drawing a diagram. Assume that this line *AB*, passing over the two fixed and stable points *A* and *B*, has hanging at its ends, as you see, two immense weights, *C* and *D*. These, drawing it with very great force, really do make it stay stretched straight if this [*AB*] is a simple line without any heaviness. But here something needs to be added. I say that if from point *E*, at the center of this [line *AB*], you suspend some small weight such as *H* here, line *AB* will yield and will tend toward point *F*. Being thus lengthened [without stretching], it will constrain the two very heavy weights, *C* and D, to rise. I demonstrate this as follows.

Around the two points, *A* and *B*, as centers, describe two quadrants, *EIG* and *ELM*; since the two radii, *AI* and *BL*, are equal to *AE* and *EB*, the advances *FI* and *IL* will be the quantity of lengthening of the parts *AF* and *FB* beyond *AE* and *EB*. Hence these determine the rises of the weights *C* and *D*, provided that weight *H* shall have had the ability to go down in [the direction *E*]*F*. This could happen if line *EF*, which is the quantity of descent of weight *H*, had a greater ratio to line *FI* (which determines the rise of the two weights *C* and *D*) than the heaviness of these weights has to the heaviness of weight *H*. But that will necessarily be the case, no matter how great the heaviness of weights *C* and *D*, or how small that of *H*. For the excess of weights *C* and *D* over weight *H* is not so great that the excess of tangent *EF* over secant *FI* is not in greater ratio.[33] We prove this as follows.

Let there be the circle whose diameter is *GAI*; and whatever the ratio of the heaviness of weights *C* and *D* to the heaviness of *H*, let line *BO* have this [ratio] to some other line, *C*. Let *D* be less than *C*, so that *BO* will have a greater ratio to *D* than to *C*. Now take *DE* as the third proportional to *OB* and *D*, and as *OE* is to *EB*, make the diameter *GI* to *IF* (prolonging *GI*). From end *F*, draw the tangent *FN*. Now since, by construction, *GI* is to *IF* as *OE* is to *EB*, then, by composition, as *OB* is to *BE*,

312

---

33. The germinal idea here resembles that of the later concept of infinitesimals of higher order. Galileo had used this notion before, in a different connection; cf. *Dialogue*, pp. 199–202 (*Opere*, VII, 225–29).

so *GF* is to *FI*. But *D* is the mean proportional between *OB* and *BE*, and *NF* is that between *GF* and *FI*. Therefore *NF* has to *FI* the same ratio that *OB* has to *D*, which ratio is greater than that of the weights *C* and *D* to weight *H*. The descent or speed of weight *H* having therefore a greater ratio to the rise or speed of the weights *C* and *D*, than the heaviness of these weights *C* and D has to the heaviness of weight *H*, it is clear that weight *H* will descend; that is, the line *AB* will depart from horizontal straightness.

And what happens to the straight line *AB*, devoid of heaviness, when there is attached at *E* any minimal weight *H*, happens also to the rope *AB* made of weighty material, without the addition of any further heavy object, since on it is suspended the weight of the material of the rope *AB* itself.

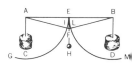

Simp. I am fully satisfied. And now Salviati, in agreement with his promise, shall explain to us the utility that may be drawn from the little chain, and afterward give us those speculations made by our Author about the force of impact.

Salv. Sufficient to this day is our having occupied ourselves in the contemplations now finished. The time is rather late, and will not, by a large margin, allow us to explain the matters you mention; so let us defer that meeting to another and more suitable time.

Sagr. I concur with you opinion. From what I have heard in my various discussions with close friends of our Academician, this matter of the force of impact is very obscure, nor have its recesses been penetrated by anyone who has treated of it. It **313** is filled with shadows, and is completely alien to men's first impressions [*prime immaginazioni*]. Among the conclusions I have heard offered, a very extravagant one sticks in my mind, which is that the force of impact is unbounded, not to say infinite. We shall therefore await Salviati's convenience. But meanwhile, tell me; what are those things I see written there after the treatise on projectiles?

Salv. These are some propositions pertaining to the center of gravity of solids which our Academician discovered in his youth, when it appeared to him that there were still some defects in what had been left written on the subject by Federico Commandino.[34] He thought that these propositions which

---

34. Federico Commandino (1509–75), *Liber de centro gravitatis solidorum* (Bologna, 1565), a work in the strict Archimedean tradition, written to supplement the ancient treatise *On Plane Equilibrium*. Commandino had been the teacher of Galileo's patron, Guidobaldo del Monte.

you see written here might supply that which Commandino's book left to be desired, and he applied himself to this study at the instance of the illustrious Marquis Guidobaldo del Monte, a very great mathematician of his time as shown by his various published works. Our Author gave a copy of these to that gentleman, intending to pursue the subject for other solids not touched on by Commandino. But some time later, he ran across the book of Luca Valerio, a prince of geometers, and saw that this resolved the entire subject without omitting anything; hence he went no further, though his own advances were made along quite a different road from that taken by Valerio.[35]

Sagr. It will be good, then, in the time between our meetings just concluded and those in the future, for you to leave this book in my hands. Thus I may look at in the meantime, and study one by one the propositions written there.

Salv. Very gladly do I yield to your request, and hope that you will take pleasure in these propositions.

[*The Fourth Day Ends*]

35. Cf. notes 20 to First Day and 37 to Second Day.

Appendix

In which are contained theorems and related demonstrations concerning the center of gravity of solids, written earlier by the Author[1]

Opere, I

187

POSTULATE

We assume that, of equal weights similarly arranged on different balances, if the center of gravity of one composite [of weights] divides its balance in a certain ratio, then the center of gravity of the other composite also divides its balance in the same ratio.

LEMMA

Let line AB *be bisected at* C, *and the half* AC *be divided at* E *so that the ratio of* BE *to* EA *is that of* AE *to* EC. *I say that* BE *is double* EA.

A   E   C   B

Indeed, since *EA* is to *EC* as *BE* is to *EA*, we shall have, by composition and permutation [of ratios], *AE* to *EC* as *BA* is to *AC*; but as *AE* is to *EC* (that is, as *BA* is to *AC*), *BE* is to *EA*; whence *BE* is double *EA*.

These things granted, it is to be demonstrated [that]:

[PROPOSITION 1]

If any number of magnitudes equally exceed one another, the

---

1. These theorems date, in part at least, from the period 1585–87. The last proposition and its lemma appear to have been written first, having been submitted by Galileo with an application for a position at the University of Bologna in 1587. Early in the next year he corresponded with Christopher Clavius and Guidobaldo del Monte about the first proposition. The others may have been done in response to encouragement from the latter and from Michael Coignet (1544–1623) at that time. A plan to publish this work in 1613 was postponed; cf. note 37 to Second Day. In the original printing the lemmas, theorems, and corollaries were not numbered, and they were not always clearly distinguished typographically; both have been done here for ease of reference.

excesses being equal to the least of them, and they are so arranged on a balance as to hang at equal distances, the center of gravity of all these divides the balance so that the part on the side of the smaller [magnitudes] is double the other part.

188

Thus, on balance AB, let hang at equal distances any number of magnitudes F, G, H, K, N, such as described above, of which the least is N; let the points of suspension be A, C, D, E, B, and let X be the center of gravity of all the magnitudes thus arranged. It is to be shown that the part of the balance BX, on the side of the lesser magnitudes, is double XA, the other part.

Bisect the balance at point D, which lies either at some point of suspension, or necessarily falls midway between two suspension points. The remaining distances between suspension [points], A and [C, C and] D, are to be bisected at points M and I, and all the magnitudes are to be divided into parts equal to N. Then the number of parts of F will be equal to the number of magnitudes that hang from the balance, while the parts of G will be one fewer, and so on for the rest. Thus the parts of F are N, O, R, S, T; those of G [are] N, O, R, S; those of H [are] N, O, R; and finally the parts of K are N and O. All the parts marked N are then equal to [those in] F; all the parts marked O will be equal to G; those marked R will be equal to H; those marked S will be equal to K; and finally the magnitude T is equal to N.

Since all the magnitudes marked N are equal to one another, their point of balance will be at D, which bisects the balance AB. For the same reason, the point of balance for all the magnitudes marked O is at I; of those marked R, it is at C; those marked S have their point of balance at M, while finally T is hung at A. Thus along the balance AD, [considered as separated from DB], there are hung, at the equal distances D, I, C, M, A, magnitudes that equally exceed one another and whose excess is equal to the least thereof. But [of these] the greatest [magnitude], composed of all the N's, hangs [as if] from D, while the least (that is, T) hangs from A, and the others are all arranged in order.

And again, there is the other balance AB on which corresponding magnitudes are arranged in the same order [though reversed], equal in number and sizes to the foregoing. Wherefore we see the balances AB and AD divided in the same ratio by the centers [of gravity] of all the magnitudes compounded. But the center of gravity of the said magnitudes [so arranged] is X;[2] therefore X divides the balances BA and AD in the same ratio, in such a way that as BX is to XA, so XA is to XD. Therefore BX is double XA, by the above lemma. Q.E.D.

[PROPOSITION 2]

If to a parabolic conoid one figure is inscribed and another **189**
is circumscribed, [both] of cylinders having equal height, and the axis of the conoid is divided in such a way that the part toward the apex is double the part toward the base, the center of gravity of the inscribed figure will be closer to the base of the section than [will] the said division point, while the center of gravity of the circumscribed figure will be farther than that same point from the base of the conoid; and the distance from that point of each of the two centers will be equal to the line that is one-sixth the height of one of the cylinders of which the figures are constructed.

Let there be a parabolic conoid and the said figures, one inscribed and the other circumscribed; let the axis of the conoid be AE, divided at N so that AN is double NE. It is to be shown that the center of gravity of the inscribed figure lies in line NE, while that of the circumscribed figure lies in AN.

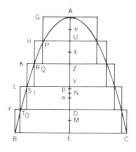

Let the figures thus arranged be cut by a plane through the axis, and let the parabola BAC be cut, the [inter]section of the cutting plane with the base of the conoid being line BC; the sections of the cylinders are rectangular figures, as appears in the diagram.

The first inscribed cylinder, of which the axis is DE, has to the cylinder of which the axis is DY the same ratio that the square [on] ID has to the square [on] SY, which is [in turn] as DA is to AY.[3] The cylinder of which the axis is DY is, moreover, to the cylinder YZ as the square on SY is to the square on RZ, which is as YA to AZ; and for the same reason the cylinder of which the axis is ZY, to that of which the axis is ZU, is as ZA is to AU. Thus the said cylinders are to one

---

2. Both Clavius and Guidobaldo (note 1, above) believed this assumption to beg the question. The latter was satisfied by Galileo's explanation, sent to him in 1588 with a redrawn diagram showing all the weights as touching horizontally; cf. p. **198**.

3. It was a well known property of the parabola that the squares on the abscissae are in the ratio of the ordinates, but cf. note 4, below.

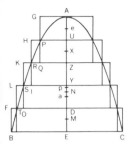

another as the lines *DA, AY, ZA, AU*; but these [lines] equally exceed one another, and the excess is equal to the least of them; hence *AZ* is the double of *AU, AY* is its triple, and *DA* its quadruple. Therefore the said cylinders are magnitudes equally exceeding one another, whose excess is equal to the least of them. Moreover, line *XM* is that along which these are hung at equal distances (indeed, each cylinder has its center of gravity at the midpoint of its own axis); whence, by the things previously demonstrated, the center of gravity of the magnitude composed of all [these] magnitudes divides the line *XM* so that the part toward *X* is double the remainder. Let it be divided thus, and let *Xa* be double *aM*; then point *a* is the center of gravity of the inscribed figure.

190

Let *AU* be bisected at point *e; eX* will be double *ME*; but *Xa* is double *aM*, whence *eE* will be triple *Ea*. Further, *AE* is triple *EN*; thus it is clear that *EN* is greater than *Ea*, and for that reason point *a*, which is the center of the inscribed figure, more nearly approaches to the base of the conoid than [does] *N*. And since as *AE* is to *EN*, so the removed part *eE* is to the removed part *Ea*, the remainder will be to the remainder (that is, *Ae* [will be] to *Na*) as *AE* is to *EN*. Therefore *aN* is one-third of *Ae*, and one-sixth of *AU*.

Further, the cylinders of the circumscribed figure will be shown in the same way to exceed one another equally, the excess being equal to the least of them, and to have their centers of gravity equidistant along line *eM*. Hence if *eM* is divided at *p* so that *ep* is double the remainder *pM*, then *p* will be the center of gravity of the whole circumscribed magnitude; and since *ep* is double *pM*, and *Ae* is less than double *EM* (for these are equal), all *AE* is less than triple *Ep*; whence *Ep* will be greater than *EN*. And since *eM* is triple *Mp*, and *ME* plus double *eA* is likewise triple *ME*, all *AE* plus *Ae* will be triple *Ep*. But *AE* is triple *EN*, so the remainder *Ae* will be triple the remainder *pN*. Therefore *Np* is one-sixth of *AU*. But these were the things to be proved. And from this it is manifest that:

[COROLLARY]

To a parabolic conoid, one figure may be inscribed and another circumscribed so that their centers of gravity may be made less distant from N *than any assigned length.*

In fact, if a line is taken six times the assigned length,

and the axes of the cylinders composing those figures are made less than the said line, then the distances between the [respective] centers of gravity of these [two] figures and the point *N* will [both] be less than the assigned line.

The same [proposition], otherwise [demonstrated]:

Let *CD* be the axis of a conoid, so divided at *O* that *CO* is double *OD*. It must be shown that the center of gravity of the inscribed figure lies in *OD*, while the center of the circumscribed [figure] lies in *CO*.

As above, the figures are intersected by a plane through the axes and through *C*. Now, cylinders *SN*, *TM*, *VI*, and *XE* are to one another as the squares on lines *SD*, *TN*, *VM*, and *XI*; and these are to one another as are lines *NC*, *CM*, *CI*, and *CE*, which moreover exceed one another equally, and this excess is equal to the least [of them], which is *CE*; and cylinder *TM* equals cylinder *QN*, while cylinder *VI* equals cylinder *PN*, and cylinder *XE* equals cylinder *LN*; therefore cylinders *SN*, *QN*, *PN*, and *LN* exceed one another equally and the excess is equal to the least of these, that is, to cylinder *LN*. But the excess of cylinder *SN* over cylinder *QN* is a ring of height *QT* (or *ND*) and of breadth *SQ*; the excess of cylinder *QN* over cylinder *PN* is a ring of breadth *QP*; and finally the excess of cylinder *PN* over cylinder *LN* is a ring of breadth *PL*. Hence the said rings *SQ*, *QP*, *PL* are equal [in volume] to one another and to cylinder *LN*. Ring *ST* is therefore equal to cylinder *XE*; ring *QV*, double ring *ST*, is equal to cylinder *VI*, which is likewise double the cylinder *XE*; and for the same reason, ring *PX* will be equal to cylinder *TM*, and cylinder *LE* [equal] to cylinder *SN*.

Therefore along the balance *KF*, which joins the midpoints of lines *EI* and *DN* and is cut into equal parts by points *H* and *G*, there are magnitudes (that is, cylinders *SN*, *TM*, *VI*, and *XE*) of which the centers of gravity are respectively *K*, *H*, *G*, and *F*. Further, we have another balance, *MK*, which is one-half *FK*, and which is divided into as many equal parts by as many points, that is, [lines] *MH*, *HN*, and *NK*; and on this there are other magnitudes equal in number and size to those found on the balance *FK*, having their centers of gravity at points *M*, *H*, *N*, *K*, and being arranged in the same order. In fact, cylinder *LE* has its center of gravity at *M* and is equal to cylinder *SN*, which has its center of gravity at *K*; ring *PX* has its center of gravity at *H* and is equal to the cylinder *TM*, of which the center of gravity is

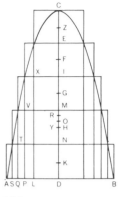

191

at H; ring QV, having its center of gravity at N, is equal to cylinder VI, of which the center is G; finally, ring ST, having its center of gravity at K, equals cylinder XE of which the center is at F. Therefore the center of gravity of [each of] the said magnitudes divides the [respective] balance in the same ratio. But their center [of gravity] is unique, and is therefore at some point common to both balances; let this be Y. Hence FY will be to YK as KY is to YM; therefore FY is double YK; and, CE being bisected at Z, ZF will be double KD, and consequently ZD will be triple DY. But CD is triple DO; therefore line DO is greater than DY, and hence the center of gravity Y of the inscribed figure is closer to the base than is the point O. And since as CD is to DO, so the removed part ZD is to the removed part DY, then the remainder CZ will also be to the remainder YO, as CD is to DO; that is, YO will be one-third of CZ, or one-sixth of CE.

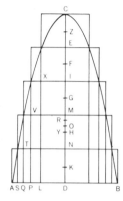

By the same procedure we may show, on the other hand, that the cylinders of the circumscribed figure exceed one another equally, that their excesses are equal to the minimum cylinder, and that their centers of gravity are situated at equal distances along balance KZ; and likewise the rings equal to the cylinders are disposed in a like manner along the balance KG, which is one-half of balance KZ, and that hence the center of gravity R of the circumscribed figure divides the balance so that ZR is to RK as KR is to RG. Therefore ZR will be double RK; but CZ will be equal to line KD, and not its double; hence all CD will be less than triple DR, and so line DR is greater than DO; or the center of gravity of the circumscribed figure is farther from base than is the point O. And since ZK is triple KR, and KD plus double ZC is triple KD, all CD plus CZ will be triple DR. But CD is triple DO; hence the remainder CZ will be triple the other remainder RO; that is, OR is one-sixth of EC. Which was the proposition.

These things first demonstrated, it will be proved that:

[PROPOSITION 3]

The center of gravity of a parabolic conoid divides its axis so that the part toward the vertex is double the part toward the base.

The parabolic conoidal [figure] whose axis is AB is divided at N so that AN is double NB. It is to be shown that the center of gravity of the conoid is point *N*. If, indeed, it is not *N*, it is below this [point] or above it. First let it be below, at *X*, and draw *LO* equal to *NX*; and let *LO* be divided anywhere at *S*; and whatever ratio *BX* plus *OS* has to *OS*, let the [volume of the] conoid have to the solid *R*.

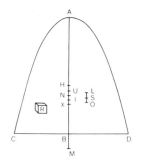

Inscribe in the conoid a figure made up of cylinders of equal height in such a way that between its center of gravity and the point *N*, [a distance] less than *LS* shall be intercepted; and let the excess by which the conoid exceeds it be less than the solid *R*. It is manifest that this can be done. Thus let the inscribed [figure] be that of which the center of gravity is *I*; now *IX* will be greater than *SO*; and since as *XB* plus *SO* is to *SO*, so the conoidal [figure] is to *R*; and further, *R* is greater than the excess by which the conoid exceeds it; the ratio of the conoid to the said excess will be greater than *BX* plus *OS* to *SO*; and by division, the inscribed figure will have a greater ratio to the said excess than *BX* has to *SO*. But *BX* has to *XI* a smaller ratio than to *SO*; therefore the inscribed figure will have to the remaining parts a much greater ratio than *BX* [has] to *XI*. Therefore the ratio of the inscribed figure to the remaining parts will be that of some other line to *XI*, which [line] must be greater than *BX*. Let it be *MX*. Thus we have *X*, the center of gravity of the conoid; but the center of gravity of the inscribed figure is *I*. Therefore the center of gravity of the remaining portions, by which the conoid exceeds the inscribed figure, will be in the line *XM*, and at that point wherein it terminates so that the ratio of the inscribed figure to the excess by which the conoid surpasses it is the same as [the ratio of] this [line] to *XI*. But it has been shown that this ratio is that of *MX* to *XI*; therefore *M* will be the center of gravity of the portions by which the conoid exceeds the inscribed figure. But that certainly cannot be; for if a plane is drawn through *M*, parallel to the base of the conoid, all the said [excessive] parts will lie on the same side of it and will not be divided by it. Therefore the center of gravity of the conoid is not below point *N*.

But neither is it above. Indeed, if this is possible, let it be [at] *H*; and as above, draw *LO* equal to *HN* and divide this anywhere at *S*; and whatever ratio *BN* plus *SO* has to *SL*, let the conoid have to *R*. Circumscribe about the conoid a figure [composed] of cylinders, as before, exceeding the conoid by a quantity less than the solid *R*, and let the line between the center of gravity of the circumscribed figure and point *N* be less than *SO*. The remainder *UH* will be

193

greater than *LS*; and since as *BN* plus *OS* is to *SL*, so the
conoid is to *R* (*R* being greater than the excess by which
the circumscribed figure exceeds the conoid), then *BN* plus
SO has a smaller ratio to *SL* than the conoid has to the said
excess. But *BU* is less than *BN* plus *SO*, while *HU* is greater
than *SL*, whence the conoid has a much greater ratio to the

194   said portions [of excess] than *BU* has to *UH*. Therefore
whatever ratio the conoid has to the said portions, some line
greater than *BU* has to *UH*. Let this be *MU*; and since the
center of gravity of the circumscribed figure is *U*, while the
center of the conoid is *H*, and as the conoid is to the remaining
portions, so *MU* is to *UH*, then *M* will be the center of
gravity of the remaining portions, which likewise is impossible.
Therefore the center of gravity of the conoid is not above
the point *N*. But it was demonstrated not to be below it;
therefore it necessarily lies at *N*. And by the same reasoning
this may be proved for a conoid cut by a plane that is not
at right angles to its axis.

The same is shown in another way, as is clear from the
following:

[PROPOSITION 4]

*The center of gravity of a parabolic conoid falls between the
center of the circumscribed figure [of cylinders] and the center
of the [similar] inscribed figure.*

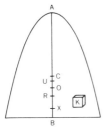

Let there be a conoid with axis *AB*; the center [of gravity]
of the circumscribed figure is *C*, while that of the inscribed
figure is *O*. I say that the center [of gravity] of the conoid
lies between points *C* and *O*. Indeed, if it does not, it lies
either above, or below, or at one of these [points]. Let it
be below, as for example at *R*; then since *R* is the center of
gravity of the whole conoid and *O* is the center of gravity
of the inscribed figure, the center of gravity of all the other
portions by which the inscribed figure is exceeded by the
conoid will lie on the extension of line *OR* beyond *R*, and
precisely at that point which terminates it in such a way
that whatever ratio the said portions have to the inscribed
[figure], that is also the ratio of line *OR* to the line intercepted
between *R* and that point. Let this ratio be that of *OR* to *RX*;
then *X* will either fall outside the conoid, or inside it, or in
its base. That it should fall outside, or in the base, is clearly

absurd. Falling inside, since *XR* is to *RO* as the inscribed figure is to the excess by which this is surpassed by the conoid, then we assume that whatever the ratio of *BR* to *RO*, such also is that of the inscribed figure to the solid *K*, which must necessarily be less than that excess.

Next, inscribe another figure which shall be exceeded by the conoid by an excess less than *K*; its center of gravity will lie between *O* and *C*. Let this be *U*; since the first figure is to *K* as *BR* is to *RO*, and since on the other hand the second figure, of which the center is *U*, is greater than the first, and is exceeded by the conoid with an excess less than *K*, we shall have that whatever the ratio of the second figure to the excess by which it is surpassed by the conoid, such also is the ratio of some line greater than *BR* to line *RU*. But the center of gravity of the conoid is *R*, while that of the inscribed figure   **195** is *U*; therefore the center of gravity of the remaining portions will lie outside the conoid, below *B*, which is impossible.

By the same procedure it will be shown that the center of gravity of this same conoid does not lie on line *CA*. Then, that it is neither of the points *C* or *O* is manifest. In fact if we suppose this, and describe other figures such that the inscribed is greater than the figure whose center [of gravity] is *O*, and that which is circumscribed is less than the figure whose center is *C*, the center of gravity of the conoid will fall outside the centers of gravity of these figures, which is impossible, as we have just concluded. It follows, then, that it lies between the center of the circumscribed figure and that of the inscribed figure. Being thus, it must necessarily lie in that point that divides the axis in such a way that the part toward the vertex is double the remainder, since indeed figures can be inscribed and circumscribed such that the lines lying between their centers of gravity and the said point may be less than any given line. Thus anyone who declared the contrary [of the above] would be led to the absurdity that the center [of gravity] of the conoid would not lie between the centers of gravity of the inscribed and circumscribed figures.

[LEMMA]

If there are three lines in [continued] proportion, and the ratio of the least to the excess by which the greatest exceeds the least is the same as that of some given line to two-thirds of the excess by which the greatest exceeds the middle [line];

*and again if the ratio of the greatest plus double the middle
[line] to triple the greatest plus triple that middle is the same
as the ratio of some [other] given line to the excess of the
greatest over the smallest; then the sum of those two given
lines is one-third of the greatest of the three proportional lines.*

Let there be three lines, *AB, BC, BF*, in [continued]
proportion, and let the ratio of *BF* to *AF* be that of *MS*
to two-thirds of *CA*; also let the ratio of *AB* plus 2*BC* to 3*AB*
plus 3*BC* be that of another [line] *SN* to *AC*. It is to be
demonstrated that *MN* is one-third of *AB*.

Since *AB, BC,* and *BF* are in continued proportion, *AC*
and *CF* are also in that same ratio; therefore, as *AB* to *BC*,
so *AC* is to *CF*, and as 3*AB* is to 3*BC*, so *AC* is to *CF*. Whatever
ratio 3*AB* plus 3*BC* has to 3*CB*, *AC* has to some smaller line
than *CF*; let this be *CO*. Then by composition and inversion
of ratios, *OA* has to *AC* the same ratio that 3*AB* plus 6*BC*
has to 3*AB* plus 3*BC*; further, *AC* has to *SN* the same ratio
as 3*AB* plus 3*BC* to *AB* plus 2*BC*; by equidistance of ratios,
therefore, *OA* has to *NS* the same ratio as 3*AB* plus 6*BC*
to *AB* plus 2*BC*. But the ratio of 3*AB* plus 6*BC* to *AB* plus
2*BC* is 3(*AB* plus 2*BC*); therefore *AO* is triple *SN*.

Next, since *OC* is to *CA* as 3*CB* is to 3*AB* plus 3*CB*, while
as *CA* is to *CF*, so 3*AB* is to 3*BC*, then by equidistance of
ratios in perturbed proportion, as *OC* is to *CF*, so 3*AB*
will be to 3*AB* plus 3*BC*; and by inversion of ratios, as *OF*
is to *FC*, so 3*BC* is to 3*AB* plus 3*BC*. Also, as *CF* is to *FB*,
so *AC* is to *CB*, and 3*AC* is to 3*BC*; therefore, by equidistance
of ratios in perturbed proportion, as *OF* is to *FB*, so 3*AC*
is to 3(*AB* plus *BC*). Hence all *OB* will be to *BF* as 6*AB* is
to 3(*AB* plus *BC*); and since *FC* has the same ratio to *CA*
that *CB* has to *BA*, then as *FC* is to *CA*, so *BC* will be to
BA; and by composition, as *FA* is to *AC*, so is the sum of
BA plus *AC* to *BA*, as likewise [are] their triples. Therefore,
as *FA* is to *AC*, so 3*BA* plus 3*BC* is to 3*AB*; whence as *FA*
is to two-thirds *AC*, so 3*BA* plus 3*BC* is to two-thirds of
3*BA*, which is 2*BA*. But as *FA* is to two-thirds *AC*, so *FB*
is to *MS*; therefore as *FB* is to *MS*, so 3*BA* plus 3*BC* is to
2*BA*. But as *OB* is to *FB*, so 6*AB* was to 3(*AB* plus *BC*).
Therefore, by equidistance of ratios, *OB* has to *MS* the same
ratio as 6*AB* to 2*BA*, whence *MS* is one-third *OB*. And
it was shown that *SN* is one-third *AO*; hence it is clear that
MN is likewise one-third *AB*. Q.E.D.

196

[PROPOSITION 5]

The center of gravity of any frustum cut from a parabolic conoid lies in the straight line that is the axis of this frustum; this being divided into three equal parts, the [said] center of gravity lies in the middle [part] and so divides this [part] that the portion toward the smaller base has, to the portion toward the larger base, the same ratio as that of the larger base to the smaller.

From a conoid whose axis is *RB*, cut a solid with axis *BE*, the cutting plane being parallel to the base. Let it be cut also by another plane, perpendicular to the base, this section giving the parabola *URC*, the sections of the cutting plane and of the base being the straight lines *LM* and *UC*. The diameter of ratios, or parallel diameter, will be *RB*, while *LM* and *UC* will be ordinately applied.[4]

197

Let the line *EB* be divided into three equal parts, of which the middle one is *QY*; this is further divided at *I* so that whatever ratio the base of diameter *UC* has to the base of diameter *LM* (that is, [the ratio] of the square of *UC* to the square of *LM*), *QI* has also to *IY*. It is to be demonstrated that the center of gravity of the frustum *ULMC* is *I*.

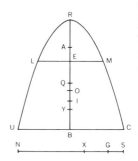

Draw *NS* equal to *BR*, and let *SX* be equal to *ER*; and to *NS* and *SX* take the third proportional *SG*; and as *NG* is to *GS*, let *BQ* be to *IO*. It does not matter whether point *O* falls above or below *LM*. And since in section *URC* the lines *LM* and *UC* are ordinately applied, as the square of *UC* is to the square of *LM*, so line *BR* will be to *RE*; and further as the square of *UC* to the square of *LM*, so is *QI* to *IY*; and as *BR* is to *RE*, so is *NS* to *SX*; therefore *QI* is to *IY* as *NS* is to *SX*. Whence as *QY* is to *YI*, so will *NS* plus *SX* be to *SX*; and as *EB* is to *YI*, so is triple *NS* plus triple *SX* to *SX*. Further, as *EB* is to *BY*, so triple the sum of *NS* and *SX* is to the sum of *NS* and *SX*; therefore as *EB* is to *BI*, so is triple *NS* plus triple *SX* to *NS* plus double *SX*. Therefore the three lines *NS*, *SX*, and *GS* are in continued proportion, and whatever the ratio of *SG* to *GN*, the same will be that of some assigned line *OI* to two-thirds of *EB* (that is, of *NX*); and whatever ratio *NS* plus double *SX* has to triple *NS* plus triple *SX*, the same will be that of some assigned line

---

4. Galileo's "diameter of ratios" in the diagram would now be called the axis of ordinates, while his "ordinates" are our abscissae.

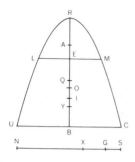

198

IB to *BE* (that is, to *NX*). Therefore, by what was demonstrated above, these [assigned] lines taken together will be one-third of *NS* (that is, of *RB*). Therefore *RB* is triple *BO*, whence *BO* will be the center of gravity of the conoid *URC*.

Now let *A* be the center of gravity of the conoid *LRM*; then the center of gravity of the frustum *ULMC* lies in line *OB*, and at the point where this terminates so that whatever ratio the frustum *ULMC* has to the portion *LRM*, the line *AO* has that same ratio to the intercept between *O* and the said point [of termination]. Since *RO* is two-thirds of *RB*, *RA* is two-thirds of *RE*, and the remainder *AO* will be two-thirds the remainder *EB*. And since as the frustum *ULMC* is to the portion *LRM*, so *NG* is to *GS*, and as *NG* is to *GS*, so is two-thirds *EB* to *OI*, and two-thirds *EB* is equal to line *AO*; then as the frustum *ULMC* is to the portion *LRM*, so *AO* is to *OI*. Therefore it is clear that the center of gravity of the frustum *ULMC* is point *I*, and the axis is so divided [by it] that the part toward the smaller base is to the part toward the larger base as double the larger base plus the smaller is to double the smaller plus the larger. Which is the proposition, but more elegantly expressed.

[LEMMA]

If any number of magnitudes are so arranged that the second adds to the first double the first, and the third adds to the second triple the first, while the fourth adds to the third quadruple the first, and so every following magnitude exceeds the preceding one by a multiple of the first magnitude according to its number in order; if, I say, such magnitudes are arranged on a balance and suspended at equal distances, then the center of equilibrium of the whole composite divides that balance so that the part toward the smaller magnitudes is triple the remainder.

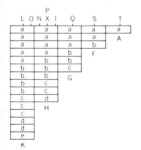

Let *LT* be the balance, and the magnitudes hanging from it, of the kind described, are *A, F, G, H, K*, of which *A* is hung first, from *T*. I say that the center of equilibrium cuts the balance *TL* so that the part toward *T* is triple the remainder. Let *TL* be triple *LI*, and *SL* triple *LP*, and *QL* [triple] *LN*, and *LP* [triple] *LO*; then *IP, PN, NO, OL* will be equal. Take at *F* a magnitude of 2*A*, and at *G* another, 3*A*; at *H*, 4*A*, and so on; and let these be the magnitudes [marked] *a* in the diagram. And do the same in magnitudes *F, G, H, K*;

indeed, let the magnitude in the remainder of F, which is b, be equal to a; and in G take $2b$, in H, $3b$, etc.; and let these be the magnitudes containing b's. And in the same way take those containing c's, d's, and e. Then all those in which a is marked are equal to [all in] K; the composite of all b's will equal H; that of the c's, G; that composed of all d's will be equal to F, and e [will equal] A itself. And since TI is double LI, I will be the point of equilibrium of magnitudes made up of all the a's; likewise, since SP is double PL, P will be the point of equilibrium of the composite of all the b's; and for the same cause, N will be the point of equilibrium **199** of the composite of all c's, O [will be that] of the composite of d's, and L, of e itself.

There is thus a certain balance TL on which at equal distances there hang certain magnitudes K, H, G, F, A; and further, there is another balance LI on which at equal distances hang a like number of magnitudes, equal to and in the same order as those described. Indeed, there is a composite of all a's that hangs from I, equal to K hanging from L; and a composite of all b's that hangs from P, equal to H hanging from P; and likewise a composite of c's that hangs from N, equal to G, and a composite of d's that hangs from O, equal to F; and e, hanging from L, is equal to A. Whence the balances are divided in the same ratio by the center of [equilibrium of] the composites of magnitudes. But there is [only] one center of the composites of the said magnitudes, and it will be a common point of the line TL and the line LI. Let this be X. And thus as TX is to XL, so LX will be to XI, and all TL to LI. But TL is triple LI, whence TX is triple XL.

[LEMMA]

If any number of magnitudes are taken, and the second adds above the first, triple the first, while the third exceeds the second by five times the first, and the fourth exceeds the third by seven times the first, and so on, each addition over the preceding being a multiple of the first according to the successive odd numbers (as the squares of lines that equally exceed one another and of which the excess is equal to the first thereof), and if these be hung at equal distances along a balance, then the center of equilibrium of all combined will divide the balance so that the part toward the lesser magnitudes is more than triple the remainder; but one distance being removed, it will be less than triple.

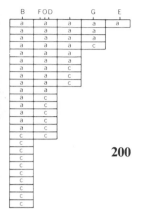

200

Let there be on the balance *BE* magnitudes such as those described, from which let then be removed some magnitudes arranged among themselves as in the preceding [lemma]; let [for example] all the *a*'s [in the present diagram be taken away]. The remainder will be the *c*'s, [still] arranged in the same order [as was the whole], but wanting the greatest [magnitude]. Let *ED* be triple *DB*, and *GF* triple *FB*; *D* will be the center of equilibrium of everything composed of the *a*'s, while *F* will be that of the *c*'s; hence the center of the compound of [both] *a*'s and *c*'s falls between *D* and *F*. And thus it is manifest that *EO* is more than triple *OB*, while *GO* is less than triple *OB*. Which was to be proved.

[PROPOSITION 6]

If any cone or portion of a cone has one figure of cylinders of equal height inscribed to it, and another circumscribed, and if its axis is divided so that the part intercepted between the point of division and the vertex is triple the remainder, then the center of gravity of the inscribed figure will be closer to the base of the cone than [will] the point of division, but the center of gravity of the circumscribed [figure] will be closer to the vertex than [will] that same point.

Let there be a cone with axis *NM*, divided at *S* so that *NS* is triple the remainder *SM*. I say that any figure as described that is inscribed in the cone has its center of gravity in the axis *NM*, and that it approaches more nearly the base of the cone than does the point *S*, while the center of gravity of one circumscribed is likewise in the axis *NM*, but closer to the vertex than is *S*.

Assume an inscribed figure of cylinders whose axes *MC, CB, BE, EA* are equal. Thus this first cylinder, of which the axis is *MC*, has, to the cylinder with axis *CB*, the same ratio as [that of] its base to the base of the other (since their altitudes are equal); and this ratio is the same as that which the square of *CN* has to the square of *NB*. It is likewise shown that the cylinder with axis *CB* has to the cylinder with axis *BE* the same ratio as that of the square of *BN* to the square of *NE*; while the cylinder around axis *BE* has to the cylinder around axis *EA* the ratio of the square of *EN* to the square of *NA*. Moreover, the lines *NC, NB, EN, NA* equally exceed one another, and their excess is equal to the least, namely *NA*. There are therefore magnitudes (i.e. the inscribed cylinders) which have successively to one another the ratio of squared lines equally exceeding one another, of which the excess is equal to the least. Thus these are arranged on the balance *TI*, with the single centers of gravity therein, and at equal distances. Hence by those things demonstrated above, it is evident that the center of gravity of all these compounded in the balance *TI* so divides it that the part toward *T* is more than triple the remainder.[5] Let this center be *O*; then *TO* is more than triple *OI*. But *TN* is triple *IM*; therefore all *MO* will be less than one-quarter of all *MN*, of which *MS* was assumed to be one-quarter. It is therefore evident that point *O* comes nearer the base of the cone than does *S*.

201

Now let the circumscribed figure consist of cylinders whose axes *MC, CB, BE, EA, AN* are equal to one another. As with the inscribed [figure], these are shown to be to one another as the squares of lines *MN, NC, BN, NE, AN*, which equally exceed one another and whose excesses equal the least, *AN*. Whence, from what went before, the center of gravity of all the cylinders thus arranged (and let this be *U*) so divides the balance *RI* that the part toward *R* (that is, *RU*) is more than triple the remainder *UI*, while *TU* will be less than triple the same. But *NT* is triple *IM*; therefore all *UM* is greater than one-quarter of all *MN*, of which *MS* was assumed to be one-quarter. And thus point *U* is closer to the vertex than is point *S*.   Q.E.D.

[PROPOSITION 7]

Given a cone, a figure can be inscribed and another circumscribed to it, made up of cylinders having equal heights, so that the line intercepted between the center of gravity of the circumscribed [figure] and that of the inscribed [figure] is less than any assigned line.

Given a cone with axis *AB*, and given further a straight line *K*; I say, let the cylinder *L* be drawn equal to that [which may be] inscribed in the cone, having an altitude of one-half the axis *AB*. Divide *AB* at *C* so that *AC* is triple *CB*; and whatever ratio *AC* has to *K*, let this cylinder *L* have to some solid, *X*. Circumscribe about the cone a figure of cylinders having equal altitudes, and inscribe another one, so that the circumscribed exceeds the inscribed by a quantity less than

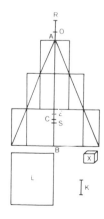

_____
5. Because *TI* omits one distance, *NA*.

the solid *X*. Let the center of gravity of the circumscribed [figure] be *E*, which falls above *C*, while the center of the inscribed one is *S*, falling below *C*. I now say that line *ES* is less than *K*.

For if it is not, put *CA* equal to *EO*; then since *OE* has to *K* the same ratio as that of *L* to *X*, the inscribed figure is not less than cylinder *L*, and the excess by which the circumscribed figure surpasses it is less than solid *X*; therefore the inscribed figure has to the said excess a greater ratio than *OE* will have to *K*. But the ratio of *OE* to *K* is not less than that of *OE* to *ES*, since *ES* cannot be assumed less than *K*; therefore the inscribed figure has a greater ratio to the excess by which the circumscribed [figure] surpasses it than *OE* has to *ES*. Hence whatever ratio the inscribed [figure] has to the said excess, some line greater than *EO* will have this to the line *ES*. Let this [line] be *ER*. Now, the center of gravity of the inscribed figure is *S*, while that of the circumscribed is *E*; hence it is evident that the remaining portions by which the circumscribed exceeds the inscribed [figure] have their center of gravity in line *RE*, and at that point where it is terminated so that whatever ratio the inscribed [figure] has to those portions, the line intercepted between *E* and that point has to line *ES*. But *RE* has this ratio to *ES*; hence the center of gravity of the remaining portions by which the circumscribed figure exceeds the inscribed will be *R*; which is impossible, since indeed the plane through *R* [drawn] parallel to the base of the cone does not cut these portions. Therefore it is false that line *ES* is not less than *K*, and hence it will be less.

Moreover, in a way not dissimilar, this may be demonstrated to hold for pyramids.

From this it is manifest that:

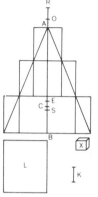

202

[COROLLARY]

About a given cone, a figure can be circumscribed, and [within it] another inscribed, of cylinders having equal altitudes, such that the lines between their centers of gravity and the point which divides the axis of the cone so that the part toward the vertex is triple the remainder are less than any given line.

For indeed, as was demonstrated, the said point dividing the axis in the said way is always found between the centers of gravity of the circumscribed and inscribed [figures]; and it is possible for the line between those same centers to be less than any assigned line; so that which is intercepted between either of the two centers and the point that thus divides the axis must be much less than this assigned line.

The center of gravity of any cone or pyramid so divides the axis that the part toward the vertex is triple the remainder toward the base.

Given the cone with axis *AB*, divided so that *AC* is triple the remainder *CB*, it is to be shown that *C* is the center of gravity of the cone. For if it is not, the center of the cone will be either above or below point *C*. First let it be below, at *E*, and draw line *LP* equal to *CE*, and divide this anywhere at *N*; and whatever ratio *BE* plus *PN* shall have to *PN*, let this cone have to some solid, *X*. Inscribe in the cone a solid figure made up of cylinders of equal height; the center of gravity of this shall be less distant from point *C* than [the length of] line *LN*, and the excess by which the cone exceeds [this figure] will be less than solid *X*. It is clear from what has been demonstrated that these things can be done. Let this solid figure which we assume have its center of gravity at *I*. Then line *IE* will be greater than *NP*, since *LP* is equal to *CE*; and *IC* [is] less than *LN*; and since *BE* plus *NP* is to *NP* as the cone is to *X*, and moreover the excess by which the cone exceeds the inscribed figure is less than solid *X*, the cone will have a greater ratio to the said excess than that of *BE* plus *NP* to *NP*; and by division, the inscribed figure has a greater ratio to the excess by which the cone exceeds it than *BE* has to *NP*. Moreover, *BE* has to *EI* a still smaller ratio than it has to *NP*, since *IE* is greater than *NP*, whence the inscribed figure has a much greater ratio to the excess by which the cone surpasses it than *BE* has to *EI*.

Therefore whatever ratio the inscribed [figure] has to the said excess, some greater line *BE* has to line *EI*. Let this be *ME*; since *ME* is to *EI* as the inscribed figure is to the excess by which the cone surpasses it, and [if] *E* is the center of gravity of the cone, while *I* is the center of gravity of the [figure] inscribed, then *M* will be the center of gravity of the remaining portions by which the cone exceeds the inscribed figure in it; which is impossible. Therefore the center of gravity of the cone is not below point *C*.

203

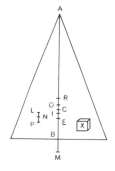

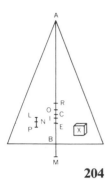

But neither is it above. For, if possible, let it be R; again take the line LP, cut anywhere at N. Whatever ratio BC plus NP has to NL, let the cone have to X, and likewise circumscribe about the cone a figure that exceeds it by a lesser quantity than the solid X; the line intercepted between its center of gravity and C shall be less than NP. Now let there be circumscribed [a figure] having center of gravity O; the remainder OR will be greater than NL. And since as BC plus PN is to NL, so the cone is to X, but the excess by which the circumscribed [figure] surpasses the cone is less than X, and BO is less than BC plus PN, while OR is greater than LN, the cone will have a greater ratio to the remaining portions by which it is exceeded by the circumscribed figure than BO has to OR. Let MO have that ratio to OR; then MO will be greater than BC, and M will be the center of gravity of the portions by which the cone is exceeded by the circumscribed figure; which is contradictory. Therefore the center of gravity of this cone is not above the point C, but neither is it below, as was shown; therefore it is C itself. And the same may be demonstrated in the above way for any pyramid.

204

[LEMMA]⁶

If there are four lines in [continued] proportion, and whatever ratio the least of these has to the excess by which the greatest exceeds the least, that same [ratio] is had by some [assumed] line to 3/4 of the excess by which the greatest exceeds the second [line]; and whatever ratio a line equal to the greatest plus double the second plus triple the third has to a line equal to four times the sum of the greatest, the second, and the third together, that same ratio is had by [another] assumed line to the excess by which the greatest exceeds the second; and these two [assumed] lines taken together will be one-quarter of the greatest of the original lines.

Let there be four lines in continued proportion, AB, BC. BD, BE; and whatever ratio BE has to EA, let FG have t· three-quarters of AC; and further, whatever ratio a line equal to AB plus $2BC$ plus $3BD$ has to a line equal to four times the sum of AB, BC, and BD, let HG have to AC. It is to be shown that HF is one-quarter of AB.

205

---

6. A manuscript copy submitted in 1587 (note 1, above) exhibits some variants from the printed text, but none of a substantial character.

Since *AB, BC, BD,* and *BE* are proportional, then *AC, CD,* and *DE* will be in that same ratio; and as four times the sum of *AB, BC,* and *BD* is to *AB* plus 2*BC* plus 3*BD*, so the quadruple of *AC* plus *CD* plus *DE* (that is, 4*AE*) is to *AC* plus 2*CD* plus 3*DE*; and thus is *AC* to *HG*. Therefore as 3*AE* is to *AC* plus 2*CD* plus 3*DE*, so is three-quarters of *AC* to *HG*. Moreover, as 3*AE* is to 3*EB*, so is three-quarters of *AC* to *GF*. Hence, by the converse of [Euclid] V, 24, as 3*AE* is to *AC* plus 2*CD* plus 3*DB*, so is three-quarters of *AC* to *HF*; and as 4*AE* is to *AC* plus 2*CD* plus 3*DB* (that is, to *AB* plus *CB* plus *BD*), so *AC* is to *HF*. And permuting, as 4*AE* is to *AC*, so *AB* plus *CB* plus *BD* is to *HF*. Further, as *AC* is to *AE*, so *AB* is to *AB* plus *CB* plus *BD*. Hence, by equidistance of ratios in perturbed proportion, as 4*AE* is to *AE*, so *AB* is to *HF*. Whence it is clear that *HF* is one-quarter of *AB*.

[PROPOSITION 9]

Any frustum of a pyramid or cone cut by a plane parallel to its base has its center of gravity in the axis, and this so divides it that the part toward the smaller base is to the remainder **206** *as three times the greater base plus double the mean proportional between the greater and smaller bases plus the smaller base is to triple the smaller base plus the said double of the mean proportional distance plus the greater base.*

From a cone or pyramid with axis *AD*, cut a frustum by a plane parallel to the base having axis *UD*; and whatever ratio triple the larger base, plus double the mean proportional [of both bases] plus the smaller [base], has to triple the smaller, plus double the [above] mean proportional plus the greatest, let *UO* have to *OD*. It is to be shown that *O* is the center of gravity of the frustum.

Let *UM* be one-quarter of *UD*. Draw line *HK* equal to *AD*, and let *KX* equal *AU*; let *XL* be the third proportional to *HX* and *KX*, while *XS* is the fourth proportional. Whatever ratio *HS* has to *SX*, let *MD* have to a line from *O* in the direction of *A*, and let this be *ON*. Now since the larger base is to the mean proportional between the larger and the smaller as *DA* is to *AU* (that is, as *HX* is to *XK*), and the said mean proportional is to the smaller as *KX* is to *XL*, then the larger, the mean proportional, and the smaller base will be in the ratio of lines *HX, XK,* and *XL*.

Thus as triple the larger base plus double the mean pro-

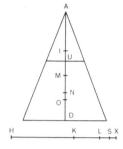

207 portional plus the smaller is to triple the smaller plus double the mean proportional plus the larger (that is, as *UO* is to *OD*), so is triple *HX* plus double *XK* plus *XL* to triple *XL* plus double *XK* plus *XH*. And, by composition and inverting, *OD* will be to *DU* as *HX* plus double *XK* plus triple *XL* is to four times the sum of *HX*, *XK*, and *XL*. Therefore there are four lines in continued proportion, *HX, XK, XL,* and *XS*; and whatever ratio *XS* has to *SH*, some assumed line *NO* has to three-quarters of *DU* (that is, to three-quarters of *HK*). Further, whatever ratio *HX* plus double *XK* plus triple *XL* has to four times the sum of *HX, XK,* and *XL*, some assumed line *OD* has to *DU* (that is, to *HK*). Hence, by what was demonstrated, *DN* will be one-quarter of *HX* (that is, of *AD*), whence point *N* will be the center of gravity of the cone or pyramid having axis *AD*.

Let *I* be the center of gravity of the pyramid or cone having axis *AU*. It is then clear that the center of gravity of the frustum lies in line *IN* extended beyond *N*, and at that point of it which, with point N, intercepts a line to which *IN* has the ratio that the frustum cut off has to the pyramid or cone having axis *AU*. Thus it remains to be shown that *IN* has to *NO* the **208** same ratio that the frustum has to the cone whose axis is *AU*. But as the cone with axis *DA* is to the cone with axis *AU*, so is the cube of *DA* to the cube of *AU*, that is, as the cube of *HX* to the cube of *XK*; and this is the ratio of *HX* to *XS*. Whence, dividing, as *HS* is to *SX*, so the frustum having axis *DU* will be to the cone or pyramid having axis *UA*. And as *HS* is to *SX*, so also *MD* is to *ON*; whence the frustum is to the pyramid having axis *AU* as *MD* is to *NO*. And since *AN* is three-quarters of *AD*, and *AI* is three-quarters of *AU*, the remainder *IN* will be three-quarters of the remainder *UD*, wherefore *IN* will be equal to *MD*. It was demonstrated that *MD* is to *NO* as the frustum is to the cone *AU*; therefore it is clear that *IN* has also this same ratio to *NO*. Whence the proposition is clear.

Finis[7]

---

7. The end of the original printed edition.

Added Day

On the Force of Percussion[1]

Interlocutors: Salviati, Sagredo and Aproino

Sagr. Your absence during this past fortnight, Salviati, has given me an opportunity to look at the propositions concerning centers of gravity in solids, as well as to read carefully the demonstrations of those many new propositions on natural and violent motions; and since there are among these not a few that are difficult to apprehend, it has been a great help to me to confer with this gentleman whom you see here.

Salv. I was about to ask you concerning the gentleman's presence, and about the absence of our good Simplicio.

Sagr. I imagine—indeed, I think it certain—that the reason for Simplicio's absence is the obscurity to him of some demonstrations of various problems relating to motion, and still more, that of those about centers of gravity. I speak of those [demonstrations] which, through their long chains of assorted propositions of [Euclid's] *Elements of Geometry*, become incomprehensible to people who do not have those elements thoroughly in hand.

The gentleman you see is Signor Paolo Aproino, a nobleman of Treviso, who was a pupil of our Academician when he taught at Padua; and not only his pupil, but his very close friend, with whom he held long and continual conversations, together with others [of like interests]. Outstanding among

---

1. Although the word *percussio* is literally translated in the title above, it is rendered by "impact" in the text as the more usual English term. Galileo first wrote on these problems in 1594 as a brief appendix to his *Mechanics*. The composition of this dialogue probably began about March, 1635, shortly after his old pupil, Aproino, saw at Venice the manuscript of the First Day.

322 these was the most noble Signor Daniello Antonini[2] of Udine, a man of surpassing intellect and superhuman worth who died gloriously in defence of his country and its serene ruler, receiving honors worthy of his merit from the great Venetian Republic. [With him, Aproino] took part in a large number of experiments that were made at the house of our Academician, concerning a variety of problems. Now, about ten days ago this gentleman came to Venice, and he visited me, as is his custom; and learning that I had here these treatises by our friend, he wanted to look them over with me. Hearing about our appointment to meet and talk over the mysterious problem of impact, he told me that he had discussed this many times with the Academician, though always questioningly and inconclusively, and he told me that he was present at the performance of divers experiments relating to various problems, of which some were made with regard to the force of impact and its explanation. He was just now on the point of mentioning, among others, one which he says is most ingenious and subtle.

Salv. I consider it my great good fortune to meet Signor Aproino and to know him personally, as I already knew him by reputation and the many reports of our Academician. It will be a great pleasure for me to be able to hear at least a part of these various experiments made at our friend's house on different propositions, and in the presence of minds as acute as those of Aproino and Antonini, gentlemen of whom I have heard our friend speak on many occasions with praise and admiration. Now since we are here to reason specifically about impact, you, my dear Aproino, may tell us what was drawn from the experiments, in this matter; promising, however, to speak on some other occasion about others made concerning other problems. For I know that such are not lacking to you, from our Academician's assurance that you were always no less curious than careful as an experimentalist [*sperimentatore*].

Apr. If I were to try with proper gratitude to repay the debt to which your excellency's courtesy obliges me, I should have to spend so many words that little or no time would be left in this day to speak of the matter here undertaken.

---

2. Antonini (1588–1616) became a correspondent of Galileo's after studying with him at Padua about 1608–10. This coupling of his name with that of Aproino suggests that the experiments described belonged to that period, as do the most precise experiments of which Galileo left any manuscript records.

Sagr. No, no, Aproino; let us start right in with learned discussion, leaving ceremonious compliments to the courtiers. For what it is worth, I shall stand pledge between you two that mutual satisfaction will be given by words that are few, but candid and sincere.   **323**

Apr. I hardly expect to say anything not already known to Salviati, so the entire burden of discourse ought to be borne on his shoulders. Yet to give him a start at least, if for no other reason, I shall mention the first steps and the first experiment that our friend essayed in order to get to the heart of this admirable problem of impact.

What is sought is the means of finding and measuring its great force, and if possible simultaneously of resolving the essence [of impact] into its principles and prime causes; for this effect seems, in acquiring its great power, to proceed very differently from the manner in which multiplication of force proceeds in all other mechanical machines; I say "mechanical" to exclude the immense force of gunpowder [*fuoco*, fire].[3] In machines, it is very conclusively perceived that speed in a weak mover compensates the power [*gagliardia*] of a strong resistent [which is] moved but slowly. Now since it is seen that in the operation of impact, too, the movement of the striking body conjoined with its speed acts against the movement of the resistent and the much or little that it is required to be moved, it was the Academician's first idea to try to find out what part in the effect and operation of impact belonged, for example, to the weight of a hammer, and what [part belonged to] the greater or lesser speed with which it was moved. He wanted if possible to find one measure that would measure both of these, and would assign the energy of each;[4] and to arrive at this knowledge, he imagined what seems to me to be an ingenious experiment.

He took a very sturdy rod, about three braccia long, pivoted like the beam of a balance, and he suspended at the ends of these balance-arms two equal weights, very heavy. One of these consisted of copper containers; that is, of two buckets, one of which hung at the said extremity of the beam and was filled with water. From the handles of this bucket

3. Cf. p. **278** and note 14 to Fourth Day.

4. Galileo's approach related the problem to compound ratios; see Introduction and Glossary. His discussion is accordingly mainly one of momentum rather than of force in its modern sense. This concentrates attention on velocity rather than on acceleration, but see pp. **330**, **332**, **344**, and notes 12, 14, below.

hung two cords, about two braccia each in length, to which
was attached by its handles another like bucket, but empty;
this hung plumb beneath the bucket already described as
324   filled with water. At the end of the other balance-arm he hung
a counterweight of stone or some other heavy material,
which exactly balanced the weight of the whole assembly
of buckets, water, and ropes. The bottom of the upper bucket
had been pierced by a hole the size of an egg or a little smaller,
which hole could be opened and closed.

Our first conjecture was that when the balance rested in
equilibrium, the whole apparatus having been prepared as
described, and then the [hole in the] upper bucket was unstop-
pered and the water allowed to flow, this would go swiftly
down to strike in the lower bucket; and we conceived that
the adjoining of this impact must add to the [static] moment
on that side, so that in order to restore equilibrium it would
be necessary to add more weight to that of the counterpoise
on the other arm. This addition would evidently restore
and offset the new force of impact of the water, so that we
could say that its momentum was equivalent to the weight
of the ten or twelve pounds that it would have been necessary
[as we imagined] to add to the counterweight.

Sagr. This scheme seems to me really ingenious, and I
am eagerly waiting to hear how the experiment succeeded.

Apr. The outcome was no less wonderful than it was unexpected by us. For the hole being suddenly opened, and
the water commencing to run out, the balance did indeed
tilt toward the side with the counterweight; but the water
had hardly begun to strike against the bottom of the lower
bucket when the counterweight ceased to descend, and
commenced to rise with very tranquil motion, restoring
itself to equilibrium while water was still flowing;[5] and upon
reaching equilibrium it balanced and came to rest without
passing a hairbreadth beyond.

Sagr. This result certainly comes as a surprise to me. The
outcome differed from what I had expected, and from which
I hoped to learn the amount of the force of impact. Nevertheless it seems to me that we can obtain most of the desired
information. Let us say that the force and moment of impact
is equivalent to the moment and weight of whatever amount

5. The experimenters expected some constant effect as long as the flow
of water continued, enabling them to re-establish equilibrium by adding
weight to the counterpoise.

of falling water is found to be suspended in the air between the upper and lower buckets, which quantity of water does not weigh at all against either upper or lower bucket. Not against the upper, for the parts of water are not attached together, so they cannot exert force and draw down on those above, as would some viscous liquid, such as pitch or lime, for example. Nor [does it weigh] against the lower [bucket], because the falling water goes with continually accelerated motion, so its upper parts cannot weigh down on or press against its lower ones. Hence it follows that all the water contained in the jet is as if it were not in the balance. Indeed, that is more than evident; for if that [intermediate] water exerted any weight against the buckets, that weight together with the impact would greatly incline the buckets downward, raising the counterweight; and this is seen not to happen. This is again exactly confirmed if we imagine all the water suddenly to freeze; for then the jet, made into solid ice, would weigh with all the rest of the structure, while cessation of the motion would remove all impact.

Apr. Your reasoning conforms exactly with ours— immediately after the experiment we had witnessed. To us also, it seemed possible to conclude that the speed alone, acquired by the fall of that amount of water from a height of two braccia, without [taking into account] the weight of this water, operated to press down exactly as much as did the weight of the water, without [taking into account] the impetus of the impact. Hence if one could measure and weigh the quantity of water hanging in air between the containers, one might safely assert that the impact has the same power to act by pressing down as would be that of a weight equal to the ten or twelve pounds of falling water.

Salv. This clever contrivance much pleases me, and it appears to me that without straying from that path, in which some ambiguity is introduced by the difficulty of measuring the amount of this falling water,[6] we might by a not unlike experiment smooth the road to the complete understanding which we desire.

Imagine, for instance, one of those great weights (which I believe are called pile drivers [*berte*]) that are used to drive stout poles into the ground by allowing them to fall from some height onto such poles. Let us put the weight of such a

6. Notes survive in which Galileo made calculations concerning the volume of this jet of water.

325

pile driver at 100 pounds, and let the height from which this falls be four braccia, while the entrance of the pole into hard ground, when driven by a single [such] impact, shall be four inches. Next, suppose that we want to achieve the same pressure and entrance of four inches without using impact, and we find that this can be done by a weight of 1000 pounds, which, operating by its heaviness alone, without any preceding motion, we may call "dead weight." I ask whether, without

326   error or fallacy, we may affirm that the force and energy of a weight of 100 pounds, combined with its speed acquired in falling from a height of four braccia, is equivalent to the dead weight of 1000 pounds. That is, does the force [*virtù*] of this speed alone signify as much as the pressure of 900 pounds of dead weight, which is the remainder after subtracting from 1000 [pounds] the 100 of the pile driver?

I see that you both hesitate to reply, perhaps because I have not explained my question properly. Then let us merely ask briefly whether, from the experiment described, we may assert that the pressure of this dead weight will always produce the same effect on a resistance as the weight of 100 pounds falling from a height of four braccia. To make things perfectly clear, [say that] the pile driver, falling from the same height but striking on a more resistant pole, will drive it no more than two inches. Now, can we be sure of this same effect from the pressing down alone of the dead weight of 1000 pounds? I mean, will that drive the pole two inches?

Apr. I think, at least on first hearing this, that it would not be rejected by anyone.

Salv. And you, Sagredo, do you raise any question about this?

Sagr. Not at the moment, no; but my having experienced a thousand times the ease with which one is deceived prevents my being so bold as to feel no trepidation.

Salv. Even you, whose great perspicacity I have known on many occasions, now show yourself as leaning toward the wrong side; hence I believe that it would be hard to find even one or two men in a thousand who would not be snared into so plausible a fallacy. But what will astonish you still more will be to see this fallacy to be hidden beneath so thin a veil that the slightest breeze would serve to uncover and reveal it, though it is now concealed and hidden.

First, then, let the pile driver in question fall on the pole as before, driving this four inches down, and let it be true that to accomplish this with dead weight would require exactly 1000 pounds. Next, let us raise this same pile driver to the same height, so that it falls a second time on the same pole, but drives it only two inches, by reason of the pole's having encountered harder ground. Must we suppose that it would be driven as much by the pressure of that same dead weight of 1000 pounds?

Apr. So it seems to me.

Sagr. Alas, Paolo, for us; this must be emphatically denied. For if in the first placement, the dead weight of 1000 pounds drove the pole only four inches and no more, why will you have it that by merely being removed and replaced, it will drive the pole two more inches? Why did it not do this before it was removed, while it was still pressing? Do you suppose that just taking it off and gently replacing it makes it do that which it could not do before?

Apr. I can only blush and admit that I was in danger of drowning in a glass of water.

Salv. Do not reproach yourself, Aproino, for I can assure you that you have plenty of company in remaining fastened by knots that are in fact quite easy to untie. No doubt every fallacy would be inherently easy to discover, if people went about untangling it and resolving it into its principles, for it cannot be but that something connected with it, or close to it, would plainly reveal its falsity. Our Academician had a certain special genius in such cases for reducing with a few words to absurdity and contradiction conclusions that are palpably false, and which nevertheless have hitherto been believed to be true. I have collected many conclusions in physics that had always passed for true, which were later shown by him to be false by means of brief and quite simple reasoning.

Sagr. Truly this is one of them, and if the others are like this, it will be good that at some time you will share them with us. But meanwhile let us continue with the question we have undertaken; we are searching for a way (if there is one) in which to give a rule and assign a just and known measure to the force of impact. It seems to me that this cannot be had through the experience proposed; for as sensible experiment shows us, repeated blows of the pile driver on the pole do drive it further and further, and it is clear that each succeeding blow does act, which is not true of the dead weight. Having acted when it made its first pres-

sure, it does not go on and produce the effect of the second
[blow] when replaced; that is, [it does not] again drive the
pole. Indeed, it is clearly seen that for this second entrance
we need a weight of more than 1000 pounds; and if we want
with dead weights to equal the entrances of the third, fourth,
and fifth blow, and so on, we shall need the heaviness of
328 continually greater and greater dead weights. Now, which
of these can we take as a constant and secure measure of the
force of that blow which, considered by itself, seems to be
always the same?

Salv. This is one of the prime marvels that I believe must
doubtless have held in perplexity and hesitation all speculative minds. Who, indeed, will not find it novel to hear
that the measure of the force of impact must be taken not
from that which strikes, but rather from that which receives
the impact? As to the experiment cited, it seems to me that
from this one may deduce the force of impact to be infinite
—or rather, let us say indeterminate, or undeterminable,
being now greater and now less, according as it is applied
to a greater or lesser resistance.

Sagr. Already I seem to understand that the truth may
be that the force of impact is immense, or infinite. For in
the above experiment, given that the first blow will drive
the pole four inches and the second, three, and continuing
ever to encounter firmer ground, the third blow will drive
it two inches, the fourth an inch and one-half, the ensuing
ones a single inch, one-half, one-fourth, and so on; it seems
that unless the resistance of the pole is to become infinite
through this firming of the ground, then repeated blows
will always budge the pole, but always through shorter and
shorter distances. But since the distance may become as small
as you please, and is always divisible and subdivisible, entrance [of the pole] will continue; and this effect having
to be made by the dead weight, each [movement] will require
more weight than the preceding. Hence it may be that in
order to equal the force of the latest blows, a weight immensely greater and greater will be required.

Salv. So I should certainly think.

Apr. Then there cannot be any resistance so great as to
remain firm and obstinate against the power of any impact,
however light?[7]

7. The compact phrasing here is meant to convey the double idea that
(1) no resistance exists that can withstand a blow of unlimited strength,

Salv. I think not, unless what is struck is completely immovable; that is, unless its resistance is infinite.

Sagr. These statements seem remarkable, and so to speak, prodigious. It appears that in this effect [and in this] alone, art may overpower and defraud nature—something that at first glance it [mistakenly] appears that other mechanical instruments can do, very heavy weights being raised with small force by the power of the lever, screw, pulleys, and the **329** rest. But in this effect of impact, a few blows of a hammer weighing no more than ten or twelve pounds may flatten a cube of copper that is not broken or mashed by resting a big marble steeple or even a very high tower upon the hammer. This seems to me to defeat all the physical reasoning by which one might try to remove the wonder from it. Therefore, Salviati, take the clue in your hand and lead us from this complicated maze.

Salv. From what you two have to say, it appears that the principal knot of the difficulty lies in puzzlement why the action of impact, which seems infinite, may arise in a different way from that of other machines which overcome immense resistances with very small forces. But I do not despair of explaining how in this, too, one proceeds in the same manner. I shall try to clear up the process; and though it seems to me quite complicated, perhaps, as a result of your questions and objections, my remarks may become more subtle and acute, and sufficient at least to loosen the knot, if not to untie it.

It is evident that the property [*facultà*] of force in the mover and [that] of resistance in the moved is not single and simple, but is compounded from two actions, by which their energy must be measured. One of these is the weight, of the mover as well as of the resistent; the other is the speed with which the one must move and the other be moved. Thus, if the moved must be moved with the speed of the mover—that is, if the spaces traversed by both in a given time are equal— it will be impossible for the heaviness of the mover to be less than that of the moved, but rather it must be somewhat greater; for in exact equality [of weight] resides equilibrium and rest, as seen in the balance of equal arms. But if with a lesser weight we wish to raise a greater, it will be necessary

---

and (2) any impact, however small, has some effect on any given resistent. Cf. note 26 to Fourth Day; pp. **337**, **341**, below; and Fragment 4 at end.

to arrange the machine in such a way that the smaller moving weight goes in the same time through a greater space than does the other weight; that is to say, the former is moved more swiftly than the latter. And thus we are taught by experience that in the steelyard, for example, in order for the counterweight to raise a weight ten or fifteen times as heavy, the distance along the beam from the center round which it turns must be ten or fifteen times as great as the distance

330 between that same center and the point of suspension of the other weight; and this is the same as to say that the speed of the mover is ten or fifteen times as great as that of the moved. Since this is found to happen in all the other instruments, we may take it as established that the weights and speeds are inversely proportional. Let us say in general, then, that the momentum of the less heavy body balances the momentum of the more heavy when the speed of the lesser has the same ratio to the speed of the greater as the heaviness of the greater has to that of the lesser—to which, any small advantage being allowed, equilibrium is overcome and motion is introduced.

This settled, I say that not only in impact does the action [*operazione*] seem infinite as to the overcoming of whatever great resistance, but that this also shows itself in every other mechanical device. For it is clear that a tiny weight of one pound, descending, will raise a weight of 100 or 1000 or as much more as you please, if we place it 100 or 1000 times as far from the center on the arm of the steelyard as the other, great, weight; that is, if we make the space through which the former shall descend to be 100 or 1000 or more times as great as the space through which the other is to rise, so that the speed of the former is 100 or 1000 times the speed of the latter. Yet I wish, by means of a more striking example, to make it palpable to you that any little weight, descending, makes any immense or very heavy bulk ascend.

Suppose a vast weight to be attached to a rope fastened to a firm high place, around which as center you are to imagine to be described the circumference of a circle that passes through the center of gravity of the suspended bulk. This center of gravity, you know, will be vertically beneath the suspending rope; or, to put it better, will be in that straight line which goes from the point of suspension to the common center of all heavy things; that is, the center of the earth. Next, imagine a fine thread to which any weight, as small as you please, is attached in such a way that its center of gravity always remains in the previously mentioned circumference; and suppose that this little weight just touches and rests against the vast bulk. Do you not believe that this new weight, added at the side, will push the greater one somewhat, separating its center of gravity from the previously mentioned vertical line in which it originally lay? Yet it will unquestionably move along the circumference mentioned, and being moved, it will separate from the horizontal line tangent   **331** to the lowest point of the circumference in which the center of gravity of this vast bulk was situated. As to the space, the arc passed through by the heavy weight will be the same as that passed through by the tiny weight which was supported against the vast one. Yet the rise of the center of the great weight will not thereby equal the descent of the center of the tiny weight, because the latter descends through a place or space much more tilted than that of the ascent of the other center, which is made in a certain way from the tangent of the circle along an angle less than the most acute [rectilineal] angle.[8] Here, if I were dealing with people less versed in geometry than you are, I should demonstrate how a moveable leaving [along a circle] from the lowest point of tangency [with the horizontal], its [vertical] rise from [*della*] the horizontal line to some point in the circumference outside [*separato dal*] the tangent may be smaller in any desired ratio than its [vertical] drop along an equal arc [*asse*] taken at any other place not containing the point of tangency; but surely you have no doubt as to this.[9]

Now, if the simple touching of the tiny weight against the great bulk can move and raise that, what will it do when, drawn back and allowed to run along the circumference, it comes to strike there?

Apr. Truly, it seems to me that there is no room left for doubt that the force of impact is infinite, from what the experiment adduced explains about it. But this information does not suffice my mind for the clearing away of many dark shadows which hold it so obscured that I do not see how

8. The "mixed angle" of note 33 to Second Day; cf. Euclid, *Elements* III.16.

9. *Asse* (axis) was probably a scribal error for *arco*; the idea is that the vertical drop for any arc not touching the lowest point is greater than the drop for an equal arc touching that point, while the latter may be made as small as one pleases by shortening the arc. Many dubious readings are found in this posthumously printed work; cf. note 10, below.

this business of impacts proceeds; at least, not so that I could reply to every question that might be asked of me.

Salv. Before going further, I want to reveal to you a certain equivocation that is lurking in ambush. This lets us believe that, in the previous example, all blows on the pole were equal (or the same), being made by the same pile driver raised always to the same height. But this does not follow. To understand this, imagine striking with your hand against a ball that comes falling from above, and tell me: if, when this arrived upon your hand, you were to have your hand sinking along the same line and with the same speed as the ball, what shock would you feel? Surely none. But if, upon the arrival of the ball, you yielded only in part, by dropping your hand with less speed than that of the ball,

332   you would indeed receive an impact—not as with the whole speed of the ball, but only as with the excess of its speed over that of the dropping of your hand. Thus if the ball should descend with ten degrees of speed, and your hand yielded with eight, the blow would be made as by two degrees of speed of the ball. The hand yielding with four [degrees], the blow would be as six; and the yielding being as one, the blow would be as nine; the entire impact of the speed of ten degrees would be [only] that which struck the hand that did not yield.

Now apply this reasoning to the pile driver, when the pole yields to the impetus of the pile driver four inches the first time, and two [inches] the second, and a single inch the third. These impacts come out unequal, the first being weaker than the second, and the second than the third, according as the yielding of four inches retires[10] more from the [initial] speed of the first blow than the second [yielding of only two inches], and the second [impact] is weaker than the third, which takes away twice as much as the second from the same [initial] speed. Hence, if the great yielding of the pole to the first shock, and its lesser yielding to the second and still less to the third, and so on continually, is the reason that the first blow is less effective [*valido*] than the

10. Reading *retrae* for *detrae* of the printed text. In pursuance of his previous argument, Galileo reasons that even though the terminal speed of fall (initial speed of impact) is the same in each case, we should call the effective blow, or impact, weaker in the earlier strokes, because the pole offers less resistance. A quite different adumbration of Newton's third law of motion was already present in Galileo's first work; cf. *On Motion*, pp. 64, 109 (*Opere*, I, 297, 336).

second, and this than the third, what wonder is it if a lesser quantity of dead weight is needed for the first driving of four inches, and more is needed for the second, of two inches, and still more for the third, and always more and more continually, in proportion as the drivings go diminishing with diminutions of the yielding of the pole, which amounts to saying with the increase of the resistance?

From what I have said, it seems to me that one may easily gather how difficult it is to determine anything about the force of impact made upon a resistent that varies its yielding, such as this pole that becomes indeterminately more and more resisting. Hence I think it necessary to give thought to something that receives the impacts and always opposes them with the same resistance. Now, to establish such a resistent, I want you to imagine a solid weight of, say, 1000 pounds, placed on a plane that sustains it. Next, I want you to think of a rope tied to this weight and led over a pulley fixed high above. Here it is evident that when force is applied by pulling down on the end of the rope, it will always meet with quite equal resistance in raising the weight; that is, the   **333** opposition of 1000 pounds of weight. For if from the end of the rope there were suspended another weight, equal to the first, equilibrium would be established; and being raised up without support from anything below, they would remain still; nor would this second weight descend and raise the first unless given some excess of weight. And if we rest the first weight on the said plane that sustains it, we can use other weights of varying heaviness (though each of them less than the weight sustained at rest) to test what the forces of different impacts are. [This is done] by tying such weights to the end of the rope and then letting them fall from a given height, observing what happens at the other end to that great solid that feels the pull of the falling weight, which pull will be to that large weight as a blow that would drive it upward.

Here, in the first place, it seems to me to follow that however small the falling weight, it should undoubtedly overcome the resistance of the heavy weight and lift it up. This consequence seems to me to be conclusively drawn from our certainty that a smaller weight will prevail over another, however much greater, whenever the speed of the lesser shall have, to the speed of the greater, a greater ratio than the weight of the greater has to the weight of the smaller; and this [always] happens in the present instance, since the speed of the falling weight infinitely surpasses the speed of the other, whose speed is nil when it is sustained at rest. But the heaviness of the falling solid is not nil in relation to that of the other, since we did not assume the latter to be infinite, or the former to be nil; hence the force of this percussent will overcome the resistance of that on which it makes its impact.

Next we shall seek to find out how great is the space through which the impact received will raise it, and whether perhaps this [distance] will correspond to that of other mechanical instruments. Thus it is seen in the steelyard, for example, that the rise of the heavy weight will be that part of the fall of the counterweight, which the weight of the counterweight is of the greater weight. So in our case we should have to see, supposing the weight of the big resting solid to be 1000 times that of the falling weight—which falls, let us say, from a height of one braccio—whether this raises the other [weight] one one-hundredth of a braccio; if so, it would appear to be following the rule for the other mechanical instruments. Let us imagine making the first experiment by dropping from **334** some height, say one braccio, a weight equal to the other, which we have placed on a [supporting] plane, these weights being tied to the opposite ends of the same rope. What shall we believe to be the effect of the pull of the falling weight, with regard to the moving and raising of the other, which was at rest? I should be glad to hear your opinion.

Apr. Since you look at me, as if you were waiting for my reply, it appears to me that the two weights being equally heavy, and the one which falls having in addition the impetus of its speed, the other must be raised by it far beyond equilibration, inasmuch as the mere weight of the other was sufficient to hold it in balance. Hence, in my opinion, it will rise through much more than a space of one braccio, which is the measure of the descent of the falling weight.

Salv. And what do you say, Sagredo?

Sagr. The reasoning seems conclusive to me at first glance; but, as I said a while ago, many experiences have taught me how easily one may be deceived, and accordingly how necessary it is to go circumspectly before boldly pronouncing and affirming anything. Hence I shall say, still dubiously, that it is true that the weight of 100 pounds of the falling heavy body will suffice to raise the other, which also weighs 100 pounds, as far as to equilibrium, even without its being

endowed and supplied with speed; [to do this,] the excess
of a mere half-ounce will suffice. But I also think that that
equilibration will be made very slowly, and hence that when
the falling body acts with great speed, it will necessarily raise
its companion on high with like speed. Now, there seems
to me no doubt that greater force is needed to drive a heavy
body upward with great speed than to push it very slowly;[11]
so it might happen that the advantage of the speed acquired
by the falling body in free fall through one braccio would
be consumed, and so to speak spent, in driving the other
with equal speed to a like height. Hence I am inclined to
believe that these two movements, upward and downward,
would end in rest immediately after the rising weight had
gone up one braccio, which would mean two braccia of   **335**
fall for the other, counting the first braccio of free fall as
executed by that one alone.

Salv. I truly lean toward the same belief. For though the
falling weight is an aggregate of heaviness and speed, the
operation of its heaviness in raising the other [weight] is nil, this
being opposed by the resistance of equal heaviness in that
other, which clearly would not be moved without the addition
of some small weight. Therefore the operation is entirely that
of the speed, which can confer nothing but speed.[12] Being
unable to confer other [speed] than what it has, and having
nothing other than that which it acquired in the descent
of one braccio after leaving from rest, it will drive the other
upward through a like space and with a like speed, in agreement with what can be discerned in various experiences;
namely, that the falling weight, leaving from rest, is everywhere found to have that impetus which suffices to restore
it to the original height.

Sagr. I recall that this is clearly shown by a weight hanging
from a thread fixed above. Removed from the vertical by
any arc less than a quadrant, and set free, this weight descends

11. Galileo's emphasis on speed as such underlay the essential difference
between his mechanics and that of medieval, as that of Cartesian, writers;
cf. note 32 to Fourth Day. But see also note 15, below.

12. This inference was probably suggested by the use of compound ratios
in physics (note 4, above). The Aristotelian position was very different,
defining greater "force" or "power" in terms of the imparting of greater
speed; cf. *Physica* vii, § 5, especially at line 250a. Medieval physicists followed
that lead; thus the theory of proportions in motion developed by Thomas
Bradwardine (1290?–1349) was intended to justify precisely this passage
in Aristotle.

and passes beyond the vertical, rising through an arc equal
to that of its descent. From this it is evident that the ascent
derives entirely from the speed acquired in descent, inasmuch
as in [any] rising upward, the weight of the moving body
can have played no part. Indeed, that weight, resisting ascent,
goes despoiling the moveable of the speed with which it was
endowed by the descent.

Salv. If the example of what is done by the heavy solid
on the thread, of which I remember that we spoke in our
336   discussions of days past, squared and fitted as well with the
case we are now dealing with as it fits with the facts [*alla
verità*], your reasoning would be very cogent. But I find
no trifling discrepancy between these two operations; I
mean between that of the heavy solid hanging from the
thread, which released from a height and descending along
the circumference of a circle, acquires impetus to transport
itself to another equal height, and this other operation of
the falling body tied to the end of a rope in order to lift another
one equal to itself in weight. For that which descends along
the circle continues to acquire speed as far as the vertical
[position], favored by its own weight, which impedes its
ascent as soon as the vertical is passed, ascent being a motion
contrary to its heaviness. Thus [in return] for the impetus
acquired in natural descent, it is no small repayment to be
carried along by violent motion or through a height. But
in the other case, the falling weight comes upon its equal
placed at rest, not only with its acquired speed but with
its heaviness as well; and this [heaviness], being maintained,
by itself alone removes all resistance on the part of its companion [weight] to being lifted.[13] Hence the [previously] acquired
speed meets with no opposition from any weight that resists
rising; and just as impetus conferred downward on a heavy
body would encounter no cause in that [body] for annihilation
or retardation [of that impetus], so none is encountered in
that rising weight whose [effective] heaviness remains nil,
being counterpoised by the other, descending, weight.

Here, it seems to me, precisely the same thing takes place

13. Here Galileo begins to speak of an inertial motion in the modern
sense, using balanced weights for the study of impact rather than the usual
and intuitive analysis in terms of frictionless bodies striking while supported
on a hard flat surface. It was on the latter basis that Descartes deduced, in
contradiction with the ensuing discussion, that a smaller body could never
budge a larger one, however great the speed of the smaller.

which happens to a heavy and perfectly round moveable placed on a very smooth plane, somewhat inclined; this will descend naturally by itself, acquiring ever greater speed. But if, on the other hand, anyone should wish to drive it upward from the lower part [of the plane], he would have to confer impetus on it, and this would be ever diminished and finally annihilated [in the rise]. If the plane were not inclined, but horizontal, then this round solid placed on it would do whatever we wish; that is, if we place it at rest, it will remain at rest, and given an impetus in any direction, it will move in that direction, maintaining always the same speed that it shall have received from our hand and having no action [by which] to increase or diminish this, there being neither rise nor drop in that plane. And in this same way the two equal weights, hanging from the ends of the rope, will be at rest when placed in balance, and if impetus downward shall **337** be given to one, it will always conserve this equably. Here it is to be noted that all these things would follow if there were removed all external and accidental impediments, as of roughness and heaviness of rope or pulleys, of friction in the turning of these about the axle, and whatever others there may be of these.

But since we are considering the speed acquired by one of these weights in descent from some height while the other remains at rest, it will be good to determine what and how much must be the speed with which both would be moved after the [initial] fall of the one, this descending and the other ascending. From what is already demonstrated, we know that a heavy body which falls freely on departing from rest perpetually acquires a greater and greater degree of speed; hence in our case, the greatest degree of speed of the heavy body, while it descends freely, is that which it is found to have at the point at which it commences to lift its companion. Now it is evident that this degree of speed will not go on increasing when its cause of increase is taken away, this being the weight of the descending body itself; for its weight no longer acts when its propensity to descend is taken away by the repugnance to rising of its companion of equal weight. Hence the maximum degree of speed will be conserved, and the motion will be converted from one of acceleration to uniform motion.[14]

14. Reduction of accelerated motion to some equivalent uniform motion was essential before development of the calculus; cf. Third Day, Bk. II,

What the future speed will then be is manifest from the things demonstrated and seen in the [discussions of the] past days. That is, the future speed will be such that, in another time equal to that of the [initial free] descent, double the space of [free] fall would be passed.

Sagr. Then Apronio has philosophized better than I have. Thus far I am well satisfied by your reasoning, and admit what you have told me as most true. But I still do not feel that I have learned enough to remove the great wonder I feel at seeing very great resistances overcome by the force of impact of the striking body when its weight is not great and its speed not excessive. It increases my bafflement to hear you affirm that there is no resistance short of the infinite that will resist a blow without yielding, and moreover that there is no way of assigning a definite measure to [the force of] such a blow. So it is our wish that you attempt to shed light in this darkness.

338   *Salv.* No demonstration can be applied to a proposition unless what is given is one and certain; and since we wish to philosophize about the force of a striking body and the resistance of one which receives the impact, we must choose a percussent whose force shall be always the same, such as that of the same heavy body falling always from the same height; and likewise let us establish a recipient of the blow that will always offer the same resistance. To have this, and keep to the above example of the two heavy bodies hanging from the ends of the same rope, I shall have the percussent be the small weight that is allowed to fall, and the other shall be a weight as much greater [than this] as you please, in the raising of which the impetus of the small falling weight is to be exercised. It is manifest that the resistance of the larger body is the same at all times and all places, as would not be the case with the resistance of a nail, or of the pole, in which resistance increases continually with penetration, but in some unknown ratio because of the various accidental events involved, such as hardness of wood or ground, and so on, even though the nail and the pole remain always the same. It is further necessary to remember some true conclusions of which we spoke in past days in the treatise on motion. The

---

Theorem 1. No clear physics of force was likely to emerge under the theory of proportion alone, in which force can appear only as some kind of relation rather than as an entity. As will be seen in Salviati's next speech, "force" is simply made to cancel out.

first of these is that heavy bodies, in falling from a high point to a horizontal plane beneath, acquire equal degrees of speed whether their descent is made vertically or upon any of diversely inclined planes.

For example, *AB* being a horizontal plane upon which the vertical *CB* is dropped from point *C*, and other planes, diversely inclined, *CA, CD*, and *CE*, fall from the same *C*, we must understand that the degrees of speed of bodies falling from the high point *C* along any of the lines going from *C* to end at the horizontal are all equal. In the second place, it is assumed that the impetus acquired at *A* by the body falling from the point *C* is such that it is exactly needed to drive the same falling body (or another one equal to it) up to the same height, from which we may understand that such force is required to raise that same heavy body from the horizontal to height *C*, whether it is driven from point *A, D, E*, or *B*. Let us recall in the third place that the times of descent along the designated planes have the same ratio as the lengths of these planes, so that if the plane *AC*, for example, were double the length of *CE* and quadruple that of *CB*, the time of descent along *CA* would be double the time of descent along *CE* and four times that along *CB*. Further, let us recall that in order to pull the same weight over diverse inclined planes, lesser force will always suffice to move it over one which is more inclined [to the vertical] than over one less inclined, according as the length of the latter is less than the length of the former.

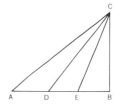

339

Now, these truths being supposed, let us take the plane *AC* to be, say, ten times as long as the vertical *CB*, and let there be placed on *AC* a solid, *S*, weighing 100 pounds. It is manifest that if a cord is attached to this solid, riding over a pulley placed above the point *C*, and to the other end of this cord a weight of ten pounds is attached, which shall be the weight of *P*, then that weight *P* will descend with any small addition of force, drawing the weight *S* along the plane *AC*. Here one must note that the space through which the greater weight moves over the plane beneath it is equal to the space through which the small descending weight is moved; from this, someone might question the general truth applying to all mechanical propositions, which is that a small force does not overcome and move a great resistance unless the motion of the former exceeds the motion of the latter in inverse ratio of their weights. But in the present instance

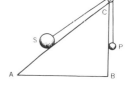

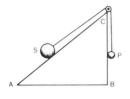

the descent of the small weight, which is vertical, must be compared [only] with the vertical rise of the great solid *S*, observing how much this is lifted vertically from the horizontal; that is, one must consider how much *S* rises in the vertical *BC*.

Having made various meditations, gentlemen, about the setting forth of that which remains to be said by me, which is the crux of the present matter, I affirm the following conclusion, which will then be explained and demonstrated.

340

PROPOSITION

If the effect made by an impact of the same weight falling from the same height shall be to drive a resistent of constant resistance through some space; and [if] to produce a similar effect there is needed a determined quantity of dead weight [merely] pressing, without impact, I say that if the original percussent, [acting] upon some greater resistent, with the given impact shall drive it (for example) through one-half the space that the other was driven, then in order to accomplish this second driving, the pressure of the said dead weight will not suffice, but there will be required another one, twice as heavy. And similarly in all other ratios, when a shorter [constantly resisted] drive is made by the same percussent, then inversely by that much there will be required, to do the same, a greater pressing quantity of dead weight.

In the earlier example of the pole, the resistance is to be understood to be such that it cannot be overcome by less than one hundred pounds of dead weight pressing, and [it is understood that] the weight of the percussent is only ten pounds, falling from a height of, say, four braccia, and driving the pole four inches. Here, in the first place, it is evident that the weight of ten pounds falling vertically will be sufficient to raise a weight of one hundred pounds along a plane so inclined that its length is ten times its height, according to what has been said above; and that as much force is needed to raise ten pounds of weight vertically as to raise one hundred on a plane whose length is ten times its vertical elevation. Hence if the impetus acquired by the falling body through such a vertical space is applied to raise another that is equal to it in resistance, it will raise it a like space; but the resistance of the vertically falling body of ten pounds is equal to that of the body of one hundred pounds rising along a plane of length ten times its vertical height. Therefore, let the weight of ten pounds fall through any height vertically, and its acquired impetus, applied to the weight of one hundred pounds, will drive this through as much space on the inclined plane as corresponds to the vertical height as great as one-tenth part of this inclined space. And it is already concluded above that the force able to drive a weight on an inclined plane is sufficient to drive it through the vertical corres-   **341** ponding to the height of this inclined plane—which vertical, in the present instance, is one-tenth the space passed along the incline, which is equal to the space of fall of the first weight, of ten pounds.

Thus it is manifest that the fall of the weight of ten pounds made vertically is sufficient to raise the weight of one hundred pounds, also vertically, but only through a space that is one-tenth the descent of the falling body of ten pounds.[15] But that force which can raise a weight of one hundred pounds is equal to the force with which the same weight of one hundred pounds presses down, and this was its power to drive the pole when placed upon it and pressing it. Behold, therefore, the explanation how the fall of ten pounds of weight is able to drive a resistance equivalent to that which a weight of one hundred pounds has to being raised, while the driving will be no more than one-tenth the descent of the percussent. And if we now assume the resistance of the pole to be doubled or tripled, so that to overcome it there is needed the pressure of two hundred or three hundred pounds of dead weight, then repeating the reasoning, we shall find that the impetus of the ten pounds falling vertically will be able to drive the pole the second and the third time, as it did the first time; and as [far as] the tenth part of its fall the first time, so the twentieth the second time, and the third time, the thirtieth of this descent. And thus, multiplying the resistance *in infinitum*, the same blow will always be able to overcome it, but by driving the resisting body always through less and less space, in inverse [*alterna*] proportion; from which it seems that we may reasonably assert the force of impact to be infinite.[16]

But we must also consider that in another way, the force

15. Here the neglect of time (or speed), unusual for Galileo, results in his adoption of a conservation principle in terms of vertical displacements alone, akin to the medieval and Cartesian approaches (cf. note 13, above) but not restricted to connected motions as in the simple machines.

16. Cf. notes 26 to Fourth Day, note 7, above, and Fragment 4 at end.

of pressing without impact is also infinite, inasmuch as if it overcomes the resistance of the pole, it will drive it not merely through some space through which the blow will have driven it, but will continue to drive it *in infinitum*.[17]

Sagr. Truly, I perceive that your attack travels very directly to the investigation of the true cause of the present problem; but since it appears to me that impact may be created in many ways, and applied to a great variety of resistances, I believe it is necessary to go on and explain some [of these] at least, the understanding of which might open our minds to the understanding of all.

342   *Salv.* You say well, and I have already prepared myself to give examples. For one thing, we shall say that at times it may happen that the operation of the percussent is revealed not on the thing struck, but in the percussent itself. Thus, a blow being struck on a fixed anvil with a lead hammer, the effect will happen to the hammer, which will be flattened, rather than to the anvil, which will not descend. Not unlike this is the effect of the mallet on the sculptor's chisel; for the mallet being of soft untempered iron and striking repeatedly on the chisel of hard tempered steel, it is not the chisel that is damaged, but the mallet that becomes dented and lacerated. Again, in another way, the effect is reflected solely in the percussent; thus we see not infrequently that if one continues to drive a nail into very hard wood, the hammer [finally] rebounds without driving the nail forward at all, and we say in this case that the blow did not "take." Not very different is the bouncing of an inflated ball on a hard pavement, or of any other body so disposed, which indeed yields to the impact, but returns to its first shape as by arching, and such a rebound occurs not only when that which strikes yields and then recovers, but also when the same occurs in that upon which it strikes; and in such a manner a ball bounces when it is of very hard and unyielding material, but falls on the tightly stretched membrane of a drum.

Also perceived with great wonder is that effect produced when a blow is added to pressure without impact, making a compound of the two. We see this in mangles or olive presses and the like, when by the simple pushing of several

17. The seeming incongruity between an infinite force of impact and the finite force of dead weight (in Galileo's sense) is here removed by him through showing how either finite or infinite strength may be attributed to impact in one way or to steady pressure in another way; see also Fragment 1, at end.

men the screw has been made to go down as far as they can manage. By drawing back a step from the bar and then striking swiftly against it, they move the screw more and more, and get it to such a point that the shock of the force of four or six men will achieve what mere pushing by a dozen or a score could not do. In this case it is required that the bar be very thick and of very hard wood, so that it bends little or not at all; for if it should give, the blow would be spent in bending it.

[Fragments]¹⁸

[1] In every moveable that is to be moved by force, it seems **343** that there are two distinct species of resistance. One relates to that internal resistance which makes us say that a body weighing a thousand pounds is harder to raise than one of a hundred; the other relates to the space through which motion must be made, as a stone requires greater force to be thrown one hundred paces than fifty, and so on. To these different resistances correspond proportionably the two different movers—the one that moves [a thing] by pressing without striking, and the other that acts by striking. The mover that operates without impact moves only a resistance which is less, though [it may be] only insensibly [less], than the power [*virtù*] or the pressing heaviness; but that will move it through an infinite distance, accompanying it always with its same force. That which moves by striking, moves any resistance, though [this may be] immense; but [moves it only] through a limited distance.

Hence I consider these two propositions true: that the percussent moves an infinite resistance through a finite and limited interval, while the pressing [force] moves a finite and limited resistance through an infinite interval; hence to the percussent, the interval is proportionable, and not the resistance, while to the pressing [force] the resistance, and not the interval [is proportionable]. These things make me doubt whether Sagredo's question has an answer, as one that seeks to equate things that are incommensurable; for

18. These fragments were collected and published at the end of this dialogue by the editors of the 1718 edition of Galileo's works, probably from manuscripts no longer extant. It is possible that despite the editor's assertion, they were not all in Galileo's own hand; particularly Fragment 2 seems suspect. Numbers are here assigned for convenience of reference; none were given in the original printed version.

such, I believe, are the actions of impact and of pressing.[19]
Thus, in this particular case [illustrated], any immense

344 resistance that may exist in the wedge *BA* will be moved by
any percussent *C*, but [only] through a limited interval,
as between points *B* and *A*, while the pressing [force] *D* will
not drive just any resistance existing in the wedge *BA*, but
only a limited resistance, and one not greater than the
weight *D*. That, however, will be driven not [only] through
the limited interval between points *B* and *A*, but *in infinitum*, provided that the resistance in the movable body *AB*
remains always equal, as must be assumed, nothing to the
contrary having been mentioned in the inquiry.

[2] The momentum of a body in the act of impact is nothing
but a composite and aggregate of infinitely many momenta,
each of them equal only to a single moment [*al solo momento*],[20] either internal and natural *per se*, as is that [moment]
of its own absolute weight which it eternally exercises when
placed on any resistant body, or else extrinsic and violent,
as is that [momentum] of the moving power [*forza*]. Such
momenta go accumulating during the time of [naturally
accelerated] motion of the heavy body from instant to
instant with equal increments, and are stored therein, in
exactly the way that the speed of a falling body goes increasing;
for as in the infinitely many instants of a time, however short,
a heavy body goes ever passing through new and equal
degrees of speed, always retaining those acquired in the
previously elapsed time, so also in the moveable those momenta (either natural or violent, conferred on it by nature or
by art) go conserving [themselves] and compounding from
instant to instant, etc.

19. This concept of incommensurability in the Euclidean sense, carried
over by Galileo into physics, disappeared in the algebraic treatment of
nearly every later writer. The authenticity of this fragment appears to me
incontestable, and its content suggests some probable reasons for which
Galileo finally decided to withhold the Added Day from publication.

20. Because the word *momento* is used in both senses of "static moment"
and "momentum" (see Glossary), not alternately, but in a mixed way un-
characteristic of Galileo, this fragment is hard to translate. It fits much better
with Torricelli's later modification of Galileo's thought than with his own
writings, and it may be apocryphal. The idea here is that moments, like
degrees of speed, are uniformly added with time, so that the momentum
of a body on impact resembles its terminal speed in free fall as being a finite
aggregate of infinitely many unquantifiable parts; cf. *parti non quante* in
Glossary.

[3] The force of impact is equivalent to [*di*] infinite [static] moment, provided that it is applied in one momentum and in one instant by the striking heavy body, upon unyielding material, as will be demonstrated.

[4] The yielding of a material struck by a heavy body moved at any speed cannot take place instantaneously, because otherwise there would be instantaneous movement through some **345** finite space, which is demonstrably impossible. If therefore the yielding in the place struck takes time, time is also required for the application of those momenta acquired in the motion of the percussent, which time is sufficient to extinguish and dissipate in part that aggregate of the aforesaid momenta. These, if they were exercised against the resisting body in an instant (as would happen if the materials of the thing struck and of the percussent did not yield at all), would absolutely have an effect and an action far greater, in moving it and overcoming it, than if applied in a time, however short. I say "greater effect" because they will have some effect against the thing struck, however tiny the blow or however swift the yielding; but this effect may perhaps be imperceptible to our senses, even though it really exists, as we shall demonstrate in the proper place. Yet that is also clearly revealed by experience, since, if with quite a small hammer one shall strike with uniform impacts against the end of a very large beam that is lying on the ground, then after a great many impacts, the beam will eventually be seen to have been moved through some perceptile space—a most evident sign that every impact acted separately on its own, in driving the beam.[21] For if the first impact had no part in the effect, then all those which followed, as in the place of the first, would achieve nothing at all; which is contrary to experience, to sense, and to the proof that will be given, etc.

[5] The force of impact is infinite in [equivalent static] moment, because there is no resistance, however large, that is not overcome by the force of the tiniest impact.

[6] He who shuts the bronze doors of San Giovanni will try in vain to close them with one single simple push; but with a

---

21. Mersenne, in the preface to his French translation of Galileo's *Mechanics* (Paris, 1634) alluded to an experiment of this sort carried out by Galileo, though there is no trace of it in his published works.

continual impulse he goes impressing on that very heavy movable body such a force that when it comes to strike and knock against the jamb, it makes the whole church tremble. From this one sees how there is impressed in moveables—and the more, the heavier these are—and how there is multiplied and conserved in them the force that has been communicated to them over some time, etc.

346   A similar effect is seen in a great bell, which is not set in strong and impetuous motion with a single pull of its rope, nor with four, or six [pulls], but [is] with a great many. These being long repeated, the final [pulls] add force to that acquired from the preceding pulls; and the thicker and heavier the bell shall be, the more force and impetus it acquires, this being communicated to it in a longer time and by a larger number of pulls than are required for a small bell, into which impetus is readily put, but from which it is also readily taken away, this [small bell] not drinking in, so to speak, as much force as the larger one.

A similar thing happens also in ships, which are not set in full course by the first tugs at the oars, or by the first impulses of wind; but, by continual rowing or continual impression of force made by the wind on the sails, they acquire very great impetus, capable of breaking the vessels themselves, when, carried by this [impetus] they strike a reef.

[7] A weak but long bow of a balestra will sometimes make a greater throw than another [bow] much stronger but not as long; for the former, accompanying the ball for a longer time, goes on continually impressing force on it, while the latter soon abandons it.

[*The   End*]

Reference
Matter

Bibliography of
Principal Editions and Translations of
the *Two New Sciences*

1. *Discorsi e dimostrazioni matematiche, intorno a due nuove scienze attenenti alla Mecanica & i Movimenti Locali . . . con una Appendice del centro di gravità d'alcuni Solidi.* Leyden, 1638.
2. *Les nouvelles pensées de Galilée.* Paris, 1639. Partial translation and paraphrase by Marin Mersenne.
3. *Discorsi . . . .* Bologna, 1655. In *Opere del Galileo*, Bologna, 1655–56. Contains the addition dictated to Viviani.
4. *Mathematical Discourses and Demonstrations, Touching Two New Sciences . . . Englished from the Originall Latine and Italian*, by Thomas Salusbury, Esq. London, 1665 (but actually printed in 1662 and published in 1666). In Thomas Salusbury, *Mathematical Collections and Translations: The Second Tome . . . The First Part.* London, 1665. English translation from the 1655 edition. Facsimile reprint (of both tomes) issued in 1967 by Dawson's of Pall Mall (London) and Zeitlin and Ver Brugge (Los Angeles).
5. *Principio della Quinta Giornata del Galileo. Da aggiugnersi all'altre quattro de' Discorsi . . .* In Vincenzio Viviani, *Quinto Libro degli Elementi d'Euclide, ovvero Scienza universale delle proporzioni . . . .* Florence, 1674. This so-called Fifth Day of the *Discorsi* was reprinted several times in a small two-volume Euclid, Florence, 1690, 1718, etc.
6. *Discursus et demonstrationes mathematicae, circa duas novas scientias . . . .* Leyden, 1699. Latin translation from the edition of 1638. Issued with the Latin translation of Galileo's *Dialogue* by Mathias Bernegger (1st ed., 1635) under the general title, *Galilaei Galilaei Lyncei Dialogi, tam eos quos . edidit De Systemate Mundi quam quos De Motu Locali.* Leyden, 1699–1700. Possibly this translation was made under Galileo's own direction for a proposed Latin edition of his works projected by the Elzevirs.
7. *Discorsi . . . .* In *Opere di Galileo Galilei . . .*, vol. 2 (of 3). Florence, 1718; contains the so-called Fifth Day (see 5,

above) and the added dialogue on percussion, called the Sixth Day. Subsequent collected editions generally contain both additional "days."

8. *Mathematical Discourses Concerning Two New Sciences . . . Done into English from the Italian, by Tho. Weston, late Master, and now publish'd by John Weston, present Master, of the Academy at Greenwich.* London, 1730. English translation from the edition of 1638, apparently with the Salusbury and the Latin translations (4 and 6, above) at hand. A reissue, differing only as to title page, appeared in 1734.

9. *Discorsi* . . . . In *Opere di Galileo Galilei* . . . , vol. 3 (of 4). Padua, 1744.

10. *Discorsi* . . . . In *Opere di Galileo Galilei* . . . , vols. 8–9 (of 13). Milan, 1811.

11. *Discorsi* . . . . In *Opere di Galileo Galilei* . . . , vol. 1 (of 2). Biblioteca Enciclopedica Italiana, vol. 20. Milan, 1832.

12. *Discorsi* . . . . In *Le Opere di Galileo Galilei, Prima Edizione Completa* . . . , edited by Eugenio Alberì, vol. 13 (of 16). Florence, 1855.

13. *Unterredungen und mathematische Demonstrationen über zwei neue Wissenzweige . . . ubersetzt und herausgegeben von Arthur von Oettingen,* 3 vols. Ostwald's Klassiker der exakten Wissenschaften. Leipzig, 1890–91. The first edition with critical notes; several reprint editions.

14. *Discorsi.* . . . In *Le Opere di Galileo Galilei,* edited by Antonio Favaro, vol. 8 (of 20). Edizione Nazionale. Florence, 1898. With the additional days, notes and fragments. Reprinted with some additions 1933, 1965 (largely destroyed by flood), and 1968.

15. *Dialogues Concerning Two New Sciences . . . Translated from the Italian and Latin into English by Henry Crew and Alfonso De Salvio, with an Introduction by Antonio Favaro.* New York, 1914, reissued 1933; reprinted, Northwestern University Press, Evanston and Chicago, 1939, 1946, 1950; reprinted frequently by Dover Books, New York; included in *Great Books of the Western World,* vol. 28.

16. *Besedy i matematicheskie dokazatelstva, kasaiushchiesia dvuch novykh otraslei nauki* . . . . Moscow, 1934. Russian translation by S. N. Dolgov. Reprinted in *Galileo Galilei, Izbranny Trudy,* vol. 1 (of 2). Moscow, 1964.

17. *Discorsi* . . . . In *Galileo Galilei, Opere,* edited by Pietro Pagnini, vol. 4 (of 5). Florence, 1935; reprinted 1964.

18. *Discorsi* . . . . In *Opere,* edited by Sebastiano Timpanaro, vol. 2 (of 2). Milan, 1936.

19. *Dialogos acerca de Dos Nuevas Ciencias.* Buenos Aires, 1945. Spanish translation by Jose San Roman Villasante. The most accurate modern translation to have been published.

20. *Discorsi . . . .* In *Galileo Galilei, Opere Scelte*, edited by Ferdinando Flora. Milan and Naples, 1953.

21. *Discorsi . . . .* Edited by Adriano Carugo and Ludovico Geymonat. Torino, 1958. Italian translations of Latin passages; includes notes and fragments from *Opere* (14, above).

22. Japanese translation by Aoki Seizo, 2 vols. Tokyo, 1961.

23. *Dialoguri aspura stiintelor noi.* Bucharest, 1962. Romanian translation. This may be an abridged version; details lacking.

24. *Discours et demonstrations mathematiques concernant Deux Sciences Nouvelles.* Paris, 1970. French translation by Maurice Clavelin.

Index

abstraction, 12, 58, 78–80, 112–13, 131, 135n, 162, 217, 223, 225, 227

Academician (Author; our friend; i.e., Galileo), xii, 7, 15, 22, 34, 74, 97, 142, 143, 152, 155, 160, 161, 162, 169, 171, 174, 175, 215, 217, 218, 221, 223, 229, 230, 232, 235, 236, 242, 245, 250, 259, 260, 281, 282, 283, 287

acceleration: 53n, 215; along curves, 164; and medium, 94–95; and weight, 90; cause of, xxvii–xxviii, 77, 157–59, 158n, 196, 198, 297; continuity of, xxv, 154, 157, 233; force of, 198; indefinite, 94, 297; law of, ix, xvii, 77, 166, 223; limited, 95; natural, xxvii, 77, 78, 147–48, 153 ff., 169, 197, 230, 232, 233; reduction of, 172, 293. *See also* deceleration; double-distance rule; motion; odd-number rule; speeds; times-squared law; uniform acceleration

accidents, external, 15, 77, 87, 162, 217, 223, 225, 227, 297, 298. *See also* motion, impediments to; resistance

act and potency, 42, 42n, 43, 44, 53

actual. *See* act and potency; potential

air: as cause of motion, 98; as habitat, 128–29; compressed, 76, 82–85; conflict with water, 74–75; escape of, 23, 73, 82, 83; expansion into void, 20, 84; lifting of, 84n; particles of, 27; penetration of, 20, 24; pressure of, 25n, 75n, 129n; rarefaction of, 24; removal of, 78; resistance to speed, 78, 88n, 90–91, 95, 163, 225–28; speed in, 70, 79, 81, 224, 226 ff.; vibration of, 99; weight in, 76; weight of, 76, 80–85. *See also* medium; waves

Albertus Magnus (1206–80), xxxv

ambit and ambient, 85

angle of contact, 291

animals: aquatic, 73, 127, 129; fall of, 14, 123; proportions of, 127; size of, 14, 123, 127; strength of, 128. *See also* fish

antipathy, 75

antiperistasis, 98, 159

Antiphon (fl. 400 B.C.), xxvi

Antonini, Daniello (1588–1616), 282, 282n

anvil, 302

Apollonius of Perga (262–190 B.C.), 215, 218, 218n

Aproino, Paolo (d. 1638): xiv, 282n; as interlocutor, xiv, 281

Aquinas, Thomas (1225?–1274), xxxv

Archimedes (287?–212 B.C.): xxiv, xxvi, xxxii, 38, 48, 110, 121n, 140, 141, 215, 223; and physics, xx, 110, 151n; method of exhaustion, 139–41, 139n; mirror of, 48, 48n; postulates of, xxvi, 110, 110n, 139; principle of, *see* buoyancy

—works: *Conoids and Spheroids*, 139n; *Floating Bodies*, xxvi; *Plane Equilibrium*, xxvi, 110, 110n, 223, 259n; *Quadrature of Parabola*, 141n, 223; *Spiral Lines*, xxvi, xxxv, 139, 139n, 148n, 149n

architects, 224

Ariosto, Lodovico (1475–1533), 127n

Aristotle (384–322 B.C.): and cause, xxviii; axioms of, 21, 27n, 42n; experiments of, 66, 69, 81–82; on mathematics, xxvi–xxvii; on speed of fall, 65–71, 79; on the void, 20, 20n, 34, 55, 65–71; on weight of air, 80–83; on Zeno's paradoxes, xxv–xxvi; vocabulary of, xxvi–xxviii; wheel of, 29 ff., 56–57. *See also* Peripatetics; philosophers; physics, Aristotelian

—works: *De caelo*, xiv, 21n, 42n; *Metaphysics*, xxvii, xxxii; *Physics*, 20n, 21n, 39n, 42n, 148n, 229n, 295n. *See also Questions of Mechanics*

arquebus, 95, 228

Arrighetti, Andrea (1592–1672), 125n

arrows, 225

artillery, 49, 224, 229n, 245

artisans, 11, 142, 143, 182

astronomers, 8, 169

atomists, 34, 34n, 65, 66, 71

atoms, xxxvii, 26, 27n, 28, 33, 34n, 47, 54, 92. *See also* indivisibles; *minima naturalia*

218 Apollonius
169 unif accel, math & expt
* 153 m - natural fall
171-75 proof of principle assumed